Methods in Molecular Biology

Series Editor
John M. Walker
School of Life and Medical Sciences
University of Hertfordshire
Hatfield, Hertfordshire, UK

For further volumes:
http://www.springer.com/series/7651

For over 35 years, biological scientists have come to rely on the research protocols and methodologies in the critically acclaimed *Methods in Molecular Biology* series. The series was the first to introduce the step-by-step protocols approach that has become the standard in all biomedical protocol publishing. Each protocol is provided in readily-reproducible step-by-step fashion, opening with an introductory overview, a list of the materials and reagents needed to complete the experiment, and followed by a detailed procedure that is supported with a helpful notes section offering tips and tricks of the trade as well as troubleshooting advice. These hallmark features were introduced by series editor Dr. John Walker and constitute the key ingredient in each and every volume of the *Methods in Molecular Biology* series. Tested and trusted, comprehensive and reliable, all protocols from the series are indexed in PubMed.

Dictyostelium discoideum

Methods and Protocols

Edited by

Alan R. Kimmel

Laboratory of Cellular and Developmental Biology, NIDDK, NIH, Bethesda, MD, USA

Humana Press

Editor
Alan R. Kimmel
Laboratory of Cellular
and Developmental Biology
NIDDK, NIH
Bethesda, MD, USA

ISSN 1064-3745 ISSN 1940-6029 (electronic)
Methods in Molecular Biology
ISBN 978-1-0716-3893-4 ISBN 978-1-0716-3894-1 (eBook)
https://doi.org/10.1007/978-1-0716-3894-1

Cover Caption: A false color image of the view of terminal development of Dictyostelium discoideum. The blue shows terminally differentiated spores and the red shows terminally differentiated stalks. Image courtesy of Dr. Pundrik Jaiswal and Dr. Alan R. Kimmel: Laboratory of Cellular and Developmental Biology, The National Institute of Diabetes, and Digestive and Kidney Disease, The National Institute of Health, Bethesda, MD, USA

This Humana imprint is published by the registered company Springer Science+Business Media, LLC, part of Springer Nature.
The registered company address is: 1 New York Plaza, New York, NY 10004, U.S.A.

Preface

Dictyostelium discoideum and related Dictyostelids are a monophyletic group of facultative multicellular, soil-dwelling eukaryotes. In the wild, *Dictyostelium* propagate as unicellular ameboid cells through mitotic cell division, using bacteria as a primary nutrient supply. Growth-phase *Dictyostelium* are highly chemotactic to many compounds (e.g., folates, cAMP, etc.) that are secreted by bacteria. *Dictyostelium* are also professional phagocytes and thus rapidly and very efficiently engulf and digest surrounding bacteria as a food source; many strains are additionally able to obtain nutrients through fluid-phase macropinocytosis. In many aspects, *Dictyostelium* serves as a model for macrophage function in aspects of innate immunity. *Dictyostelium* and macrophages share functionally identical G protein-coupled chemoattractant receptors and regulated downstream signaling circuits. Phagocytic and lysosomal degradative pathways for bacteria are likewise similar. Accordingly, the first chapters focus to issues of growth, membrane function and damage, and nutrient/bacterial processing.

When bacteria within a defined area are fully cleared by feeding *Dictyostelium*, available nutrients become depleted. Starved *Dictyostelium* undergo growth arrest and, at a sufficient cell-density threshold, transition to multicellular development. Here again *Dictyostelium* process parallel mammalian cell functions. Nutrient sensing-response requires the reciprocal actions of kinases mTORC1 and AMPK, where their mis-regulations can form the bases for loss of cell growth arrest in several cellular over-growth disease phenotypes in humans. As in other metazoa, inhibition of mTORC1 through nutrient depletion leads to autophagic activation; in *Dictyostelium* autophagy is required to recycle organelles and macromolecules to provide energy and metabolic precursors, for these otherwise starved cells. Autophagy activation and function, at the junction between growth and development, is the next described chapter.

Multicell development involves the secretion of cAMP as a chemoattractant and the expression of the G protein-coupled cAMP chemoattractant receptor CAR1 and its linked downstream machinery. Cells within a localized region begin to secrete cAMP, recruiting proximal cells to migrate inwardly toward the center of cAMP signaling. The migrating cells also relay the cAMP signal outwardly, enlisting more distal cells in a collective directional migratory stream. Unidirectional movement is imposed through time-directed response/adaptation/re-response state variations for cAMP synthesis, cAMP degradation, and cAMP chemotaxis. Downstream of CAR1, multiple intracellular signaling arms converge at the actin cytoskeleton to direct cell polarity and chemotactic movement, involving Ras, PI3K/PTEN, PIP_3, PIP_2, IP_3, cGMP, and others, and again where similar pathways are required for innate immune function in neutrophils and other migratory cells. Related chapters focus broadly to protein localization, interaction, and activation, before a chapter that examines cell migration and contact guidance for quantification of cytoskeletal dynamics. These chapters are followed by ones focused to more specific regulatory pathways.

Multicellular formation of *Dictyostelium* then follows the collective cell migration toward a territory center and allo-recognized aggregation and cell adhesion. Multicell aggregates undergo cytodifferentiation and morphogenetic movement leading to a terminally differentiated organism. Thus, the *Dictyostelium* life cycle is highly complex. Although the genome is relatively small, ~34 mb, it is very compact and encodes >12,000 genes; more than half are developmentally activated. The volume concludes with chapters that describe global mutational screening, transcriptome analyses, and proteomics that apply to all stages of the *Dictyostelium* life cycle and which may further generalize analyses for human disease models. [This work was supported by the Intramural Research Program of the National Institutes of Health (NIH), the National Institute of Diabetes and Digestive and Kidney Diseases (NIDDK)].

Bethesda, MD, USA ***Alan R. Kimmel***

Contents

Preface .... *v*
Contributors .... *ix*

1 Discovery and Quantitative Dissection of Cytokinesis Mechanisms Using *Dictyostelium discoideum* 1
Jennifer M. K. Nguyen, Yinan Liu, Ly Nguyen, Venkataramana K. Sidhaye, and Douglas N. Robinson

2 Superresolution Expansion Microscopy in *Dictyostelium* Amoebae 29
Marianne Grafe, Valentin Pitzen, Irene Meyer, and Ralph Gräf

3 Inflicting, Monitoring, Visualizing, and Quantitating Various Sterile Membrane Damages and the Repair Response in *Dictyostelium discoideum* 45
Lyudmil Raykov, Davide D'Amico, A. T. López-Jiménez, and Thierry Soldati

4 Assaying Lysosomal Enzyme Activity in *Dictyostelium discoideum* 55
William D. Kim, Adam F. DiGiacinto, and Robert J. Huber

5 Procedures to Measure *Dictyostelium* Phagocytosis and Macropinocytosis 81
Lou W. Kim and Cong Gu

6 Quantitative Analysis of the Inactivation Process of Internalized Bacteria in *Dictyostelium* Cells 89
Lou W. Kim and Victor Osorio-Castillo

7 Methods to Assess Autophagic Activity in *Dictyostelium* 97
Laura Antón-Esteban, Alba Tornero-Écija, Olivier Vincent, and Ricardo Escalante

8 Controlling the Subcellular Localization of Signaling Proteins Using Chemically Induced Dimerization and Optogenetics 107
Mariam Beshay, Yu Deng, and Chris Janetopoulos

9 APEX2-Mediated Proximity Protein Labeling in *Dictyostelium* 119
Jamie A. Takashima, Helena A. Woroniecka, and Pascale G. Charest

10 GRminusRD: A Sensitive Assay to Detect Activation Processes at the Plasma Membrane in Living Cells 133
Peter J. M. van Haastert, Ineke Keizer-Gunnink, and Arjan Kortholt

11 Analyzing the Micro-topography Guidance of *Dictyostelium* Cells Using Microfabricated Structures 149
Gen Honda and Satoshi Sawai

12 Evaluating the Activity of Ras/Rap GTPases in *Dictyostelium* 163
Stephen F. Smith and Pascale G. Charest

13 Analysis of cGMP Signaling in *Dictyostelium* 177
Peter J. M. van Haastert, Ineke Keizer-Gunnink, and Arjan Kortholt

14 Quantitative Monitoring of GPCR-Mediated Spatiotemporal IP_3 Dynamics Using Confocal Fluorescence Microscopy 195
Xuehua Xu, HyunGee Ha, Joseph Brzostowski, and Tian Jin

15 Mutation Screening of *Dictyostelium* Restriction Enzyme-Mediated Integration (REMI) Libraries 209
George Heslop-Harrison, Anthony Goddard, and Robin S. B. Williams

16 Single Cell Transcriptome Analysis During Development in *Dictyostelium* 223
Vlatka Antolović and Jonathan R. Chubb

17 Proteomic Analysis of *Dictyostelium discoideum* by Mass Spectrometry 247
Ramesh Rijal and Richard H. Gomer

Index *257*

Contributors

VLATKA ANTOLOVIĆ • *UCL Laboratory for Molecular Cell Biology, University College London, London, UK*

LAURA ANTÓN-ESTEBAN • *Instituto de Investigaciones Biomédicas Sols-Morreale, Madrid, Spain*

MARIAM BESHAY • *Department of Biological Sciences, University of the Sciences, Philadelphia, PA, USA*

JOSEPH BRZOSTOWSKI • *Imaging Core Facility, Laboratory of Immunogenetics, National Institute of Allergy and Infectious Diseases (NIAID), National Institutes of Health (NIH), Rockville, MD, USA*

PASCALE G. CHAREST • *Department of Chemistry and Biochemistry, University of Arizona, Tucson, AZ, USA; Department of Molecular and Cellular Biology, University of Arizona, Tucson, AZ, USA*

JONATHAN R. CHUBB • *UCL Laboratory for Molecular Cell Biology, University College London, London, UK*

DAVIDE D'AMICO • *Département de Biochimie, Faculté des Sciences, Université de Genève, Sciences II, Geneva, Switzerland*

YU DENG • *Department of Cell Biology and Center for Cell Dynamics, School of Medicine, Johns Hopkins University, Baltimore, MD, USA; Department of Chemical and Biomolecular Engineering, Whiting School of Engineering, Johns Hopkins University, Baltimore, MD, USA*

ADAM F. DIGIACINTO • *Department of Biology, Trent University, Peterborough, ON, Canada*

RICARDO ESCALANTE • *Instituto de Investigaciones Biomédicas Sols-Morreale, Madrid, Spain*

ANTHONY GODDARD • *Centre for Biomedical Sciences, School of Biological Sciences, Royal Holloway University of London, Egham, UK*

RICHARD H. GOMER • *Department of Biology, Texas A&M University, College Station, TX, USA*

MARIANNE GRAFE • *Department of Cell Biology, University of Potsdam, Potsdam-Golm, Germany*

RALPH GRÄF • *Department of Cell Biology, University of Potsdam, Potsdam-Golm, Germany*

CONG GU • *Biological Sciences, Biomolecular Sciences Institute, Florida International University, Miami, FL, USA*

HYUNGEE HA • *Chemotaxis Signaling Section, Laboratory of Immunogenetics, National Institute of Allergy and Infectious Diseases (NIAID), National Institutes of Health (NIH), Rockville, MD, USA*

GEORGE HESLOP-HARRISON • *Centre for Biomedical Sciences, School of Biological Sciences, Royal Holloway University of London, Egham, UK*

GEN HONDA • *Komaba Institute for Science, Graduate School of Arts and Sciences, The University of Tokyo, Meguro, Tokyo, Japan*

ROBERT J. HUBER • *Environmental and Life Sciences Graduate Program, Trent University, Peterborough, ON, Canada; Department of Biology, Trent University, Peterborough, ON, Canada*

CHRIS JANETOPOULOS • *Department of Cell Biology, School of Medicine, Johns Hopkins University, Baltimore, MD, USA; Total Experience Learning, Albright College, Reading, PA, USA*
TIAN JIN • *Chemotaxis Signaling Section, Laboratory of Immunogenetics, National Institute of Allergy and Infectious Diseases (NIAID), National Institutes of Health (NIH), Rockville, MD, USA*
INEKE KEIZER-GUNNINK • *Department of Cell Biochemistry, University of Groningen, Groningen, The Netherlands*
LOU W. KIM • *Biological Sciences, Biomolecular Sciences Institute, Florida International University, Miami, FL, USA*
WILLIAM D. KIM • *Environmental and Life Sciences Graduate Program, Trent University, Peterborough, ON, Canada*
ARJAN KORTHOLT • *Department of Cell Biochemistry, University of Groningen, Groningen, The Netherlands*
YINAN LIU • *Department of Cell Biology, Johns Hopkins School of Medicine, Baltimore, MD, USA; Department of Pharmacology of Molecular Sciences, Johns Hopkins School of Medicine, Baltimore, MD, USA*
A. . T. LÓPEZ-JIMÉNEZ • *Département de Biochimie, Faculté des Sciences, Université de Genève, Sciences II, Geneva, Switzerland; Department of Infection Biology, London School of Hygiene and Tropical Medicine, London, UK*
IRENE MEYER • *Department of Cell Biology, University of Potsdam, Potsdam-Golm, Germany*
JENNIFER M. K. NGUYEN • *Department of Cell Biology, Johns Hopkins School of Medicine, Baltimore, MD, USA; Department of Pharmacology of Molecular Sciences, Johns Hopkins School of Medicine, Baltimore, MD, USA*
LY NGUYEN • *Department of Cell Biology, Johns Hopkins School of Medicine, Baltimore, MD, USA*
VICTOR OSORIO-CASTILLO • *Biological Sciences, Biomolecular Sciences Institute, Florida International University, Miami, FL, USA*
VALENTIN PITZEN • *Department of Cell Biology, University of Potsdam, Potsdam-Golm, Germany*
LYUDMIL RAYKOV • *Département de Biochimie, Faculté des Sciences, Université de Genève, Sciences II, Geneva, Switzerland*
RAMESH RIJAL • *Department of Biology, Texas A&M University, College Station, TX, USA*
DOUGLAS N. ROBINSON • *Department of Cell Biology, Johns Hopkins School of Medicine, Baltimore, MD, USA; Department of Pharmacology of Molecular Sciences, Johns Hopkins School of Medicine, Baltimore, MD, USA; Department of Medicine, Johns Hopkins School of Medicine, Baltimore, MD, USA*
SATOSHI SAWAI • *Department of Basic Science, Graduate School of Arts and Sciences, University of Tokyo, Tokyo, Japan; Research Center for Complex Systems Biology, University of Tokyo, Meguro-ku, Tokyo, Japan; Universal Biological Institute, University of Tokyo, Bunkyo-ku, Tokyo, Japan; Department of Biology, Graduate School of Science, University of Tokyo, Bunkyo-ku, Tokyo, Japan*
VENKATARAMANA K. SIDHAYE • *Department of Medicine, Johns Hopkins School of Medicine, Baltimore, MD, USA*
STEPHEN F. SMITH • *Department of Chemistry and Biochemistry, University of Arizona, Tucson, AZ, USA*
THIERRY SOLDATI • *Département de Biochimie, Faculté des Sciences, Université de Genève, Sciences II, Geneva, Switzerland*

JAMIE A. TAKASHIMA • *Department of Chemistry and Biochemistry, University of Arizona, Tucson, AZ, USA; Division of Chemistry and Chemical Engineering, California Institute of Technology, Pasadena, CA, USA*

ALBA TORNERO-ÉCIJA • *Instituto de Investigaciones Biomédicas Sols-Morreale, Madrid, Spain*

PETER J. M. VAN HAASTERT • *Department of Cell Biochemistry, University of Groningen, Groningen, The Netherlands*

OLIVIER VINCENT • *Instituto de Investigaciones Biomédicas Sols-Morreale, Madrid, Spain*

ROBIN S. B. WILLIAMS • *Centre for Biomedical Sciences, School of Biological Sciences, Royal Holloway University of London, Egham, UK*

HELENA A. WORONIECKA • *Department of Chemistry and Biochemistry, University of Arizona, Tucson, AZ, USA*

XUEHUA XU • *Chemotaxis Signaling Section, Laboratory of Immunogenetics, National Institute of Allergy and Infectious Diseases (NIAID), National Institutes of Health (NIH), Rockville, MD, USA*

Chapter 1

Discovery and Quantitative Dissection of Cytokinesis Mechanisms Using *Dictyostelium discoideum*

Jennifer M. K. Nguyen,
Yinan Liu,
Ly Nguyen, Venkataramana K. Sidhaye, and Douglas N. Robinson

Abstract

The social amoeba *Dictyostelium discoideum* is a versatile model for understanding many different cellular processes involving cell motility including chemotaxis, phagocytosis, and cytokinesis. Cytokinesis, in particular, is a model cell-shaped change process in which a cell separates into two daughter cells. *D. discoideum* has been used extensively to identify players in cytokinesis and understand how they comprise the mechanosensory and biochemical pathways of cytokinesis. In this chapter, we describe how we use cDNA library complementation with *D. discoideum* to discover potential regulators of cytokinesis. Once identified, these regulators are further analyzed through live cell imaging, immunofluorescence imaging, fluorescence correlation and cross-correlation spectroscopy, micropipette aspiration, and fluorescence recovery after photobleaching. Collectively, these methods aid in detailing the mechanisms and signaling pathways that comprise cell division.

Key words *Dictyostelium discoideum*, cDNA library complementation, Cytoskeletal network, Mechanoresponsive, Protein dynamics

1 Introduction

Cytokinesis is the last step during cell division in which a cell dramatically changes shape to separate into two daughter cells. For this process to occur, the cell utilizes a complex signaling pathway system with biochemical and mechanical signals and regulatory feedback loops. Thus, cytokinesis is fundamentally a mechanical process and a robust one, making it ample for many essential cellular functions such as cell proliferation, cell survival, and tissue formation. Understanding the mechanisms of cytokinesis allows for insight into these fundamental processes and into diseases where these mechanisms are dysregulated, such as cancer.

Alan R. Kimmel (ed.), *Dictyostelium discoideum: Methods and Protocols*, Methods in Molecular Biology, vol. 2814,
https://doi.org/10.1007/978-1-0716-3894-1_1,

Studies on cytokinesis have occurred in various mammalian cell types and other model systems including yeast and *Drosophila* [1–2]. We focus on using the social amoeba, *Dictyostelium discoideum*, as a model for cytokinesis. With its genetic tractability, rapid growth rate, ease of handling, and genetic homology to mammalian cells, *D. discoideum* is a suitable model for studying cytokinesis at a fundamental level. Through molecular genetic techniques, individual proteins responsible for cytokinesis have been identified through *D. discoideum* such as myosin II, cortexillin I, racE, and coronin [3–8]. Nonetheless, mapping out the complex biochemical network and pathways behind cytokinesis requires methods that integrate interaction genetics. One such approach is cDNA library complementation and high copy suppression, in which genetic suppressors of a *D. discoideum* cytokinesis-deficient mutant are recovered and identified. Through this method, we uncovered various genetic suppressors of genes like cortexillin I and myosin II, including dynacortin and enlazin [9–10].

As regulators of cytokinesis are identified via genetic selection (or other methods such as mass spectrometry [11]), we follow a workflow to examine if and how these proteins affect cytokinesis. Genetic mutant cell lines (cells expressing fluorescently tagged proteins or knockdown/knockout cell lines) are generated. They are then imaged for cytokinetic phenotypes, including multinucleation and furrow ingression defects. If there are noticeable cytokinetic phenotypes, the behaviors of these regulators during cytokinesis are expanded upon by methods like fluorescence correlation spectroscopy (FCS), fluorescence recovery after photobleaching (FRAP), and micropipette aspiration (MPA). These methods offer insight into protein movement, localization, and cell mechanics. Protein-protein interactions can also be studied if multiple cytokinetic regulators are observed in mutant cell lines. The previously mentioned methods can be used with multiple regulators along with another assay, fluorescence cross-correlation spectroscopy (FCCS), which specifically involves examining the interactions between proteins tagged with two differently colored fluorophores.

In this chapter, we detail the workflow of identifying potential regulators of cytokinesis and examining how they contribute to the cell division process. We first describe *D. discoideum* cell culture and the different approaches in creating mutant cell lines. From there, we expand on imaging these cells and using them in these quantitative assays: FRAP, FCS, FCCS, and MPA. The materials needed for each assay are also described.

2 Materials

2.1 D. discoideum Cell Culture and Mutant Cell Line Generation

2.1.1 General Cell Handling

1. 1.5× HL-5 medium: For 2 L of medium, mix 30.0 g proteose peptone, 0.255 g $Na_2HPO_4{\cdot}7H_2O$, 7.8 g Bacto™-yeast extract, 0.255 g KH_2PO_4, and 6.0 g glucose in deionized and double distilled water (*see* **Note 1**). Adjust pH to 6.5 and aliquot into 500 mL bottles. Autoclave and store at room temperature.
2. Hans' Enriched HL-5 medium:
   (a) ForMedium (FM): For 2 liters, mix 38 g FM with ~1500 mL of water. Dissolve by stirring rapidly. Raise volume to 2 L. Aliquot into 500 mL bottles and cover bottles with foil because FM is sensitive to light. Autoclave and store at room temperature. Add 43 mL of FM per 500 mL of 1.5× HL-5, which yields Hans' Enriched HL-5 medium (1.4× HL-5, 8% FM).
   (b) 500× Penicillin/Streptomycin (P/S): Dissolve 1.00 g Penicillin G and 1.56 g Streptomycin sulfate in 156 mL water. Filter sterilize and aliquot into 15 mL tubes. Store at −20 °C.
3. 10 cm Petri dishes.

2.1.2 Identification of Cytokinetic Mutants via cDNA Library Screening

2.1.3 Screening of Cytokinetic Mutants

1. Hemocytometer.
2. Electroporation (E-pore) buffer: 17.115 g/L sucrose and 10 mL/L 1 M potassium phosphate buffer, pH 6.5. Filter sterilize and store at room temperature.
3. Gene Pulser®/MicroPulser™ Electroporation Cuvettes, 0.4 cm gap (Bio-Rad, #1652088).
4. Gene Pulser Electroporator (Bio-Rad, model #1652076).
5. 100 mg/mL G418: Add 81.6 μL into 500 mL of Hans' Enriched HL-5 for 15 μg/mL G418. Store G418 stock solution at −20 °C.
6. 150 mL Erlenmeyer flasks.

Monitoring Growth Rate of D. discoideum

1. Hans' Enriched HL-5 medium (*see* Subheading 2.1.1, **items 1** and **2**)
2. 150 mL Erlenmeyer flasks
3. Hemocytometer

Multinucleation

1. Hans' Enriched HL-5 medium (*see* Subheading 2.1.1, **items 1** and **2**).
2. Coverslips (Fisher Scientific).
3. 10% paraformaldehyde: Dissolve 1 g PFA powder in 65 °C water. Add 1 M NaOH dropwise and vortex to dissolve PFA.
4. 100 μg/μL Hoechst dye.
5. 1× PBS: 8 g/L NaCl, 0.2 g/L KCl, 1.44 g/L Na_2HPO_4-$7H_2O$, and 0.24 g/L KH_2PO_4 in ~800 mL of water. Adjust pH to 7.4 with HCl or NaOH and bring volume up to 1 L.
6. 1× PBST: 0.1% (v/v) Tween 20 in 1× PBS.
7. Triton X-100.
8. Fixation solution: 8 mL Hans' Enriched HL-5 medium, 2 μL Triton X-100, 2 mL 10% PFA, and 2 mL water.
9. Anti-fade mounting media.

2.1.4 Development of Cell Lines Expressing Fluorescently-Labelled Proteins

1. PCR reagents:
   (a) Primers are obtained from Integrated DNA Technologies (IDT).
   (b) Q5® High-Fidelity 2× Master Mix (New England Biolabs).
2. Transformation materials (*see* Subheading 2.1.3, **items 2–5**).

2.1.5 Generation of *D. discoideum* Cytokinetic Knockdown/Knockout Cell Lines

Homologous Recombination

1. Transformation materials (*see* Subheading 2.1.3, **items 2–5**).

RNA Hairpin

1. PCR reagents (*see* Subheading 2.1.4, **item 1**).
2. Expression vector for cloning.
3. Transformation materials (*see* Subheading 2.1.3, **items 2–5**).
4. RNA extraction reagent such as TRIzol® (Thermo Fisher Scientific).
5. RT-qPCR reagents such as the Verso 1-step RT-qPCR Kit with SYBR Green (Thermo Fisher Scientific).

CRISPR

1. *D. discoideum* strains growing at exponential growth phase in 10 mL of Hans' Enriched HL-5 media on 10 cm Petri dish.
2. SM-5 plates: 1.0 g/L Bacto™-tryptone, 1.0 g/L Bacto™-- yeast extract, 1.0 g/L glucose, 0.95 g/L KH_2PO_4, 0.5 g/L K_2HPO_4, 0.1 g $MgSO_4$-$7H_2O$, and 20 g/L Bacto™-agar. Adjust pH to 6.4 with 2.5 M NaOH to be more basic or 1 M KH_2PO_4 to be more acidic. Autoclave the mixture, allow it to

cool to approximately 50 °C, and pour 30 mL of the mixture in each 10 cm Petri dish. Store the plates at 4 °C.

3. *Klebsiella aerogenes* culture: Add approximately 0.5 mL of *K. aerogenes* in 50 mL of LB medium and grow in suspension at 37 °C overnight to be used the next day.
3. LB medium: 5 g/L Bacto™-tryptone, 2.5 g/L Bacto™-yeast extract, 2.5 g/L sodium chloride, and 100 μL/L 5 M NaOH in water. Autoclave the medium and store at room temperature.
4. Transformation materials (*see* Subheading 2.1.3, **items 2–5**).
5. All-in-one CRISPR-Cas9 pTM1285 vector was obtained from NBRP Nenkin.
6. We selected gRNA guide sequences targeting the gene of interest using Cas-Designer [12]. Customized RNA oligos were obtained from Integrated DNA Technologies (IDT).
7. Forward and reverse primers complementary to gene regions within 50 nucleotides away from *Bbs* I-cut sites/gRNA insertion sites on pTM1285 vector were customized and obtained from IDT.
8. Fermentas FastDigest *Bbs* 1 restriction enzyme, FastDigest Buffer (10×), and FastAP Thermosensitive Alkaline Phosphatase (Thermo Fisher Scientific).
9. QIAquick gel extraction kit (Qiagen).
10. Quick ligation kit (NEB).
11. T4 PNK (NEB).
12. T4 ligation buffer (NEB).

2.2 Assessing Cytokinesis via Cell Imaging

2.2.1 Live Cell Imaging/Furrow Measurements

1. Glass coverslips (Fisher Scientific).
2. Glass bottom imaging chamber.

2.2.2 Fixed Cell Imaging

1. Glass coverslips and glass slides (Fisher Scientific).
2. 4% paraformaldehyde.
3. Acetone.
4. 10× PBS: For 1 L, dissolve 80 g NaCl, 14.4 g $Na_2HPO_4{\cdot}7H_2O$, 2.0 g KCl, and 2.4 g KH_2PO_4 in ~800 mL water. Adjust pH to 7.4 with HCl or NaOH. Raise volume to 1 L. Autoclave and store at room temperature.
5. Blocking buffer: 1× PBS with 0.05% Triton X-100 and 0.5% BSA.

6. 1× PBST: Dilute 10× PBS 10-fold in water and then add Triton X-100 for final concentration of 0.05%.
7. Primary and secondary antibodies compatible with immunofluorescence.
8. Anti-fade mounting media.

2.3 Fluorescence Correlation Spectroscopy (FCS) and Fluorescence Cross-Correlation Spectroscopy (FCCS)

1. Confocal microscope with a fluorescence correlation and dual color cross-correlation capabilities.
2. A high numerical aperture, water-immersion objective.
3. Analysis software such as Zen Blue or Zen Black software for acquisition and data analysis.
4. Glass bottom imaging chambers (Thermo Fisher Scientific).
5. Coverslips (Fisher Scientific).
6. Low fluorescence media: For 1 L, pour ~600 mL water into a glass container. Dissolve 3.85 g glucose, 1.20 g Na_2HPO_4-$12H_2O$, 1.78 g Bacto™ Proteose-Peptone, 0.485 g KH2PO4, and 0.450 g Bacto™-yeast extract. Adjust pH to 6.5 using 1 M KH_2PO_4 or 1 M K_2HPO_4. Bring volume up to 1 L. Autoclave.
7. Rhodamine 6G.

2.4 Fluorescence Recovery After Photobleaching (FRAP)

1. Zeiss Axiovert 200 inverted microscope with LSM510-Meta confocal module and a 63× (NA 1.4) objective.
2. Glass bottom imaging chambers (Thermo Fisher Scientific).
3. MES starvation buffer: 50 mM MES, pH 6.8, 2 mM $MgCl_2$, and 0.2 mM $CaCl_2$ in deionized water.

2.5 Micropipette Aspiration (MPA)

This list briefly details our materials used for MPA, but an MPA system can be designed in various ways with different motorization settings.

1. Microscope: Olympus IX81 or IX71 capable of DIC and fluorescence imaging, with 10×, 40× (NA1.3), and 60× (NA1.45) objectives and a 1.6× Optovar.
2. Micromanipulator: Sutter Instrument MP225.
3. Micropipette puller: MicroData Instrument PMP-102.
4. Microforge: MicroData Instrument MFG-5 Microforge-Grinding Center.
5. Custom-built rectangular imaging chambers made of anodized aluminum with a thickness of about 1.5–2 mm.
6. Vacuum grease (Dow Corning high-vacuum grease) for attaching cover glass to imaging chamber.
7. Cover glass (No. 1.5 thickness, 22 × 60 mm, Fisher Scientific).
8. MES starvation buffer: *see* **item 3** in Subheading 2.4.

9. Micropipette needles: Micropipettes needles are made with a micropipette puller and microforge. Thin-wall borosilicate glass tubing (1 mm outer diameter, 0.75 mm inner diameter, 10 cm length, Sutter Instrument) can be used with the micropipette puller. When pulling pipettes, the inner diameter should ideally be in the 4–6 μm range. The micropipette tip should be straight and polished with the microforge. Pre-Pulled Glass Pipettes are also commercially available through World Precision Instrument.
10. 1 μm blue polystyrene microbeads (Polysciences, #15712) for calibrating the MPA system.

3 Methods

3.1 D. discoideum Cell Culture and Mutant Cell Line Generation

3.1.1 General Cell Handling

Vegetative *D. discoideum* cells are routinely grown in Hans' Enriched HL-5 (1.5× HL-5 medium supplemented with ForMedium (FM, 8% final concentration), penicillin (60 U/mL), and streptomycin sulfate (60 mg/mL)). Cells are grown at 22 °C on polystyrene Petri dishes and passaged by simple pipetting before cells reach the stationary phase (between 2×10^6 and 6×10^6 cells/mL). Cells can also be grown in suspension in Erlenmeyer flasks at 180 rpm at 22 °C.

3.1.2 Identification of Cytokinetic Mutants via cDNA Library Screening

While proteins involved or required in cytokinesis in *D. discoideum* have been identified with common molecular genetic methods, uncovering complex genetic interactions with these methods are more difficult to do because *Dictyostelium* are primarily haploid organisms that are not easily genetically crossed. To bypass this issue, we focus on using cDNA library complementation and high-copy suppression methods to probe for more complex genetic interactions during cytokinesis. Numerous genes have been identified to play a role in cytokinesis using this method [9–10, 13] by transformation of the cDNA library into cell lines with identified cytokinetic aberrations, including high nuclei/cell ratio and a lower saturation density. The generation of the cDNA library in the pLD1A15SN expression vector is detailed in reference 9. Below, we describe the transformation of the cDNA library and identification of genetic suppressors in a cell line with a cytokinesis defect resulting in multinucleation and decreased growth rate.

3.1.3 Screening of Cytokinetic Mutants

1. Grow mutant cell line on 20 10-cm Petri dish plates until they reach late log phase growth, generally between 2×10^6 and 6×10^6 cells/mL.
2. Pool cell lines and determine cell density with a hemocytometer.

3. Spin down cells at 2000 rpm for 5 min and wash with 10 mL ice-cold electroporation buffer.
4. Decant supernatant and resuspend pellet in 10 mL ice-cold electroporation buffer. Keep cells on ice.
5. Spin down cells at 2000 rpm for 5 min and resuspend cells at a density of 10^7 cells/350 μL of electroporation buffer.
6. Transform 1 μg of plasmid cDNA library per 350 μL of cells using a Gene Pulser. Add DNA into cuvette with cells. Keep cuvettes on ice.
7. Pulse cells one time in 0.4-cm cuvettes at 3 μF and 1.3 kV for ORF 7–3 cells (HS1000 cells: (Ax3: Replicase orf+)) (*see* **Note 2**). Pour cells from the cuvette into a petri dish.
8. Allow cells to grow in enriched Hans' HL-5 media for 24 h at 22 °C. Replace media with Hans' enriched HL-5 with G418 (typically 10–15 μg/mL but determined using kill curves and/or optimization of transformation efficiency with the empty plasmid). Change media every 2–3 days until clones are harvested.
9. When clones reach near confluency on 10 cm plates, harvest cells. About 1000–1500 clones/μg of DNA can be recovered from ORF+-derived strains. These cells can be combined into pools as large as 30,000–35,000 clones per pool as was originally done, but we find that smaller pool sizes (&3000 clones/pool) allow for much more rapid identification of genetic interactors (*see* **Note 3**).
10. To begin genetic selection, grow cells in suspension in at 1×10^5 cells/mL in 10 mL cultures in 125 mL Erlenmeyer flasks at 180 rpm at 22 °C. Allow cells to grow to $5–6 \times 10^6$ cells/mL before splitting cells into new flasks at 1×10^5 cells/mL. Culture cells over 3–4 growth cycles for smaller pool sizes.
11. Monitor cell growth rates (as detailed in the next section). During log phase growth, aliquot cells from suspension cultures for nuclei/cell ratio determination (as described in the Multinucleation section below).
12. Identify cultures with wild-type cell morphology (deemed as winners) and harvest DNA.
13. Transform harvested DNA into STBL2 *E. coli* cells (*see* **Note 4**). Perform DNA clean-up and appropriate restriction endonuclease digestion on isolated bacterial clones for DNA sequencing. Identify genes from DNA sequencing via sequence database (i.e., Dictybase BLAST).
14. To confirm that the suppression phenotype can be recapitulated, transform recovered cDNA into original mutant cell line and monitor cell growth rates and nuclei/cell ratio.

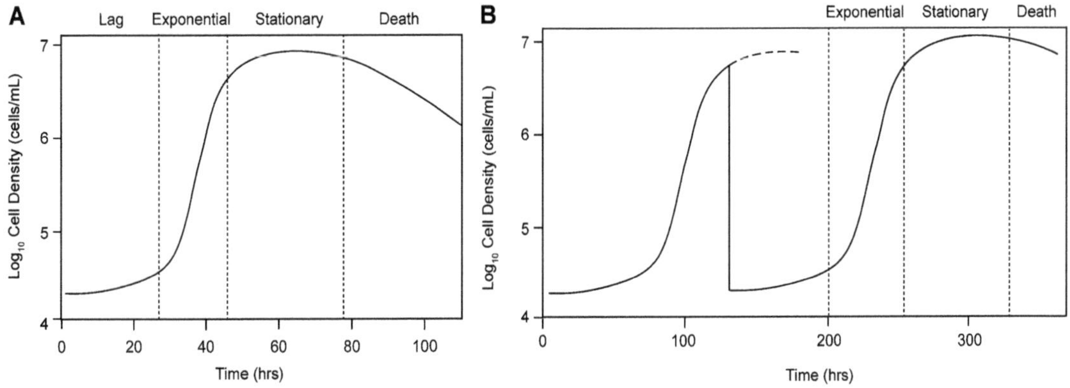

Fig. 1 Relative *D. discoideum* growth rates can be determined by plotting cell densities versus time. (**a**) Sample *D. discoideum* growth curve. (**b**) Example growth curves over multiple cycles of growth. Cells are split within the exponential phase of growth before being subjected to a subsequent cycle of growth

Monitoring Growth Rate of D. discoideum

1. Seed cells in 10 mL of Hans' Enriched HL-5 medium in 150 mL Erlenmeyer flasks at a starting concentration of 1×10^5 cells/mL (*see* **Note 5**).
2. Shake cells in 22 °C incubator at 180 rpm. Cell concentration is measured and recorded roughly every 24 h using a hemocytometer. It is important to note that because of the time lag required for the winner cells to populate the cultures, we normally acquire and analyze data from second and third cycles of suspension culture.
3. When cells reach the later point of exponential growth phase (about 5–6 × 10^6 cells/mL), split down cultures to a concentration of 0.5–1 × 10^5 cells/mL (*see* **Note 6**). Continue to measure roughly every 24 h, recording the exact times cell density is measured.
4. Relative growth rates are determined by plotting cell densities versus time (Fig. 1a). The resulting exponential phase curve is fitted to a single exponential equation using a data analysis software such as KaleidaGraph (Synergy Software). If cell concentrations are measured throughout multiple cycles of suspension culture, growth rates can be obtained from each cycle (Fig. 1b). Growth rate (k) can be determined for each growth curve through the exponential growth equation:

$$n(t) = n_0 e^{kt}$$

where $n(t)$ = population at time t, n_0 = initial population concentration, and k = relative growth rate.

5. Relative growth rates between cell lines can be compared to their respective control lines to determine if there are any growth defects.

Multinucleation

1. Seed cells at approximately 70% confluency on coverslips in Hans' Enriched HL-5 media and allowed to adhere for 15–30 min.
2. Remove media and fix cells with ice-cold fixation solution for 10 min.
3. Wash cells quickly with 1× PBS and stain nuclei with 10 μg/uL Hoechst for 10 min. Afterward, wash cells with 1×PBST three times, 10 min each.
4. Before visualization, mount cells on coverslips using an anti-fade of choice (*see* **Note** 7).
5. Image nuclei using a fluorescence microscope and count nuclei through an imaging software such as ImageJ.

3.1.4 Development of Cell Lines Expressing Fluorescently Labeled Proteins

Generating *D. discoideum* cell lines expressing fluorescently labeled cytokinetic proteins allows for further examination into their potential role in cytokinesis. With fluorescent tags, protein localization throughout cytokinesis can be assessed as shown in Fig. 2 [14]. Protein-protein interactions of cytokinetic players can also be examined by expressing fluorescent proteins in different background strains (e.g., GFP-cortexillin I in a *myosin II*-null strain). In addition, behavioral changes based on protein expression levels can also be assessed (e.g., overexpression, rescue cell lines).

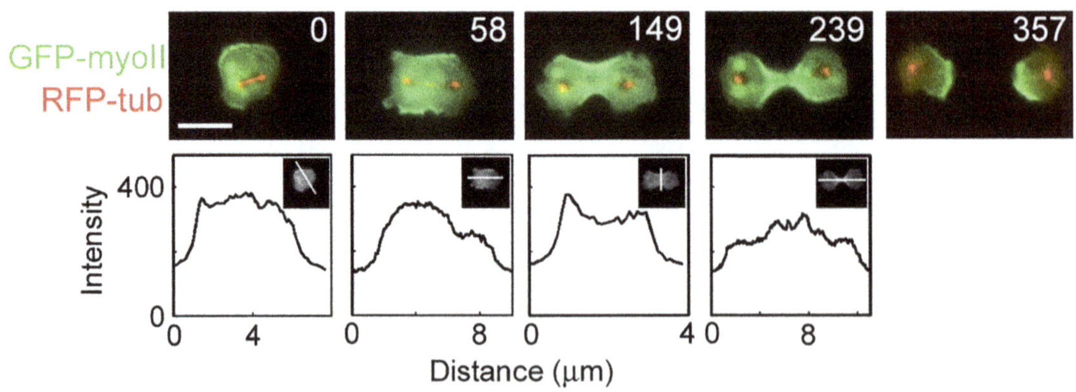

Fig. 2 Fluorescently labelled proteins can be used to assess protein localization during cytokinesis. Proteins with fluorescent tags such as GFP-myosin II and RFP-tubulin (a mitotic spindle marker) can be visualized during cell division over time. GFP-myosin II concentrates in the cleavage furrow and bridge as the cell divides as indicated by the time lapse images and corresponding line scans. (Reproduced from ref. 14 with permission from Elsevier)

D. discoideum readily take up plasmid DNA [15]. Transformants are isolated via drug selection and expression of recombinant protein is confirmed via fluorescence microscopy or Western analysis. The generation of these cell lines is detailed below.

3.1.5 Cloning and Transformation of Recombinant Protein

1. Amplify gene of interest via PCR. The amplified gene of interest should contain restriction enzyme sites for insertion into expression vectors.
2. Insert gene of interest into expression vectors containing sequences for fluorescent proteins (e.g., GFP, mCherry, etc.). We use a variety of expression vectors including pLD1A15SN, pDM181, pDRH, and pBIG. A list of expression vectors can be found on the *D. discoideum* database, Dictybase.
3. Transform cells by electroporation (*see* Subheading 3.1.3, **steps 1–8**, for more details on transformation), and select for transformants with supplemented Hans' Enriched HL-5 medium containing G418 (10–15 μg/mL) or hygromycin B (35–40 μg/mL) (*see* **Note 8**).
4. Change out drug media every 2–3 days. Transformants will appear as colonies on the petri dish. The amount of time needed for colonies to appear can vary depending on which genes are being expressed (*see* **Note 9**).

3.1.6 Generation of D. discoideum *Cytokinetic Knockdown/Knockout Cell Lines*

In addition to generating cell lines with fluorescently labeled proteins, knocking down or knocking out potential cytokinetic proteins creates another avenue for examining their role in cytokinesis. Our studies and others [4–11] detail gene knockout and knockdown in *D. discoideum* with various methods, including homologous recombination and RNAi. As regards to RNAi technology, both short hairpin (25 nt) and long RNA hairpin (300–500 nt) have been used in *D. discoideum* previously. More recent findings utilize CRISPR technology in *D. discoideum* to knockout genes using a pTM1285 vector, mapped in Fig. 3 [16–17]. We detail various methods of gene knockdown/knockout utilized in our studies below.

3.1.7 Homologous Recombination

In *D. discoideum*, gene knockout can occur through homologous recombination by simply transforming a plasmid containing a drug selection marker with flanking regions from the gene of interest, ideally from the 5′ and 3′ UTRs [3]. *See* Subheading 3.1.3 for general transformation and isolation of mutant cells.

RNA Long Hairpin

1. Determine a target region for knockdown. Both coding sequence DNA and 3′ UTR can be targeted. While it is tempting to target the untranslated regions (UTRs) to allow for rescue experiments, *Dictyostelium* UTRs have very high AT

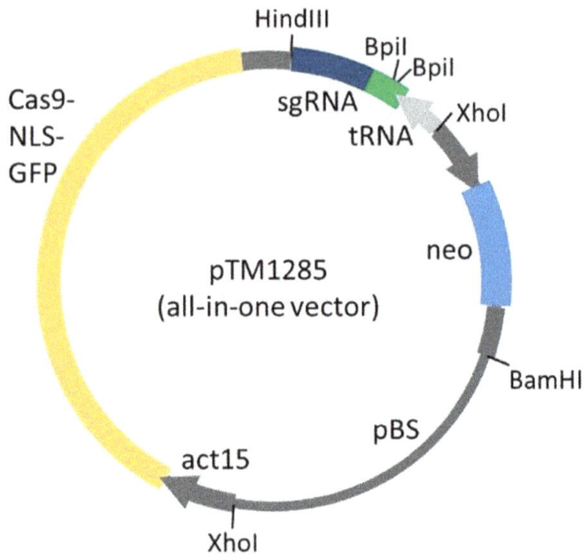

Fig. 3 Map of the pTM1285 plasmid. (Reproduced from ref. 17)

content, which leads to a very low complexity. To increase specificity, we have had to focus on coding sequences to identify regions where extensive DNA sequence searching reveals no, or very little, homology across >15 bp to any other region in the genome.

2. Once the target region is determined, use PCR to amplify both the anti-sense and sense strand DNA templates for cloning into an expression vector. We routinely use pLD1A15SN, a G418-resistant expression vector that is readily titratable, to achieve different levels of knockdown.
3. Determine the spacer sequence. For long hairpin structure, the spacer sequence is generally between 100 and 300 nucleotides. We routinely use the sequence immediately upstream of targeting coding sequence as the spacer to simplify cloning procedure. However, because this stretch is included in the antisense portion, we also apply the same standard that it should not include any stretch >15 bp with homology to any other region of the genome. Only two PCR runs are needed to amplify both antisense and sense strands (Fig. 4).
4. Introduce the hairpin RNA plasmid into cells via transformation. For transformation by electroporation, *see* Subheading 3.1.3, **steps 1–8**, for more details. For pLD1A15SN vectors expressing hairpin RNA, drug select for transformed cells with G418 (*see* **Note 10**). Empty vector-containing cells also need to be generated to serve as the control comparator.
5. To determine the level of knockdown, extract RNA using commercial extraction reagents (such as TRIzol®) from cells

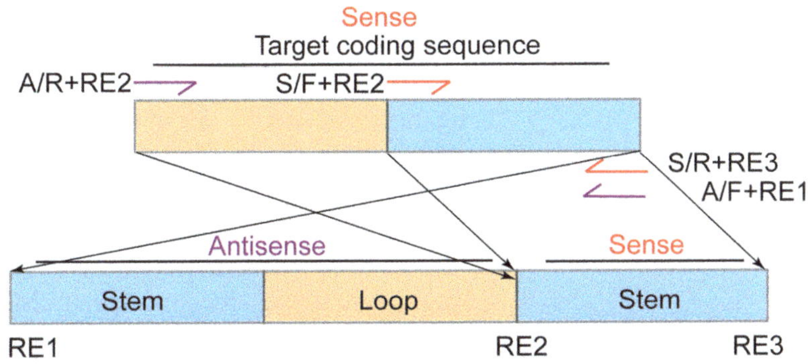

Fig. 4 Hairpin RNA can be generated to knock down genes in *D. discoideum*. Hairpin RNA for *D. discoideum* can be created by amplifying the target coding sequence containing the gene of interest (blue regions) and an upstream spacer region (beige regions). Restriction enzyme recognition sites (RE2, RE3) are added to sense strand primers (sense forward (S/F) and sense reverse (S/R)) to amplify the sequence encoding the sense fraction of the RNA hairpin. Likewise, restriction enzyme recognition sites (RE1, RE2) are added to the antisense strand primers (antisense forward (A/F) and antisense reverse (A/R)) to amplify the antisense fraction of the RNA hairpin plus spacer. The stem is generated by combining the antisense and sense strands with the antisense spacer loop separating the regions containing the gene of interest

expressing the RNA hairpin vector and cells expressing the empty vector. Perform qRT-PCR with the extracted RNA to quantify the relative amount of RNA for the gene of interest in each cell line. Primers for qRT-PCR should be targeting a sequence of 100–200 base pairs outside the sequence included in the RNA hairpin. A housekeeping gene such as GAPDH could be used as a normalization factor during knockdown level quantification.

CRISPR

1. To begin the process of inserting the gRNA into the pTM1285 vector (Fig. 3), digest 1 μg of pTM1285 plasmid with 1 μL FastDigest *Bbs* 1, 1 μL FastAP, 2 μL 1× FastDigest Buffer, and water to a final reaction volume to 20 μL total. Incubate the reaction for 30 min at 37 °C. After the digestion is completed, load the entire reaction volume on a 1% DNA gel. Gel-purify the plasmid using a QIAquick Gel Extraction Kit and elute it in EB buffer.
2. Phosphorylate 1 μL of complementary gRNA oligos by 0.5 μL T4 PNK in a 10 μL reaction volume containing 1 μL 10× T4 ligation buffer and water. Anneal the pair of oligos in a thermocycler for 30 min at 37 °C, followed by 5 min at 95 °C, and then ramped down to 25 °C at 5 °C/min.

3. Ligate 50 ng of the vector and 150 ng of the annealed gRNA oligos with 1 μL Quick Ligase in an 11 μL reaction volume containing 2× Quick Ligation Buffer and water. Incubate the ligation reaction for 30 min at room temperature.
4. Transform 2 μL of the ligation reaction from **step 3** into 50 μL of STBL2 competent cells. Heat shock the competent cells for 28 sec at 42 °C and plate on ampicillin-containing LB plates (*see* **Note 11**). Incubate plates at 30 °C for 16 h and isolate individual clones for mini-prep. Mini-prepped recombinant pTM1285 plasmids should be confirmed using pTM1285 forward and reverse primers. The correct pTM1285 with the corresponding gRNA can be maxi-prepped for future use.
5. With the correct pTM1285 plasmid, transform 1–10 μg of plasmid into *Dictyostelium* cells by electroporation and drug select with G418. *See* Subheading 3.1.3, **steps 1–8**, for more details on transformation.
6. After drug selection, plate any remaining live cells on SM-5 plates with cultured *K. aerogenes*, and incubate at room temperature for 3–4 days until visible, isolated plaques form (*see* **Notes 12–13).** Transfer individual plaques separately into 48-well plates containing Hans' Enriched HL-5 media without G418. Check for colony growth after approximately 2–3 days and transfer them to 10 cm Petri dishes. Since the cells should be expressing only Cas9 and gRNA transiently, it is important to test a subpopulation of these cells for G418 sensitivity to ensure the absence of pTM1285 plasmid. Retention of the plasmid can potentially induce off-target gene modifications. The tentative KO cell lines can be identified using gene sequencing and Western analysis.

3.2 Assessing Cytokinesis by Imaging Cells

Time-resolved microscopic imaging of live *D. discoideum* is a powerful way to study cytokinesis and identify if and how it is affected in mutant cell lines. The morphological changes and dynamics during cytokinesis reflect multiple aspects of the cell, including mechanics, actomyosin contraction, adhesion, polar protrusive forces, and abscission machineries. Cells can simply be imaged during cell division, allowing for the observation of furrow ingression and bridge formation (Fig. 5). Cytokinesis dynamics and alterations to furrow ingression can be observed, quantified, and even modeled [18–19]. With fluorescently tagged proteins, fluorescence microscopy can be useful for assessing protein localization during cell division. If proteins are difficult to visualize live and antibodies for a protein of interest are available, *D. discoideum* can be fixed for immunofluorescence (IF) imaging. Fluorescence imaging is also particularly helpful for identifying multinucleated mutants, a sign of a cytokinetic defect, using nuclear stains such as DAPI and

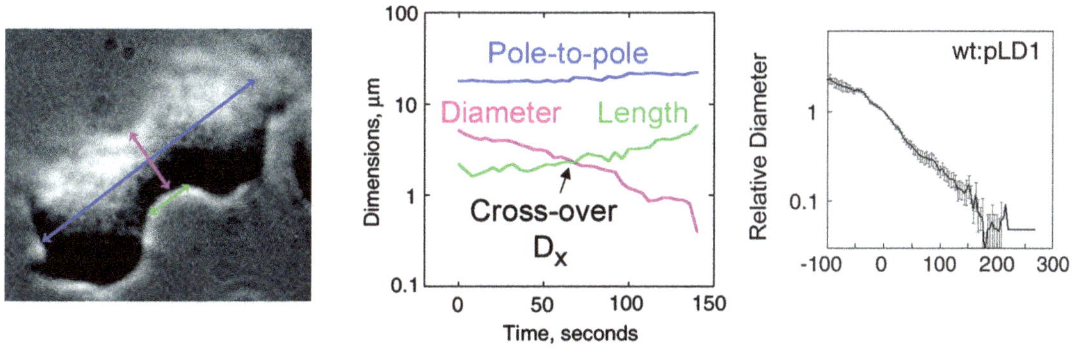

Fig. 5 *D. discoideum* cytokinesis furrow measurements can be observed and quantified. During *D. discoideum* cell division, the diameter of the division furrow, which later becomes the intercellular bridge, is measured (indicated by purple line with arrows over cell). The furrow length is also measured (indicated by green line with arrows over cell). Both values are plotted over time to find the point where diameter equals length (crossover). In the end, at each time point, relative diameter is calculated by dividing measured furrow diameter by crossover value and plotted over time. Different strains and different genetic makeup could result in apparent variations in relative diameter dynamics over time. (Reproduced from reference 18 with permission from the National Academy of Sciences. Copyright (2005) National Academy of Sciences, USA)

Hoechst (detailed in the Multinucleation section in Subheading 3.1.3). Below we detail how we use imaging to assess cytokinesis in *D. discoideum*.

3.2.1 Imaging Cell Division and Furrow Measurements

1. Seed cells on a glass bottom imaging chamber in Hans' Enriched HL-5 media and allowed to adhere for 15–30 min.
2. Put the imaging chamber on the microscope stage. Begin imaging cells using the DIC channel with the 40× objective. Scan the field of view for dividing cells. There is not an efficient cell synchronization method for imaging many dividing cells at once, but it is not necessarily required to synchronize cells beforehand. The doubling time of *D. discoideum* is 8–12 h when grown in culture media under drug selection, whereas mitosis through cytokines usually takes around 15 min. That means 2–3% of the cells are dividing at a given time during exponential growth phase. During exponential growth phase when observed under 40× objective, usually 80–120 cells can be seen within the same field of view. Thus, it is not difficult to spot one dividing cell after scanning two different fields of view, and when starting with 15 min time lapse video acquisition, it is very common to find a second and third divider within a single movie acquisition.
3. After spotting a cell undergoing mitosis, capture time lapse videos until the cell finishes dividing. A time lapse video of 15 min with 2 s intervals is recommended when imaging cell division using differential interference contrast (DIC)

microscopy. Mitosis usually starts with rounding up of the cell, followed by elongation and furrow ingression, and ends with intercellular bridge thinning and abscission. It is easier to spot cells during furrow ingression stage since there is a significant morphological difference, although rounded-up cells are also relatively easy to find.

4. Analyze the time lapse videos with imaging software such as ImageJ. Cell lines exhibiting cytokinesis failure may display furrow regression and failure in abscission (*see* **Note 14**). A variety of parameters can be considered to assess cytokinesis with these videos. Some to consider include success/failure of division, symmetry of daughter cells, total time required to divide, distance between poles, and furrow ingression and intercellular bridge-thinning dynamics. The ratio of success/failure of division can be used as a phenotype for a mutant cell line and may be related to the function of the modified gene. The symmetry of the daughter cells reflects the cell's ability to correctly position contractile machinery, which is especially challenging in multicellular cells. Total time required to finish cell division and the distance between two poles are additional parameters to consider in order to spot differences in elasticity and polar protrusive forces.
5. Finally, the kinetics of furrow ingression and intercellular bridge thinning dynamics reflect a cell's cortical viscoelasticity, active contractile forces, adhesion, protrusive forces, passive fluid mechanics, strain stiffening of the network, and the resistive stresses generated through the cortex and cytoplasm from the emerging daughter cells. To assess the furrow ingression dynamics and intercellular bridge thinning, quantify the diameter and length of furrow and intercellular bridge over time [18]. Find the crossover distance when the diameter equals length, and then normalize the bridge diameter to cross over distance to get relative furrow diameter.

3.2.2 Fixed Cell Imaging

1. Grow cells on coverslips in Hans' Enriched HL-5 media and allow to adhere for at least 1 h.
2. Gently aspirate off media and fix cells in ice-cold 4% paraformaldehyde for 10 min, and then permeabilize with −10 °C acetone for 3 min.
3. Wash cells with 1× PBS three times.
4. Block cells in 1× PBS + 0.05% Triton X-100 + 0.5% BSA for 45 min at room temperature. Follow with three washes in 1× PBST (1× PBS + 0.05% Triton X-100).
5. Incubate with primary antibody in PBST overnight at 4 °C. Wash cells with PBST five times.

6. Incubate with secondary antibody in PBST for 60 min at room temperature, away from light. Wash cells with PBST five times.
7. Mount coverslips onto glass slides. Allow sections to cure for at least 24 h at room temperature in the dark.
8. Proceed to image slides on a fluorescence microscope. We routinely image on confocal microscopes on 40× or 63× oil-immersion objectives with excitation lasers at 488 for GFP, 561 for mCherry, and 633 for far-red fluorescent proteins.

3.3 Fluorescence Correlation Spectroscopy (FCS) and Fluorescence Cross-Correlation Spectroscopy (FCCS)

Fluorescence correlation spectroscopy (FCS) and fluorescence cross-correlation spectroscopy (FCCS) are common methods used to measure the concentration and diffusion coefficient of one molecular species or the interaction between two molecular species [20, 21]. Both methods involve capturing sensitive fluorescence fluctuations within a small confocal volume, which can be used to extract diffusion and mobility parameters (Fig. 6). Whereas FCS utilizes autocorrelation analysis, FCCS utilizes cross correlation between two fluorescently tagged species. In *D. discoideum*, these methods have been used to quantify the mobility and concentration of proteins and to reveal protein-protein interactions within the cytoskeleton and the mechanosensitive network [11]. Below are detailed steps for FCS and FCCS experiments geared toward live *D. discoideum* cells, but the same methods can also be applied to molecules in vitro. More information on FCS/FCCS are also detailed in 20 and 21.

3.3.1 Cell Line Preparation

1. For in vivo FCS and FCCS, introduce fluorescently tagged molecules of interests into cells. One of the major concerns about FCS and FCCS is the brightness of the fluorescent

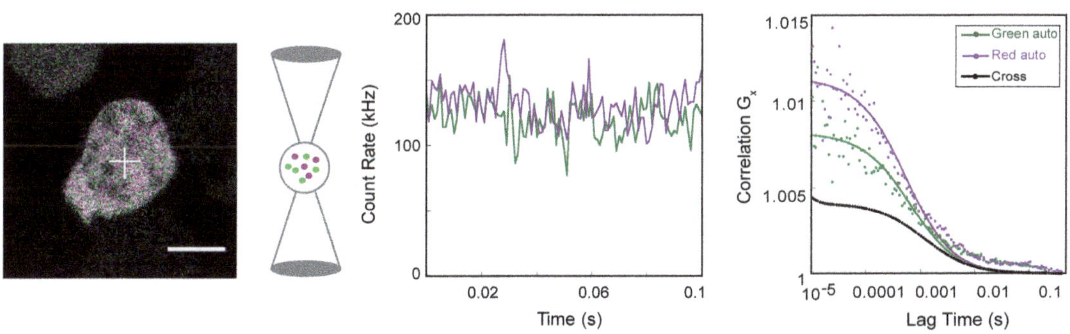

Fig. 6 Fluorescence cross-correlation spectroscopy (FCCS) can be used to probe the interaction between two molecular species. Two proteins are labeled with two different colored fluorophores and exogenously expressed in *D. discoideum* cells. Within a small confocal volume, fluctuations of both fluorophores are recorded and reported as count rates in kHz. Using auto-correlation and cross-correlation calculations, diffusion coefficients of both proteins and in vivo K_D between the two proteins can be calculated. (Reproduced from reference 11 with permission from the Company of Biologists)

probes since the effective confocal volume is relatively small. Probes should be bright enough to indicate fluctuations, but the concentration cannot be too high as it will saturate the signal, making it difficult to assess the fluctuations in signal intensity. Another factor to consider when choosing fluorescent probes is the triplet state since it directly affects the correlation calculation. For FCCS, having spectral separation between the two probes is also an important factor to be considered. Overlapping excitation and emission spectra should be minimized to avoid noise in detecting signals.

3.3.2 Microscope Setup and Calibration of System

1. Prepare a solution of rhodamine 6G with a concentration between 50 nM and 200 nM. Load into a glass imaging chamber.
2. In imaging software such as Zen, select appropriate dichroic and emission filters for the dye pair to be used. A laser power that is too low may result in count rates that are undetectable, whereas a very high laser power may cause photo damage to the cells and oversaturate the signals. Ideally, laser powers for both channels should be kept consistent.
3. Find the image plane by using line scan. Move the objective up from the load position to identify both edges of the coverslip. Stop when two spikes of signals appear. These correspond to the two edges of the coverslip. Move the focus up 200 μm into the solution.
4. To compensate for suboptimal refractive indexes of the specimen, adjust the correction collar of the objective until the maximal count rate is achieved.
5. Adjust the pinhole. We usually set the pinhole at 1 Airy unit, although a smaller size can be considered. Adjust the coarse lateral pinhole position first and then the axial position. Repeat the sequence when adjusting the fine pinhole position.
6. Acquire measurements for rhodamine 6G every 5 s for ten repetitions.
7. Select the correct model to fit the data. The equation we use to fit fluorescence fluctuations over time to obtain an autocorrelation value (G_x) for FCS is given below:

$$G_x = \frac{\langle \delta F(t) \cdot \delta F(t+\tau) \rangle}{\langle F(t) \rangle^2}$$

where t is time and $F(t)$ is the fluorescence fluctuation detected.

Fluorescence fluctuations relay information on the mobility of a species, but they may also occur to a change in intrinsic fluorescence. To correct for the change in intrinsic fluorescence, a triplet-state component must be added to the model. Set the model by putting 1 for the triplet state, the triplet relaxation time as 8 μs, and the structural parameter as 6 (a well-aligned system will have a structural parameter between 5 and 7). The diffusion time of rhodamine 6G should be between 20 and 30 μs in water. If the diffusion time largely deviates from the expected value, readjust the focal plane, the correction collar, and pinhole position to achieve fine measurements. Note that the number of components and triplet-state relaxation time might vary between different molecules of interest and different fluorescent probes used in the experiment.

The setups for FCS and FCCS are similar, but an additional laser will be used for FCCS. The second laser will excite the second fluorophore and create a second confocal volume at the same time as the first laser. Fluorescence fluctuations from both channels are recorded. Thus, three plots will be generated. Each channel will have its own autocorrelation plot and there will also be a cross-correlation plot. The model for FCCS is as follows and relays information on the cross-correlation function, G_x, a measure of how well the fluorescence fluctuations of two channels correlate over time:

$$G_x = \frac{\langle \delta F_A(t) \cdot \delta F_B(t+\tau) \rangle}{\langle F_A(t) \rangle \langle F_B(t) \rangle}$$

where t is time and $F(t)$ is the fluorescence fluctuation in either the green or red channel.

3.3.3 Obtaining Traces for Individual Cells

1. Seed cells into an imaging chamber on a coverslip at least 30 min before imaging. The temperature should be at 22–25 °C to maintain cell viability.
2. Right before imaging, change the media to water or low fluorescence media. *D. discoideum* wild-type cells are relatively motile and photosensitive; therefore acquisition should be made within a short time frame. Select the desired light path and set the acquisitions to take images every 2 s with ten repetitions.
3. Make acquisitions from the background. Select cells and regions of cells with expression levels that are significantly stronger than background but are not oversaturated. Count rates below 50 or over 1000 should be avoided.
4. Acquire images of cells before and after taking FCS/FCCS measurements. The images are taken afterward to ensure cells do not move out of the imaging volume during acquisition. For each experiment, we recommend collecting at least 20 measurements to be able to draw conclusions from the acquired data.

3.3.4 Data Analysis

1. Open measure files in analysis software such as Zen Black [22]. Manually inspect each trace. Avoid traces with noticeable downward trends, which indicate photobleaching, and large broad peaks, which indicate aggregates. Significant deviation in count rate indicating active movements should also be avoided.
2. Take the average of the remaining trace measurements, and in the analysis tab, fit the averages to the model established from the system calibration.
3. Evaluate the fitted data and compare to acquired traces. If there is a large deviation between the correlation curves and fits in general, the fitting parameters might be inappropriate. Discard fits with significant deviations.
4. From the fitted data, parameters such as the diffusion times can be derived. The diffusion time is extracted by the negative slope of the fitted data.

3.4 Fluorescence Recovery After Photobleaching (FRAP)

With cell lines expressing fluorescently tagged proteins, fluorescence recovery after photobleaching (FRAP) is a quantitative method that can be used to uncover protein turnover dynamics. As diffusion is generally much faster, these dynamics are frequently dominated by the off-rates of a protein from the structures with which it is associated. The method entails a fluorescent region of interest (ROI) in a cell being photobleached with a high intensity laser. The fluorescence in the ROI recovers over time due to the movement of fluorescent molecules into the photobleached area. The time it takes for the fluorescence to recover is measured against the fluorescence intensity to generate a curve (Fig. 7). From the data, the mobile and immobile fractions of the protein can be identified, and kinetic parameters can be calculated such as diffusion times and time to dissociate from binding interactions [23].

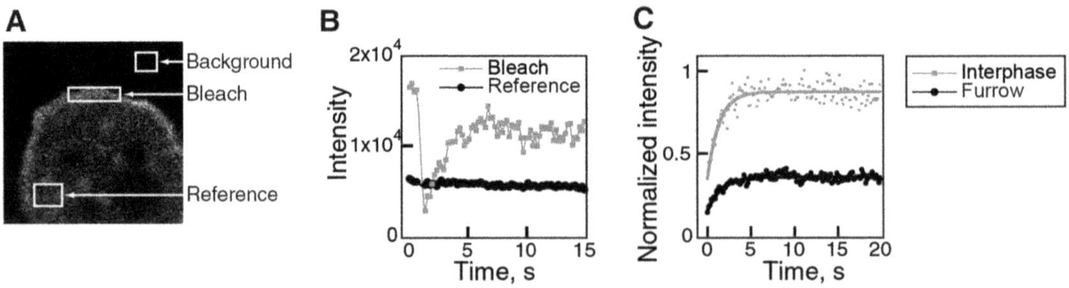

Fig. 7 Fluorescence recovery after photobleaching (FRAP) measures protein turnover dynamics. (**a**) Example image of a *D. discoideum* cell-expressing GFP-cortexillin I. The average intensities for the labeled regions are collected for FRAP calculations. (**b**) An example graph of fluorescence intensity over time for the bleached and reference regions. (**c**) After background subtraction, the intensity is fit to an exponential curve. The reference region stays consistent while the bleached region regains fluorescence intensity over time. (Reproduced from reference 25 with permission from Elsevier)

Applying this method to cytokinesis, FRAP can be used to characterize and quantify protein mobility during cell division. For example, a study utilizing FRAP in *D. discoideum* demonstrated that myosin II, a protein enriched at the cleavage furrow, is largely recruited from a cytoplasmic pool while some myosin II also flows along the cortex [24]. Interactions between different cytokinesis proteins can also be analyzed using FRAP. We measured the recovery times for various cytoskeletal proteins through FRAP including actin and proteins at the cleavage furrow such as myosin II, cortexillin I, and IQGAPs 1 and 2 [25]. These cleavage furrow proteins have slower recovery times in the cortex than in the cytoplasm, indicating that their recovery in the cortex is highly dependent on unbinding events rather than diffusion. We continued to build upon the idea of the cleavage furrow proteins participating in a meshwork by expressing these proteins in and using FRAP on mutant cell lines lacking myosin II, cortexillin I, and the IQGAPs.

Below, we detail the experimental setup for FRAP in *D. discoideum* and how to analyze the data output.

3.4.1 Experimental Setup

1. Grow cells expressing a fluorescently tagged protein in a glass bottom imaging chamber for 1 h.
2. Replace culture medium with MES starvation media before imaging.
3. Conduct experiment using a confocal microscope with a 63× objective. Take images of cells before photobleaching for later calculations.
4. Pick a small region of the cell and photobleach with a laser of the excitation wavelength of the fluorescent protein.
5. Measure time of recovery until the region is saturated (150 frames, 45–150 ms/frame depending on protein).

3.4.2 Data Analysis

1. For each frame, measure the average intensity of the bleached region, a reference (unbleached) region, and the background with analysis software such as ImageJ. Subtract the average background intensity from the average intensities of the bleached region and the reference region.
2. To correct for photobleaching, calculate the reference theoretical intensity (RTI) by fitting the background subtracted reference intensity with the following equation:

$$\mathrm{RTI}(t) = A - B \cdot \mathrm{e}^{-C \cdot t}$$

   where A, B, and C are fitting parameters.
3. Normalize the background subtracted bleached region intensity to RTI.

4. Calculate the normalized intensity (NI) by normalizing values in **step 2** to the pre-bleach intensity (the average of 4 pre-bleach images) to the following equation:

$$\mathrm{NI}(t) = m_1\left(1 - m_2 \cdot \mathrm{e}^{-kt}\right)$$

where m_1, m_2 are fitting parameters and k is the recovery rate.

5. The recovery time (τ) and the immobile fraction, F_i, can be measured with the following equations:

$$\tau = \frac{1}{k}$$

$$F_i = \frac{1 - m_1}{1 - m_1 + m_2}$$

3.5 Micropipette Aspiration (MPA)

While cytokinesis is a process that involves the biochemical interactions between various molecules, it also depends on the generation of force and the reaction of mechanosensitive components in the cell to this force [26–27]. Methods such as atomic force microscopy (AFM) [28], optical traps [29], and cell compression [30] can be used to study how cells respond to force, but they come with several caveats. AFM requires the cells to be attached to a surface, which can be problematic as *Dictyostelium* cells tend to be fairly weakly adhered to substrates. Optical traps lead to localized increases in heat from the laser light. Generating sufficient forces with an optical trap to deform a *Dictyostelium* cell can lead to lethality due to the heat. Cell compression is a fast, high-throughput method for applying mechanical stresses to cells, which can be very useful for many types of studies [31]. Micropipette aspiration (MPA) remains one of our preferred techniques as it allows for a well-controlled, specific amount of force in the nanoNewton (nN) range to be applied to a well-defined surface area (up to tens of μm^2) of the cell. The forces applied through MPA mimics the forces generated by a cell during cytokinesis, making it useful for observing how cytokinetic proteins sense and respond to mechanical perturbations. Proteins that do respond, which are mechanoresponsive, can change protein localization when a force is applied to the cell. In a previous study, we identified such mechanoresponsive cytokinetic proteins including myosin II and cortexillin I [32]. In addition, we used MPA as a tool for uncovering and expanding on the pathways leading to mechanosensory responses. Mitotic regulator proteins such as kif12 (*Dictyostelium* kinesin-6) and INCENP are also mechanoresponsive and are dependent on myosin II because they are non-mechanoresponsive in knockout cell lines lacking myosin II [33].

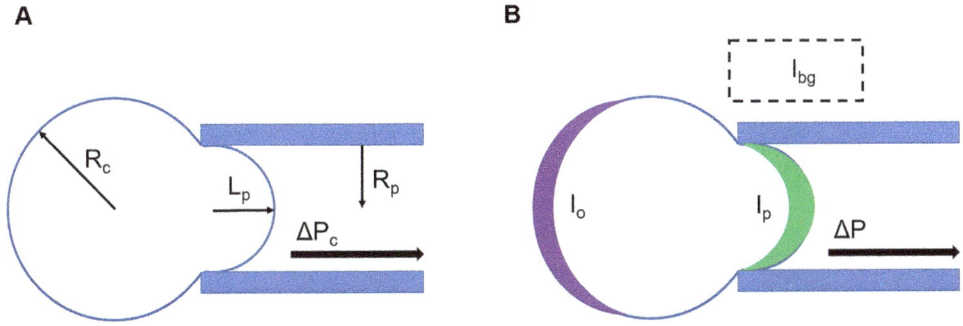

Fig. 8 Micropipette aspiration (MPA) is a tool for measuring cortical tension and mechanosensitivity. (**a**) A pipette aspirates on a cell with radius, R_c, until the critical pressure (ΔP_c) is reached, which is the point at which the length of the cell deformation into the pipette, L_p, equals R_p, the radius of the pipette. From this pressure, the cortical tension may be calculated. (**b**) To measure the mechanoresponse of a fluorescent protein, the mean intensities of the region inside the pipette (I_p), the cortex from the opposite side of the cell (I_o), and a background region outside of the cell (I_{bg}) are collected. The cell is aspirated on by a pipette with a fixed pressure (ΔP) set prior to contacting the cell

MPA can also be used to measure mechanical parameters in the cell, such as cortical tension. As shown in Fig. 8a, as the cell, with radius R_c, is aspirated upon by a pipette with radius, R_p, it will form a hemispherical deformation inside the pipette at the equilibrium pressure (ΔP_c). At the equilibrium pressure, the radius of the hemisphere, L_p, equals to R_p. With these parameters, the cortical tension can be measured with the following equation:

$$\Delta P_c = 2T_c\left(\frac{1}{R_p} - \frac{1}{R_c}\right)$$

3.5.1 Preparing the MPA System

The setup of an MPA system and data analysis have been previously detailed in refs. 34 and 35, but briefly, we proceed with the following steps for MPA.

3.5.2 Loading and Locating the Micropipette

1. Before loading micropipette tip, loosen micropipette holder and push out drops of water, leaving a drop hanging out of the opening of the micropipette tip. This will ensure that air bubbles will not be introduced into the micropipette holder when loading the tip.
2. Fill the micropipette with MES starvation buffer or growth medium. Leave an excess droplet at the opening of the micropipette.
3. Load the micropipette midway into the micropipette holder. Insert the micropipette into the holder droplet to avoid introducing an air bubble. Tighten the micropipette holder.
4. Place imaging chamber with cells onto the microscope stage. Select the 40× magnification and set up the DIC channel.

Move the micropipette into the light path and lower the tip until it is immersed. Adjust valve so that water manometer is connected to the micropipette.

5. Using the micromanipulator, move the micropipette close to the surface where cells are and position the tip into the center of the imaging field.

Calibrating the MPA System

1. To calibrate the system, locate the micropipette and connect it to the manometer.
2. Position the two water tanks so that they are the same level (home position).
3. Add ~20–50 μL of diluted bead solution (1 μm polystyrene beads diluted in 1:500 filtered-sterilized water) to the imaging chamber.
4. Move micropipette next to a free-floating microbead. The system is balanced when the bead does not move toward or away from the micropipette opening.
5. If the bead does move, adjust the manometer by raising it if the bead is going toward the pipette opening. Lower the manometer if the bead moves away from the pipette.

3.5.3 Cortical Tension Measurements

1. After loading the micropipette and calibrating the system, set up time acquisition using the DIC channel with 5 s intervals.
2. On a healthy suspended cell, apply pressure to the cell, starting with low pressures (around 10–15 mm of water pressure).
3. Increase the pressure by 5 mm every five frames until cell is deformed into pipette until $L_p = R_p$. Finer step sizes are required to home in on the final pressure at which point $L_p = R_p$.
4. Collect at least three measurements per cell. Wait for 5 min between each measurement for the cell to regain stability or aspirate on a different region of the cell.

3.5.4 Protein Accumulation Quantification

1. Put cells in MES starvation buffer on imaging slide and place sample on microscope imaging stage.
2. Calibrate system.
3. Set up image acquisition with DIC channel and channels for fluorescent proteins. Set up time intervals for 15–30 s.
4. Increase pressure in micropipette prior to contacting the cell. Keep pressure fixed and aspirate the cell.
5. Using an image analysis software such as ImageJ (https://imagej.nih.gov/ij), collect the mean gray values from the cell region inside the pipette (I_p), the cortex from the opposite side

of the cell (I_o), and a background region outside of the cell (I_{bg}) (Fig. 8b). Calculate the mechanosensitive protein response with the following equation:

$$\text{Normalized } I = (I_p - I_{bg})/(I_o - I_{bg})$$

4 Notes

1. The water used in the materials and methods mentioned is always deionized and double distilled.
2. The voltage settings can vary based on strain of *D. discoideum*. The electroporation settings (3.0 μF and 1.3 kV) are optimized for our lab's ORF+7–3 (Ax3:Rep orf+ (HS1000)) cells (from reference 9). We use 1.1 kV for 11–5.1 cells [9].
3. For a larger-scale screen, more pools of 30,000–35,000 clones can be analyzed. We have analyzed up to 240,000–280,000 independent clones [9].
4. STBL2 *Escherichia coli* are generally used for cloning because *D. discoideum* DNA sequences are AT-rich, making them unstable in many other *E. coli* strains.
5. For cells in drug media or mutants that may have growth defects, it is important to start suspension culture at a higher initial cell density (at least ~1×10^5 cells/mL).
6. Initial cell densities from the second or third cycle can be lower than those of the first cycle, roughly $0.5–1 \times 10^5$ cells/mL, as cell growth becomes more consistent.
7. After being mounted, coverslips can be sealed with nail polish to avoid dehydration and stored at −20 °C for later visualization.
8. Transforming with multiple plasmids is possible if each individual expression vector confers resistance to different selection markers.
9. The amount of time for colonies to appear can vary based on which genes are being transformed. Colonies usually form after 5–7 days but some can appear after 4 weeks.
10. G418 concentrations for drug selecting RNAi plasmids typically fall between 7.5 μg/mL and 15 μg/mL. For genes that are critical to cell survival and growth, a high concentration of G418 selection for 1 or 2 days, followed by a sudden drop and a slow ramp-up of drug concentration, is sometimes helpful for transformed cells to adjust and grow into colonies. The time frame and drug concentration of selection regime have to be empirically determined.

11. Ampicillin-containing LB plates should not be in 30 °C for more than 20 h to avoid the breakdown of ampicillin and growth of non-transformants.
12. SM-5 plates should be stored at 4 °C and used within 30 days to avoid dehydration.
13. During drug selection, dead floating cells can be gently washed with Hans' Enriched HL-5 media containing 15 μg of G418 once every 2 days. It is important to eliminate as many floating cells as possible before plating transformants on SM-5 plates to avoid growth of non-transformants.
14. Additional markers including tubulins or nuclear proteins should be considered in order to differentiate from pseudopod protrusion and retraction.

Acknowledgments

The authors would like to thank members of the Robinson lab for helpful discussion and comments on the paper. This work is supported by the National Institutes of Health (F31HL145910 to JMN, GM66817 to DNR, and HL126888 to VS and DNR) and DARPA (HR0011-16-C-0139 to DNR).

References

1. Glotzer M (2005) The molecular requirements for cytokinesis. Science 307:1735–1739
2. Balasubramanian MK, Bi E, Glotzer M (2004) Comparative analysis of cytokinesis in budding yeast, fission yeast and animal cells. Curr Biol 14:R806–R818
3. Faix J, Steinmetz M, Boves H, Kammerer RA, Lottspeich F, Mintert U, Murphy J, Stock A, Aebi U, Gerisch G (1996) Cortexillins, major determinants of cell shape and size, are actin-bundling proteins with a parallel coiled-coil tail. Cell 86:631–642
4. De Lozanne A, Spudich JA (1987) Disruption of the *Dictyostelium* myosin heavy chain gene by homologous recombination. Science 236: 1086–1091
5. Larochelle DA, Vithalani KK, De Lozanne A (1996) A novel member of the rho family of small GTP-binding proteins is specifically required for cytokinesis. J Cell Biol 133: 1321–1329
6. Larochelle DA, Vithalani KK, De Lozanne A (1997) Role of *Dictyostelium* racE in cytokinesis: mutational analysis and localization studies by use of green fluorescent protein. Mol Biol Cell 8:935–944
7. Manstein DJ, Titus MA, De Lozanne A, Spudich JA (1989) Gene replacement in *Dictyostelium*: generation of myosin null mutants. EMBO J 8:923–932
8. de Hostos EL, Rehfuess C, Bradtke B, Waddell DR, Albrecht R, Murphy J, Gerisch G (1993) *Dictyostelium* mutants lacking the cytoskeletal protein coronin are defective in cytokinesis and cell motility. J Cell Biol 120:163–173
9. Robinson DN, Spudich JA (2000) Dynacortin, a genetic link between equatorial contractility and global shape control discovered by library complementation of a *Dictyostelium discoideum* cytokinesis mutant. J Cell Biol 150: 823–838
10. Octtaviani E, Effler JC, Robinson DN (2006) Enlazin, a natural fusion of two classes of canonical cytoskeletal proteins, contributes to cytokinesis dynamics. Mol Biol Cell 17:12
11. Kothari P, Srivastava V, Aggarwal V, Tchernyshyov I, Van Eyk JE, Ha T, Robinson DN (2019) Contractility kits promote assembly of the mechanoresponsive cytoskeletal network. J Cell Sci 132:jcs226704

12. Park J, Bae S, Kim JS (2015) Cas-designer: a web-based tool for choice of CRISPR-Cas9 target sites. Bioinformatics 31:4014–4016
13. Ngo T, Miao X, Robinson DN, Zhou Q (2016) An RNA-binding protein, RNP-1, protects microtubules from nocodazole and localizes to the leading edge during cytokinesis and cell migration in *Dictyostelium* cells. Acta Pharmacol Sin 37:1449–1457
14. Effler JC, Kee YS, Berk JM, Tran MN, Iglesias PA, Robinson DN (2006) Mitosis-specific mechanosensing and contractile-protein redistribution control cell shape. Curr Biol 16:1962–1967
15. Bozzaro S (2013) The model organism *Dictyostelium discoideum*. In: Eichinger L, Rivero F (eds) *Dictyostelium discoideum* protocols, Methods in molecular biology (methods and protocols), vol 983. Humana Press, Totowa, pp 17–37
16. Sekine R, Kawata T, Muramoto T (2018) CRISPR/Cas9 mediated targeting of multiple genes in *Dictyostelium*. Sci Rep 8:8471
17. Muramoto T, Iriki H, Watanabe J, Kawata T (2019) Recent advances in CRISPR/Cas9-mediated genome editing in *Dictyostelium*. Cells 8:46
18. Zhang W, Robinson DN (2005) Balance of actively generated contractile and resistive forces controls cytokinesis dynamics. Proc Natl Acad Sci 102:7186–7191
19. Poirier CC, Ng WP, Robinson DN, Iglesias PA (2012) Deconvolution of the cellular force-generating subsystems that govern cytokinesis furrow ingression. PLoS Comput Biol 8(4): e1002467
20. Kothari P, Schiffhauer ES, Robinson DN (2017) Cytokinesis from nanometers to micrometers and microseconds to minutes. Methods Cell Biol 137:307–322
21. Bulseco DA, Wolf DE (2013) Chapter 21 – Fluorescence correlation spectroscopy: molecular complexing in solution and in living cells. Methods Cell Biol 114:489–524
22. Carl Zeiss Microimaging, Inc. Guide to basic FCS experiments on Zeiss LSM 780/Confocor3/BiG systems running ZEN 2010. https://www.zeiss.com/content/dam/Microscopy/Downloads/Pdf/FAQs/zen2010-lsm780_basic_fcs_experiments.pdf. Accessed 4 May 2020
23. Ishikawa-Ankerhold H, Ankerhold R, Drummen G (2014) Fluorescence recovery after photobleaching (FRAP). eLS. https://doi.org/10.1002/9780470015902.a0003114
24. Yumura S (2001) Myosin II dynamics and cortical flow during contractile ring formation in *Dictyostelium* cells. J Cell Biol 154:137–146
25. Srivastava V, Robinson DN (2015) Mechanical stress and network structure drive protein dynamics during cytokinesis. Curr Biol 25: 663–670
26. Hiramoto Y (1963) Mechanical properties of sea urchin eggs: II. Changes in mechanical properties from fertilization to cleavage. Exp Cell Res 32:76–89
27. Rappaport R (1967) Cell division: direct measurement of maximum tension exerted by furrow of echinoderm eggs. Science 156:1241–1243
28. Matzke R, Jacobson K, Radmacher M (2001) Direct, high-resolution measurement of furrow stiffening during division of adherent cells. Nat Cell Biol 3:607–610
29. Ashkin A, Dziedzic JM, Yamane T (1987) Optical trapping and manipulation of single cells using infrared laser beams. Nature 330: 769–771
30. Fukui Y, Yumura S, Yumura TK, Mori H (1986) Agar overlay method: high-resolution immunofluorescence for the study of the contractile apparatus. Methods Enzymol 134:573–580
31. Miao C, Schiffhauer ES, Okeke EI, Robinson DN, Luo T (2017) Parallel compression is a fast low-cost assay for the high-throughput screening of mechanosensory cytoskeletal proteins in cells. ACS Appl Mater Interfaces 9: 28168–28179
32. Reichl EM, Ren Y, Morphew MK, Delannoy M, Effler JC, Girard KD, Divi S, Iglesias PA, Kuo SC, Robinson DN (2008) Interactions between myosin and actin crosslinkers control cytokinesis contractility dynamics and mechanics. Curr Biol 18:471–480
33. Kee YS, Ren Y, Dorfman D, Iijima M, Firtel R, Iglesias PA, Robinson DN (2012) A mechanosensory system governs myosin II accumulation in dividing cells. Mol Biol Cell 23:1510–1523
34. Kee YS, Robinson DN (2013) Micropipette aspiration for studying cellular mechanosensory responses and mechanics. Methods Mol Biol 983:367–382
35. Hochmuth RM (2000) Micropipette aspiration of living cells. J Biomech 33:15–22

Chapter 2

Superresolution Expansion Microscopy in *Dictyostelium* Amoebae

Marianne Grafe, Valentin Pitzen, Irene Meyer, and Ralph Gräf

Abstract

Expansion microscopy (ExM) is a superresolution technique for fixed specimens that improves resolution of a given microscopy system approximately fourfold. The gain in resolution in ExM is not achieved by improvement of the resolution of the microscope itself but by isotropic expansion of the sample. To achieve this, the sample is cross-linked to an expandable gel matrix that swells approximately fourfold by incubation in water. We have applied the method to *Dictyostelium amoebae* and discuss the pros and cons of different labeling techniques in combination with pre- and post-expansion staining protocols.

Key words Expansion microscopy, Superresolution microscopy, *Dictyostelium*, Centrosome, Lamin

1 Introduction

During the last decade, we have witnessed a revolution in light microscopy after Ernst Abbe's diffraction limit of microscopic resolution, which restricts optical resolution in fluorescence microscopy to ~250 nm in lateral dimensions and ~700 nm along the z-axis (with green light), was overcome by several commercially available imaging systems. To achieve this so-called superresolution, the most common systems employ either stimulated emission depletion (STED), structured illumination (SIM), or single molecule localization as in stochastic optical reconstruction microscopy (STORM) [1]. All these techniques have in common that they require specialized, often rather expensive equipment that is only available in imaging centers. Yet in 2015 Tillberg, Chen and Boyden published a new method to achieve superresolution with conventional, unmodified wide-field fluorescence microscopes [2]. The method is called expansion microscopy (ExM), since it

Authors Marianne Grafe and Valentin Pitzen have equally contributed to this chapter.

Alan R. Kimmel (ed.), *Dictyostelium discoideum: Methods and Protocols*, Methods in Molecular Biology, vol. 2814, https://doi.org/10.1007/978-1-0716-3894-1_2,

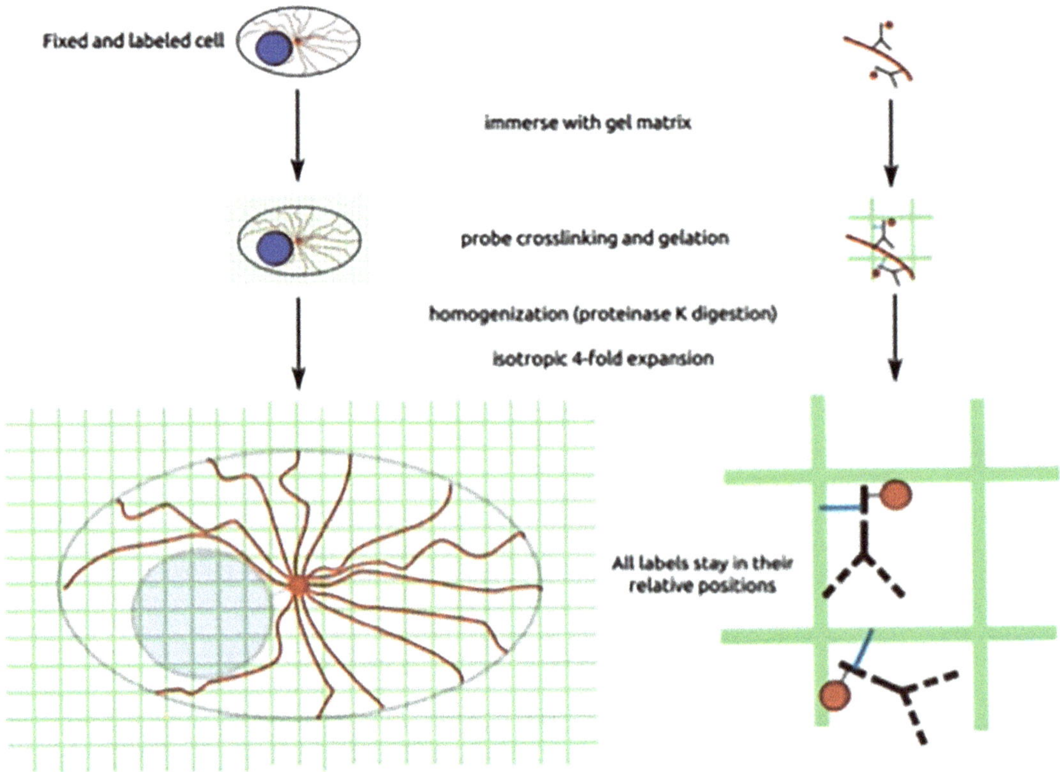

Fig. 1 Schematic representation of the classic ExM pre-expansion staining protocol. Labeled structures are shown red; fluorophores are represented by red circles. Labeled probes are shown in brown with their cross-links (blue) to the gel matrix (green)

pursues a completely different superresolution approach in increasing the size of the specimen instead of increasing the resolving power of the microscope. In the original ExM method the fixed, fluorescently labeled specimen is embedded in a hyperswellable hydrogel, usually consisting of polyacrylamide, to which the fluorophores are chemically cross-linked. After gelation and a digestion step that is required to disrupt intracellular and intercellular structures, the gel is expanded ~4-fold simply by incubation in distilled water. Because of the isotropy of expansion, the relative position of the individual cross-linked fluorophores in the three-dimensional space remains unchanged while their distance is increased by the expansion factor (Fig. 1), and structures with a lateral distance of only ~70 nm can be resolved in a conventional fluorescence microscope [2].

While the original method employed methacryloyl-modified fluorescent oligonucleotides hybridized to oligonucleotide-modified antibodies, a modified protocol also allowed deployment of conventional primary and secondary fluorescent antibodies [3]. Using this approach, we achieved a comparable resolution with in vitro assembled lamin filaments as with STED (stimulated

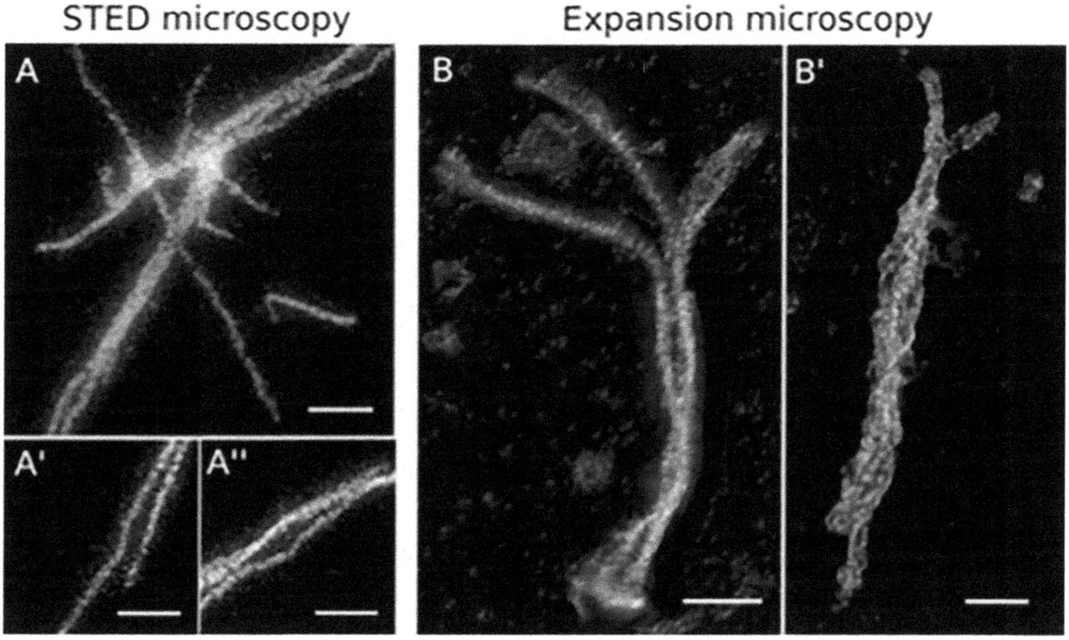

Fig. 2 ExM yields qualitatively similar results as STED microscopy. Superresolution fluorescence microscopy of HisMyc-NE81ΔNLSΔCLIM filaments. (**a–a″**) STED microscopy: assemblies formed at low protein concentration (5 μg/mL) were stained with anti-NE81/anti-rabbit-Atto 647 N. Bars = 1 μm. (**b**, **b′**) ExM with deconvolution, 3.5-fold expanded: assemblies formed at 0.3 mg/mL (**b**) and 5 μg/mL (**b′**), respectively, were stained with anti-NE81/anti-rabbit-AlexaFluor 488. Maximum intensity projections are shown. The scale bar (2 μm) refers to the original size of the specimen. (Figure and text taken from ref. [4] (CC license))

emission depletion) microscopy [4] (Fig. 2). Although principally applicable also for fluorescent proteins (FPs) [3], ExM is only rarely combined with FP-tagging, since the protease digestion (usually proteinase K), which is required to homogenize cells and tissues, also attacks the fluorescent proteins causing a significant loss of fluorescence. This can be partially circumvented by reducing protease concentration or digestion time; however, it has to be evaluated carefully whether more gentle conditions do not interfere with the isotropy of expansion. Yet, FP-tagged proteins, like also small epitope tags, can also be detected with specific, fluorescently labeled nanobodies. Very good ones are, for example, available against GFP or the Spot-tag [5, 6]. Compared to the use of primary and secondary antibodies, the nanobody approach has further advantages. First it often allows higher labeling densities, which is a general issue in superresolution microscopy as insufficient labeling density may cause continuous structures to appear discontinuous. Second, labeled nanobodies bring the fluorophores much closer to the protein of interest [7]. In case of primary/secondary antibody complexes, the distance of the fluorophore to the antigenic epitope could be up to 30 nm [8], while it is ~7.5 times lower when fluorescent nanobodies are applied [9]. Although this significantly

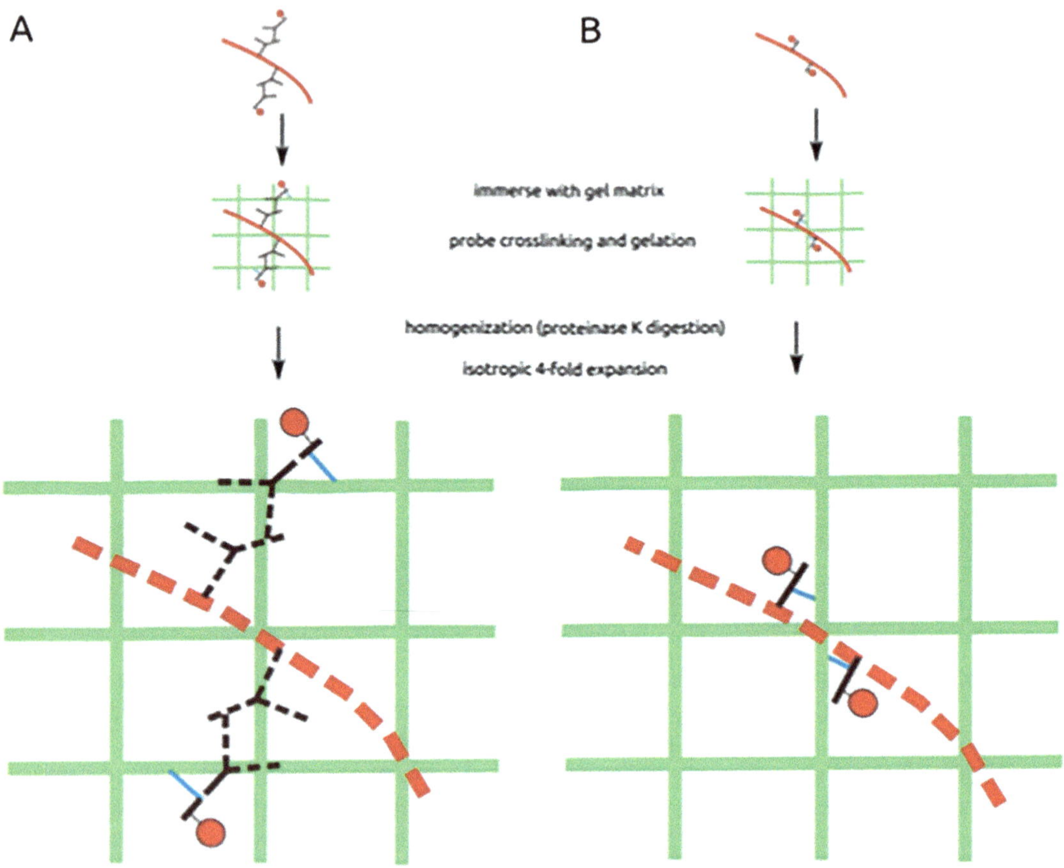

Fig. 3 Schematic comparison of pre-expansion staining with primary/secondary antibodies (**a**) versus small fluorescent probes (**b**). Note the strongly decreased distance between labeled structures (dashed red line) and fluorophores (red circles). Probes are shown in brown and cross-links to the gel matrix (green) are shown in blue

reduces resolution in all superresolution techniques, the effect is more obvious in ExM, since here fluorophore epitope distances are expanded by the same factor as the whole sample (Fig. 3). The size of primary/secondary antibody complexes can also result in a mis-estimation of molecular dimensions, especially if the structure to be labeled is dense and accessible for antibodies only from one side. This has been shown, for example, for the labeling of centrioles. These are cylindrical microtubule structures with diameter of ~200 nm and a length of ~500 nm. Gambarotto and co-workers have shown clearly that centriole size is strongly overestimated in ExM when labeled with primary and secondary antibody complexes that have access only to the outer centriole wall [10]. To circumvent this problem, they introduced a variation of the original ExM method that they called ultrastructure expansion microscopy (U-ExM). The authors modified the standard protocols for ExM [3] and the related magnified analysis of the proteome (MAP; [11])

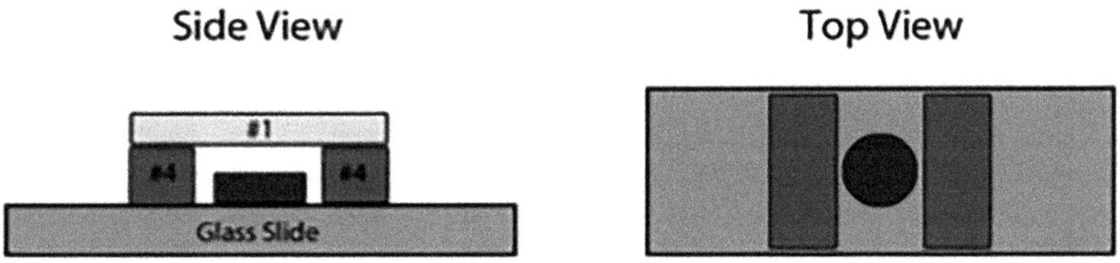

Fig. 4 Self-made gelation chamber made of glass slide with a divided coverslip on top (type #4) fixed with nail polish, agarose gel, or Twinsil. The divided coverslips need to be apart from each other at least the distance of the coverslip where the cells are fixed on (here cells are fixed on a 12 mm coverslip shown in dark gray). After addition of the gel solution, the chamber is covered with a rectangular cover slip type #1 (*see* side view)

to optimize concentrations of acrylamide and the formaldehyde cross-linker and, thus, to reduce anisotropies of expansion. Moreover, in U-ExM, the sample is homogenized by SDS/heat treatment instead of protease and expanded prior to antibody staining. Due to fourfold expansion of the sample, inaccuracies caused by the size of primary/secondary antibody complexes can be neglected, and centrioles appear with approximately the same size in U-ExM as measured with alternative methods such as dSTORM or electron microscopy [10, 12]. Recently, we have shown that correct dimensions of centrosomal structures can also be achieved with pre-expansion staining according to the original ExM protocol, if small, compact fluorescent probes such as fluorescent nanobodies or fluorescent streptavidin are used instead of primary and secondary antibody complexes [13] (Fig. 4).

The mentioned anisotropies in expansion and uneven distribution of fluorophores due to structural constraints cause the most common artifacts in ExM. To avoid these artifacts and to validate the ExM, it is always recommended to compare pre- and post-expansion samples and to know the structure of interest and its dimensions from electron microscopical data.

Besides its low requirements of specialized equipment and low costs, ExM has further advantages that makes it attractive as an alternative to the "classic" superresolution techniques. First, it works with almost any small-molecule fluorophore and is independent of its capabilities of photoconversion (as in PAL-M) or blinking behavior (as in dSTORM). Second, it also works with weaker fluorescent signals, that is, at lower signal to noise ratios, which is a serious issue, for example, in STED microscopy or SIM. This may appear surprising, since isotropic fourfold sample expansion in ExM results in a dilution of the fluorescence signal by a factor of 64. However, this usually can be tolerated as this also holds true for background signals. Third, ExM works with any fixation method. Fourth, in case of tissues, the digestion during the homogenization step followed by expansion reduces light scattering effects by the tissue. Fifth, although ExM also requires a higher number of optical slices to achieve three-dimensional reconstructions with sufficient axial resolution, it requires less computing power than SIM or dSTORM. Sixth, imaging itself is rather fast and instantly yields a superresolution image. Moreover, ExM can easily be combined with other superresolution techniques such as, for example, STED [14, 15], dSTORM [16], SIM [17] and Airyscan microscopy [13].

In this chapter, we describe the application of both pre-expansion staining with various fluorescent probes and U-ExM-post-expansion staining to *Dictyostelium amoebae* and the combination of ExM with Airyscan microscopy.

2 Materials

2.1 Strain Generation

All solutions (except HL5c-medium) should be sterile filtered using a 0.2 μm filter.

1. HL5c medium (Formedium, Hunsanton, UK): 26.55 g powdered medium per liter deionized water; autoclave 20 min at 121 °C. If necessary, add antibiotics to the required concentration (e.g., 6 μg/mL Blasticidin S or 10 μg/mL G418).
2. SPB (Soerenson phosphate buffer): 14.6 mM KH_2PO4, 2 mM Na_2HPO_4, and pH 6.0.
3. Electroporation buffer (EB): 50 mM sucrose, 10 mM KH_2PO_4, and pH 6.1 (with KOH).
4. $CaCl_2/MgCl_2$ solution: 100 mM of each salt in deionized water.
5. Blasticidin S stock 8 mg/mL, working solution 6 μg/mL (Carl Roth, Karlsruhe, Germany).
6. Penstrep (100×) stock: 100 U/mL penicillin and 100 μg/mL streptomycin (GE Healthcare, Solingen, Germany).
7. Standard uncoated tissue culture 24-well plates and 175 cm^2 flasks.
8. Conical (50 mL) tubes (BD Falcon 352098) and swinging bucket rotor (e.g., Heraeus Multifuge 1 S-R).
9. Electroporation cuvette (4 mm) and electroporator (e.g., BIO-RAD GenePulser® II).
10. Optional vectors with knock-in constructs under Blasticidin selection cassette and a mini tag, for example, Spot-tag or BioH6-tag [5, 13].
11. *Dictyostelium* cells (wild-type or respective mutants).

2.2 Cell Growth and Fixation

2.2.1 Fixations

1. Four-well plate (Nunc/Thermo Fisher, Hennigsdorf, Germany) or custom-made multiwell holder (as described in [18]).
2. Soerensen phosphate buffer (SPB): 14.6 mM KH_2 PO_4, 2 mM Na_2 HPO_4, and pH 6.0.
3. Phosphate-buffered saline (PBS): 30 mM KH_2 PO_4, 70 mM Na_2 HPO_4, 137 mM NaCl, 2.6 mM KCl, and pH 7.4.
4. Tris-buffered saline (TBS): 20 mM Tris-HCl, 150 mM NaCl, and pH 7.4.
5. Methanol fixation: Methanol, −20 °C cold.
6. Glutaraldehyde fixation: PHEM buffer (60 mM PIPES, 25 mM HEPES, 10 mM EGTA, 2 mM MgCl2, pH 6.9), 0.5% glutaraldehyde EM-grade (stock 25%), 0.5% Triton X-100 (stock 25%), SPB (*see* **step 2**), and 0.1% $NaBH_4$ (freshly prepared in SPB).
7. Formaldehyde fixation: 3.7% formalin in PBS or TBS, pH = 8.0.

2.2.2 Immunostainings

1. Phosphate-buffered saline (PBS): (*see* Subheading 2.2.1, **step 3**).
2. Primary antibodies, secondary antibodies (e.g., Alexa-Fluor conjugates, Thermo Fisher), direct conjugate (e.g., Streptavidin-AlexaFluor 488; Thermo Fisher), nanobodies (e.g., anti-Spot®V_HH-Atto 594; Chromotek, Martinsried, Germany)
3. Antibody dilution buffer: 1% Bovine serum albumin in PBS
4. 0.1 μg/mL DAPI (Carl Roth) in PBS
5. Mounting medium (120 mg/mL polyvinyl alcohol 4–88, 30% glycerol, 0.2 M Tris-HCl, pH 8.8)
6. 10 μg/mL Hoechst 33342 (Thermo Fisher) in PBS

2.3 Expansion

1. Monomer solution: 1×PBS, 2 M NaCl, 2.5% (wt/wt) acrylamide, 0.15% (wt/wt) N,N′-methylenebisacrylamide, and 8.625% (wt/wt) sodium acrylate
2. Accelerator solution: 10% TEMED (AppliChem, Darmstadt, Germany)
3. Initiator solution: 10% APS (AppliChem)

2.3.1 Solutions for ExM

1. 0.25% Glutaraldehyde
2. Digestion buffer: 50 mM Tris-HCl, pH 8.0, 1 mM EDTA, 0.5% Triton X-100, and 0.8 M guanidine HCl, freshly added 8 U/mL Proteinase K (stock 600 U/mL)
3. 50 μg/mL Hoechst 33342 in deionized water

2.3.2 Solution for U-ExM

1. Fixation solution: 1.4% formaldehyde and 2% acrylamide in PBS
2. Denaturation buffer: 200 mM SDS, 200 mM NaCl, 50 mM Tris-HCl, and pH 8.0
3. 1 mg/mL DAPI in deionized water
4. PBST (1× PBS + 0.1% Tween-20)

2.3.3 Chamber Preparation (Fig. 5)

1. Incubation/gelation chamber, for example, self-made with slide and coverslips (type #4; Menzel-Gläser, Braunschweig, Germany)
2. Rectangular coverslips (21 mm × 25 mm, thickness #1)
3. Four-well plate (Nunc/Thermo Fisher)
4. Petri dishes (Ø 90 mm, Ø 55 mm)
5. Tool for measurement (caliper)
6. 5% DABCO (Carl Roth) in deionized water

2.4 Microscopy

1. High precision cover slips (e.g., 24 × 60 mm, High Precision No. 1.5H, 170 ± 5 μm, Carl Roth, Karlsruhe, Germany)
2. Polylysine-L (Sigma-Aldrich, St. Louis, MO, USA)
3. Soft tissues (e.g., Kimtech Precision Wipes, Kimberly Clark, Koblenz, Germany)
4. Twinsil (Picodent, Wipperfürth, Germany) or low melting agarose
5. Water immersion objective (e.g., Zeiss LCI PlanNeofluar 63×/NA 1.3)

3 Methods

3.1 Strain Generation

All solutions need to be cooled (except HL5c medium) and all steps should be performed on ice. Transformation of the recombinant DNA will be done by electroporation as described elsewhere [19].

1. Cultivate *Dictyostelium* cells in HL5c medium at 21 °C in a tissue culture flask to a density of 2×10^6 cells/mL.
2. Transfer 25 mL in a 50 mL tube and chill on ice for 20 min.
3. Centrifuge at 400×*g* for 5 min and wash twice with 25 mL SPB.
4. Wash cells with 10 mL electroporation buffer and resolve in 700 μL electroporation buffer.
5. Transfer cells in a cooled electroporation cuvette (4 mm) with 20 μg of linearized plasmid DNA.
6. Perform electroporation with two pulses with an electrical capacity of 3 μF, a voltage of 1.1 kV, and a resistance of

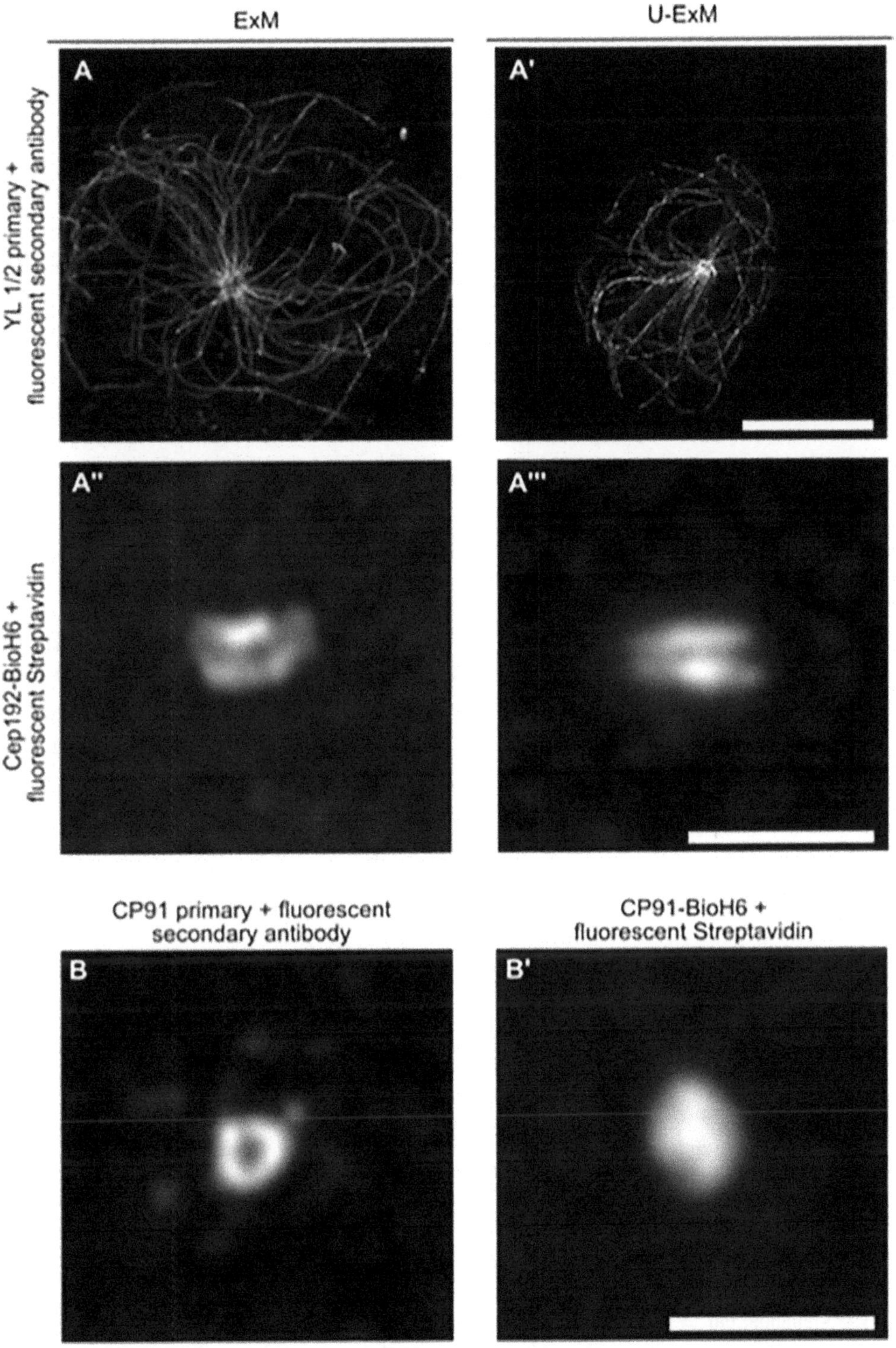

Fig. 5 Expansion microscopy results. (**a**–**a'''**) Comparison between ExM and U-ExM. Microtubules (**a**–**a'**) and centrosomes (**a''**–**a'''**) were labeled with the probes indicated in the figure. Note that microtubules appear thinner and less continuous in U-ExM versus ExM. Since the inaccuracy of the position of the fluorophore relative to the epitope of the antibody is increased by the expansion factor in ExM, the U-ExM image appears sharper and the microtubules thinner when stained with the same antibody. At the same time, this inaccuracy

1000 Ω for circular plasmids and 200 Ω for linear DNA fragments.

7. Let cells rest at room temperature for 10 min.
8. Transfer cells in a sterile 1.5 mL reaction tube and add 1 mM $MgCl_2/CaCl_2$.
9. Agitate suspension gently on a rotator for 15 min at room temperature.
10. For transformations with blasticidin S resistance, dilute cells with 25 mL HL5c-PenStrep medium and pipette to a 24-well plate.
11. Cultivate at 21 °C for 20–24 h.
12. Replace medium with blasticidin S supplemented medium (6 μg/mL).
13. Change medium carefully every 3 days and wait for colony growth.
14. Verify positive clones by widefield immunofluorescence and/or by Western blot analysis.

3.2 Cell Culture and Fixation

1. Cultivate *Dictyostelium* cells in HL5c medium to exponential growth phase (approximately density 2×10^6 cells/mL).
2. For 12 mm round coverslips, let 2×10^5 cells settle onto coverslips for 20 min. Use a multiwell plate (*see* Subheading 2.2.1) and two coverslips per experiment.
3. Carefully aspirate medium and immediately proceed with fixation (for UExM proceed with **step 1** in Subheading 3.3.2, if the used antibodies work with the mild fixative used for UExM).

3.2.1 Fixations

1. Methanol fixation: let the cells settle on ice (*see* Subheading 3.2.2). Gently add 1 mL cold (−20 °C) methanol to the well with the coverslip and incubate for 2 min at −20 °C. Carefully aspirate methanol and wash twice with PBS (*see* **Note 1**).
2. Glutaraldehyde fixation: add 1 mL fixation solution in well with coverslip and incubate for 5 min at room temperature. Wash twice with SPB. Block for 10 min with freshly prepared

Fig. 5 (continued) leads to a more continuous signal along the microtubules in ExM than in U-ExM. These differences between ExM and U-ExM can be neglected if small fluorescent probes are used as shown in the two Cep192 stainings, bar = 5 μm (**a**–**a′**) and 0.5 μm (**a″**–**a‴**). (**b**–**b′**) In ExM, the use of small fluorescent probes can also avoid artifacts due to accessibility problems at dense structures. CP91 is an inner core protein of the centrosome and is expected to exhibit a disk-like distribution when centrosomes are viewed from the top. When stained with primary and secondary antibodies, it appears as a ring instead (**b**), since antibodies cannot penetrate the inner core. The smaller streptavidin probe gets access to the inner core and the CP91 distribution appears as expected (**b′**). All images were taken at a Zeiss LSM880 Airyscan system. Bar = 0.5 μm

0.1% $NaBH_4$ (this reduces and inactivates remaining aldehyde groups). Wash twice with PBS.

3. Formaldehyde fixation: add 1 mL fixation solution in well with coverslip and incubate for 5 min at room temperature. Wash twice with PBS.

3.2.2 Immunostaining

1. Incubate cells overnight at 4 °C with primary antibody or conjugates diluted in antibody dilution buffer in a twofold higher concentration than for conventional fluorescence microscopy (*see* **Note 2**).
2. Wash three times with PBS.
3. If primary antibodies were used, incubate with secondary antibody (diluted 1:500) for 4 h.
4. Incubate with 10 μg/mL Hoechst 33342 in PBS for 10 min.
5. Wash twice with PBS.
6. Mount one coverslip with mounting medium as a control.

3.3 Expansion

3.3.1 ExM

1. Postfixation: add 200 μL 0.25% glutaraldehyde (freshly prepared in PBS) to the second coverslip and incubate for 10 min at room temperature.
2. Wash three times with PBS and proceed with gelation immediately (*see* **Note 3**).
3. Under the fume hood with nitrile gloves: place cover slip in gel chamber (Fig. 4; cells still oriented upward), and incubate for 1 min at room temperature with monomer solution.
4. Keep all solutions on ice to slow down polymerization.
5. Mix carefully in a 1.5 mL reaction tube 188 μL monomer solution +4 μL A.dest +4 μL 10% TEMED.
6. Add 4 μL 10% APS at the end; mix quickly but without bubbles.
7. Remove monomer solution carefully from coverslip (which contains harmful chemicals, dispose in an extra container!), and add 200 μL gel solution immediately and cover with a square coverslip.
8. Incubate for 1 h at 37 °C to let the gel polymerize.
9. Remove top coverslip and cut the gel into an asymmetric trapezoid with a sharp scalpel. This shape helps to remember on which site your cells are fixed in the gel (*see* **Note 4**).
10. Measure dimensions of the gel piece.
11. Put coverslip with trapezoid gel in a well of a multiwell plate (*see* Subheading 2.3), gel can stick to the coverslip, and add 500 μL digestion buffer with proteinase K.
12. Incubate 30 min to 1.5 h at 37 °C.

13. Carefully transfer the gel in a 90 mm Petri dish with approximately 50 mL deionized water (the gel is already a bit expanded and should be detached from coverslip).
14. Expand for 1–2 h and change water 2–3 times (*see* **Note 5**).
15. Measure dimensions again and calculate expansion factor.
16. Cut a part of the gel (one-third to half of it) and put in a 55 mm Petri dish.
17. Incubate with 500 μL Hoechst 33342 (50 μg/mL in deionized water) for 30 min and wash twice with deionized water (*see* **Note 6**).
18. Gels can be stored at 4 °C up to 1 week.

3.3.2 U-ExM

1. Add 1 mL UExM fixation solution to coverslip and incubate 3 h at 37 °C.
2. Wash twice with PBS and proceed with gelation (**steps 3–10** in Subheading 3.3.1 ExM).
3. Incubate with denaturation buffer for 15 min at room temperature.
4. Transfer gel in a 2 mL reaction tube with denaturation buffer and incubate 90 min at 95 °C.
5. Initial expansion: Transfer gel in a 90 mm Petri dish with approximately 50 mL deionized water and let it expand for 1 h.
6. Wash three times for 10 min with PBS and proceed with immunostaining (**steps 1–3** in Subheading 3.2.2) (*see* **Note 7**).
7. Wash three times in PBST and expand for 1 h in deionized water (final expansion).
8. Incubate with DAPI in deionized water for 20 min and wash three times with deionized water.
9. Measure dimension to calculate expansion factor and cut gel to a fitting size (*see* **Note 4**).
10. Gels can be stored at 4 °C up to 1 week.

3.4 Mounting for Microscopy

1. Carefully remove excess liquid with a soft tissue from gel.
2. Transfer gel to a high-precision coverslip coated with poly-L-lysine with cells oriented downward.
3. Remove excess liquid as much as possible, since the gel should not shift at all.
4. Mix both Twinsil chemicals freshly (1 mL each, pipette with truncated 1 mL pipette tips since chemicals are very viscous) and put around the gel piece.
5. Cover with a normal small coverslip (same dimensions).
6. As soon as the Twinsil has hardened (processing time: 5 min), microscopy can be performed (*see* **Note 8**).

4 Notes

1. The choice of fixatives depends on the cellular structure to be investigated and on dependencies of the used antibodies on certain fixation methods. In principle, ExM should be applicable to any primary fixation method. When using methanol as a fixative, be aware that methanol-fixed cells are only loosely attached to the coverslip. Therefore avoid strong agitation and aspirate liquids carefully.
2. As small fluorescent probes, we have successfully employed fluorescently labeled anti-Spot-tag nanobodies and fluorescent streptavidin. If tag-specific nanobodies should be used, we recommend generation of respective knock-in strains, in which the gene encoding the protein of interest is replaced by a modified version encoding the C-terminal tag as described recently [13]. Fluorescent streptavidin can be employed in combination with the BioH6 tag, which encodes a biotinylation signal and a 6× histidin tag [13]. The tag encodes a signal for biotinylation with endogenous biotin ligases. When using this tag, one has to bear in mind that *Dictyostelium* amoebae naturally express a couple of biotinylated mitochondrial proteins, which causes a background staining of mitochondria [20]. If the localizations of the protein of interest do not overlap with mitochondria, this background staining can be neglected.
3. Choice of gels and cross-linking methods: Polyacrylamide cross-linked with bisacrylamide is the most commonly used swellable hydrogel used in ExM. However, Lee and co-workers recently replaced acrylamide by tetra-gel, which is not based on free-radical chain-growth polymerization, and obtained a higher accuracy in fluorophore positioning during the expansion process [21].

   Glutaraldehyde cross-linking is cheap, fast, and reliable, but there are further methods for cross-linking of protein side chains to the acrylamide matrix. In another commonly used method, fixed immunostained specimens are treated with a solution containing 25 mM of the amine-reactive small-molecule MA-NHS (methacrylic acid N-hydroxysuccinimidyl ester). The MA part is incorporated in the acrylamide matrix during gelation and, hence, cross-links proteins to the acrylamide matrix [3]. For the samples we have tested, MA-NHS cross-linking was not superior to glutaraldehyde cross-linking, but this could be sample-dependent.
4. To reduce measurement inaccuracies, it is also possible to use a biopsy punch to get a defined round gel piece. Note that the orientation of the cells must be checked prior to microscopy.

Alternatively, a high-precision coverslip on both sides can be used, but unevenness on top side of the gel can occur and interfere with AiryScan measurement.

5. A post-expansion staining step with a secondary antibody in PBS is also possible in ExM and sometimes useful to increase label density. Note that the gel will shrink again. Therefore, repeat expansion in water after the staining, and put the gel in deionized water again for 1 h and change water two times.
6. Optionally, to avoid bleaching, 5% DABCO in deionized water may be added, especially to preserve DNA stainings. However, we observed shrinking effects that should be monitored carefully.
7. Caveats resulting from biased accessibility of antigenic structures for labeled antibodies: One common artifact in case of pre-expansion labeling may occur in cases in which the labeled structures are accessible for the fluorescent probe only from one side resulting in an uneven distribution of the fluorophores in the three-dimensional space. As a result, the peaks of fluorescence intensity distribution do not match the actual position of the labeled protein structure. Such mismatches are enhanced by expansion but can be widely overcome by post-expansion labeling or the use of small fluorescent probes, as it was recently shown for *Dictyostelium* centrosomes [13] (Fig. 4).

   Post-expansion staining is also often advantageous in cases, in which the fluorescent probe has only limited access to its binding site in the unexpanded sample due to sterical hindrance. Here, access may be improved after expansion. On the other hand, in some cases, the interaction sites for the fluorescent probe on the protein of interest could be inactivated by cross-linking and thus reduce or even prevent binding of the probe. This could explain the observation that some antibodies are not suitable for the U-ExM protocol.
8. Choice of microscope equipment: We routinely use a Zeiss AxioObserver widefield setup equipped with a Axiocam 506 mono camera or a Zeiss LSM880 laser scanning microscope with an Airyscan detector. Both systems have the advantage that they impose no special fluorophore requirements. The Airyscan system yields superior results, but widefield systems are often more readily accessible.

   Widefield images can often be improved by deconvolution. However, when generating a theoretical PSF (point spread function), one has to be aware that commercial programs usually do not consider acrylamide gels as an embedding medium. The use of the "water" setting instead may cause artifacts. Alternatively, fluorescent beads embedded in the expanded gel matrix can be used to create a measured PSF as described elsewhere [22].

We strongly recommend the use of water immersion lenses for ExM samples, for example, the Zeiss LCI PlanNeofluar 63×/1.3 water immersion lens, even though they have a lower numerical aperture than comparable oil immersion lenses. With the latter, the misfit between the refractive index of the immersion fluid and the sample strongly increases spherical aberration, which nullifies the gain in resolution due to higher NA. This effect could be the reason for the occasionally observed reduction in image quality with increased expansion factors [23].

In case of already thick unexpanded samples, the thickness of the post-expansion sample may increase the working distance of some lenses.

Acknowledgments

We would like to thank Cindy Marona for the expert technical assistance. This work was funded by the Deutsche Forschungsgemeinschaft (GR1642/4-3, ME3690/2-1 and INST 336/114-1 FUGG for large equipment).

References

1. Schermelleh L, Ferrand A, Huser T, Eggeling C, Sauer M, Biehlmaier O, Drummen GPC (2019) Super-resolution microscopy demystified. Nat Cell Biol 21:72–84. https://doi.org/10.1038/s41556-018-0251-8
2. Chen F, Tillberg PW, Boyden ES (2015) Optical imaging. Expansion microscopy. Science 347:543–548. https://doi.org/10.1126/science.1260088
3. Chozinski TJ, Halpern AR, Okawa H, Kim HJ, Tremel GJ, Wong RO, Vaughan JC (2016) Expansion microscopy with conventional antibodies and fluorescent proteins. Nat Methods 13:485–488. https://doi.org/10.1038/nmeth.3833
4. Grafe M, Batsios P, Meyer I, Lisin D, Baumann O, Goldberg MW, Gräf R (2019) Supramolecular Structures of the Dictyostelium Lamin NE81. Cells 8. https://doi.org/10.3390/cells8020162
5. Virant D, Traenkle B, Maier J, Kaiser PD, Bodenhöfer M, Schmees C, Vojnovic I, Pisak-Lukáts B, Endesfelder U, Rothbauer U (2018) A peptide tag-specific nanobody enables high-quality labeling for dSTORM imaging. Nat Commun 9:930. https://doi.org/10.1038/s41467-018-03191-2
6. Rothbauer U, Zolghadr K, Tillib S, Nowak D, Schermelleh L, Gahl A, Backmann N, Conrath K, Muyldermans S, Cardoso MC, Leonhardt H (2006) Targeting and tracing antigens in live cells with fluorescent nanobodies. Nat Methods 3:887–889. https://doi.org/10.1038/nmeth953
7. Platonova E, Winterflood CM, Junemann A, Albrecht D, Faix J, Ewers H (2015) Single-molecule microscopy of molecules tagged with GFP or RFP derivatives in mammalian cells using nanobody binders. Methods 88: 89–97. https://doi.org/10.1016/j.ymeth.2015.06.018
8. Harris LJ, Larson SB, Hasel KW, McPherson A (1997) Refined structure of an intact IgG2a monoclonal antibody. Biochemistry 36:1581–1597. https://doi.org/10.1021/bi962514+
9. Mikhaylova M, Cloin BMC, Finan K, van den Berg R, Teeuw J, Kijanka MM, Sokolowski M, Katrukha EA, Maidorn M, Opazo F, Moutel S, Vantard M, Perez F, van Bergen en Henegouwen PMP, Hoogenraad CC, Ewers H, Kapitein LC (2015) Resolving bundled microtubules

using anti-tubulin nanobodies. Nat Commun 6:7933. https://doi.org/10.1038/ncomms8933
10. Gambarotto D, Zwettler FU, Le Guennec M, Schmidt-Cernohorska M, Fortun D, Borgers S, Heine J, Schloetel J-G, Reuss M, Unser M, Boyden ES, Sauer M, Hamel V, Guichard P (2019) Imaging cellular ultrastructures using expansion microscopy (U-ExM). Nat Methods 16:71–74. https://doi.org/10.1038/s41592-018-0238-1
11. Ku T, Swaney J, Park J-Y, Albanese A, Murray E, Cho JH, Park Y-G, Mangena V, Chen J, Chung K (2016) Multiplexed and scalable super-resolution imaging of three-dimensional protein localization in size-adjustable tissues. Nat Biotechnol 34:973–981. https://doi.org/10.1038/nbt.3641
12. Le Guennec M, Klena N, Gambarotto D, Laporte MH, Tassin A-M, van den Hoek H, Erdmann PS, Schaffer M, Kovacik L, Borgers S, Goldie KN, Stahlberg H, Bornens M, Azimzadeh J, Engel BD, Hamel V, Guichard P (2020) A helical inner scaffold provides a structural basis for centriole cohesion. Sci Adv 6:eaaz4137. https://doi.org/10.1126/sciadv.aaz4137
13. Pitzen V, Sander S, Baumann O, Gräf R, Meyer I (2021) Cep192, a novel missing link between the centrosomal core and corona in Dictyostelium amoebae. Cells 10:2384. https://doi.org/10.3390/cells10092384
14. Gao M, Thielhorn R, Rentsch J, Honigmann A, Ewers H (2021) Expansion STED microscopy (ExSTED). Methods Cell Biol 161:15–31. https://doi.org/10.1016/bs.mcb.2020.06.001
15. Bianchini P, Pesce L, Diaspro A (2021) Expansion microscopy at the nanoscale: the nuclear pore complex as a fiducial landmark. Methods Cell Biol 161:275–295. https://doi.org/10.1016/bs.mcb.2020.07.001020-17086-8
16. Zwettler FU, Reinhard S, Gambarotto D, Bell TDM, Hamel V, Guichard P, Sauer M (2020) Molecular resolution imaging by post-labeling expansion single-molecule localization microscopy (Ex-SMLM). Nat Commun 11:3388. https://doi.org/10.1038/s41467-020-17086-8
17. Halpern AR, Alas GCM, Chozinski TJ, Paredez AR, Vaughan JC (2017) Hybrid Structured Illumination Expansion Microscopy Reveals Microbial Cytoskeleton Organization. ACS Nano 11:12677–12686. https://doi.org/10.1021/acsnano.7b07200
18. Koonce M, Tikhonenko I, Gräf R (2020) Dictyostelium Cell Fixation: Two Simple Tricks. Methods Protoc 3. https://doi.org/10.3390/mps3030047
19. Mann SKO, Devreotes PN, Eliott S, Jermyn K, Kuspa A, Fechheimer M, Furukawa R, Parent CA, Segall J, Shaulsky G, Vardy PH, Williams J, Williams KL, Firtel RA (1998) Cell biological, molecular genetic, and biochemical methods used to examine Dictyostelium. In: Celis JE (ed) Cell biology: a laboratory handbook, 2nd edn. Academic Press, San Diego, pp 431–465
20. Batsios P, Meyer I, Gräf R (2016) Proximity-Dependent Biotin Identification (BioID) in Dictyostelium Amoebae. Methods Enzymol 569:23–42. https://doi.org/10.1016/bs.mie.2015.09.007
21. Lee H, Yu C-C, Boyden ES, Zhuang X, Kosuri P (2021) Tetra-gel enables superior accuracy in combined super-resolution imaging and expansion microscopy. Sci Rep 11:16944. https://doi.org/10.1038/s41598-021-96258-y
22. Cole RW, Jinadasa T, Brown CM (2011) Measuring and interpreting point spread functions to determine confocal microscope resolution and ensure quality control. Nat Protoc 6:1929–1941. https://doi.org/10.1038/nprot.2011.407
23. Sahabandu N, Kong D, Magidson V, Nanjundappa R, Sullenberger C, Mahjoub MR, Loncarek J (2019) Expansion microscopy for the analysis of centrioles and cilia. J Microsc 276:145–159. https://doi.org/10.1111/jmi.12841

Chapter 3

Inflicting, Monitoring, Visualizing, and Quantitating Various Sterile Membrane Damages and the Repair Response in *Dictyostelium discoideum*

Lyudmil Raykov, Davide D'Amico, A. T. López-Jiménez, and Thierry Soldati

Abstract

Eukaryotic cells have been constantly challenged throughout their evolution by pathogens, mechanical stresses, or toxic compounds that induce plasma membrane (PM) or endolysosomal membrane damage. The survival of the wounded cells depends on damage detection and repair machineries that are evolutionary conserved between protozoan, plants, and animals. We use the social amoeba *Dictyostelium discoideum* as a model system to study bacteria, mechanical or sterile membrane damage that allows us to identify and monitor factors involved in PM, endolysosomal damage response (ELDR), and endolysosomal homeostasis. Importantly, the sterile damage techniques presented here homogenously affect cell populations, which allows to phenotype mutant strains and quantify various aspects of cell fitness using live cell microscopy. This is instrumental to functionally assess genes involved in the repair of damaged plasma membrane or intracellular compartments and the degradation of extensively damaged compartments. Here, we describe how to inflict sterile PM or endolysosomal membrane damage, how to monitor the cell-intrinsic response to damage, and how to proxy proton leakage from damaged acidic compartments and quantify cell viability.

Key word Digitonin, LLOMe, GPN, Endolysosome, Propidium iodide (PI), Dextran, Lysosensor, Hyperosmotic shock

1 Introduction

Maintenance of plasma membrane (PM) and endolysosomal membrane integrity is vital for the cells, which are perpetually challenged by pathogens, enzymes, protein aggregates, or chemical compounds that induce PM or endolysosomal compartments' damage [1, 2]. In mammalian cells, Ca^{2+} influx from extracellular milieu or intracellular storage and exposure of glycosylated lipids or proteins localized on the cell surface or the intra-luminal membrane leaflets of endosomes are detected by cells and trigger membrane repair

Alan R. Kimmel (ed.), *Dictyostelium discoideum: Methods and Protocols*, Methods in Molecular Biology, vol. 2814, https://doi.org/10.1007/978-1-0716-3894-1_3,

responses that result in membrane remodeling, patching, shedding, or degradation of extensively damaged compartments by autophagy [3–5]. The Endosomal Sorting Complex Required for Transport (ESCRT) machinery functions in membrane deformation away from the cytosol during the formation of endosomal intralumenal vesicles (ILV) and also plays a crucial role in virtually all membrane remodeling processes including virus budding, cytokinetic abscission, nuclear envelope reformation, phagophore closure and autophagosome formation, fusion of autophagosomes with the endolysosomal system, endolysosome membrane repair, and others [6, 7]. We have previously demonstrated that similarly to their human homologs, the ESCRT accessory protein ALIX and the ESCRT-III subunit CHIMP4B, *D. discoideum* ALIX, and Vps32 are also actively recruited to damaged PM and endolysosomal membranes [8, 9]. Here, we describe optimized induction of sterile PM or endolysosomal damage in *D. discoideum* that can be monitored with fluorescent protein fusions acting as damage reporters, such as ALIX-GFP, GFP-Vps32, or the ubiquitin E3-ligase GFP-TrafE/TrafE-GFP [8, 9].

To study PM and endolysosomal repair mechanisms, we established three sterile damaging agents: digitonin, L-leucyl-L-leucine methyl ester (LLOMe), and glycyl-L-phenylalanine 2-naphthylamide (GPN). The natural steroidal saponin digitonin is a mild detergent that at low concentrations can be used to selectively permeabilize the plasma membrane while leaving intact other organelles [10]. The lysosomotropic agent LLOMe accumulates in acidic compartments and is converted by the thiol protease cathepsin C (also known as dipeptidyl peptidase 1) to a hydrophobic polymer with membranolytic activity that creates small-size perforations in endolysosomal membranes permitting transient release of protons and other small lysosomal components of less than 10 kDa [11]. GPN is also a lysosomotropic substrate for cathepsin C that accumulates in acidic compartments, such as endolysosomes, and is considered to cause osmotic changes resulting in rupture of the compartments [12]. We trace the recovery of damaged acidic compartments using the lysosomotropic fluorescent pH tracer Lysosensor™ Green DND-189 that accumulates in acidic compartments, where its protonation relieves the fluorescence quenching. We have recently demonstrated that in WT *D. discoideum cells*, lysosensor fluorescence intensity decreases after addition of LLOMe or GPN indicating proton leakage and pH increase within the acidic compartments, which is then gradually restored, demonstrating the capacity of cells to repair LLOMe-damaged endolysosomes [9]. In addition, we control the subcellular localization of the induced damage and the extent of the endolysosomal perforations by monitoring the leakage of various sizes of fluorophore-conjugated dextrans.

As previously reported, LLOMe-induced membrane damage reduces membrane tension similarly to hyperosmotic conditions

[13]. Therefore, we designed a protocol to assess whether the recruitment of factors involved in endolysosomal damage response is membrane tension-dependent. Finally, we describe how to monitor cell viability after induction of endolysosomal sterile damage using propidium iodide (PI) as a proxy for lytic cell death. This technique is useful to identify mutant strains that fail to repair damaged endolysosomal compartments and maintain proper endolysosomal homeostasis.

2 Materials

2.1 Media, Reagents, Buffers, and Labelling Agents

All membranolytic reagents are prepared from the stock just prior to their use. Work with gloves when using all reagents listed below. Use treated ibidi slides, μ-Slide 2 Well ibiTreat CatNo, 80,286; μ-Slide 4 Well ibiTreat CatNo, 80,426; or μ-Slide 8 Well ibiTreat CatNo, 80,826 (*see* **Note 1**). Follow all waste disposal regulations when discarding waste materials.

2.2 Medium

Filtered HL5c medium (Formedium) prepared according to manufacturer's instructions in double-distilled H_2O and stored at room temperature (RT).

2.3 Microscopy Medium

LoFlo (Formedium) prepared in double-distilled H_2O according to manufacturer's instructions and stored at RT.

2.4 Reagents, Buffers, and Labelling Agents

1. LLOMe (L-leucyl-L-leucine methyl ester, Bachem) stored at −20 °C. Keep powder aliquots of LLOMe of approximately 8–10 mg at −20 °C after transferring from a principal stock in the 4 °C room. This will protect the stocks from humidity. Prior to the experiment, prepare a fresh LLOMe aliquot in the desired medium (filtered HL5c or microscopy medium), and use it promptly (within a few hours). Prepare 50 mM LLOMe by resuspending a powder aliquot in the appropriate volume of medium (8.5 mg of LLOMe powder in 0.5 mL).
2. GPN (glycyl-L-phenylalanine 2-naphthylamide, Sigma) stored at −20 °C. 1 M stocks are prepared in DMSO and stored at −20 °C. From 1 M stock, dilute 10x in DMSO (5 μL 1 M GPN in 45 μL DMSO) for 100 mM GPN. Prepare medium with 5% DMSO—for total of 200 μL, add 10 μL DMSO to 190 μL medium. Remove 4 μL from the 200 μL medium-5% DMSO and add 4 μL of the 100 mM GPN to obtain 2 mM GPN in medium-5% DMSO.
3. Digitonin (Invitrogen) stored at RT. Resuspend 130 mg of digitonin in 1 mL dH2O. Incubate at 85 °C for 3 min. Add 10 μL in 990 μL of the desired medium to obtain 1 mM and use it promptly (within a few hours).

4. 1 M Sorbitol solution stored in double-distilled H_2O at RT.
5. Dimethyl sulfoxide (DMSO), autoclaved and stored at RT.
6. Propidium iodide (PI) stored at −20 °C. Take extra caution as PI is highly toxic.
7. Lysosensor Green DND-189 (Invitrogen) stored at −20 °C.
8. Alexa 647 Dextran 10 kDa (Invitrogen) and Dextran Cascade Blue 3 kDa (Invitrogen) stored at −20 °C.

3 Methods

For microscopy (*see* **Note 2**), cells are imaged in filtered HL5c medium (*see* **Note 3**) or LoFlo medium (*see* **Note 4**). Fluorescent dyes and reagents must be kept in the dark before and after adding them to the cell cultures. For all LLOMe, GPN or digitonin imaging experiments use as positive controls, cells that express GFP-TrafE/TrafE-GFP, ALIX-GFP or GFP-Vps32 no LLOMe, GPN, or digitonin treatments represent negative controls for assessment of the phototoxicity due to fluorescent microscope lasers.

3.1 L-leucyl-L-leucine Methyl Ester (LLOMe)-Induced Sterile Damage

Sterile damage to endolysosomes in *D. discoideum* is induced using 4.5–5 mM of LLOMe from a tenfold concentrated solution added 30 s to 5 min after the start of imaging (Fig. 1). Microscopy is performed for periods of 15–180 min with imaging interval ranging from 30 s to 5 min. We found that 4.5 mM LLOMe is the optimal concentration tolerated by all *D. discoideum* strains for a time period of approximately 2 h, required for most of the imaging purposes. 5 mM LLOMe was found to be the upper limit concentration at which approximately 50% of *D. discoideum* cells die in approximately 2 h. Set the microscopy slide on the microscope, remove the cover, adjust the focus and the desired positions, and at

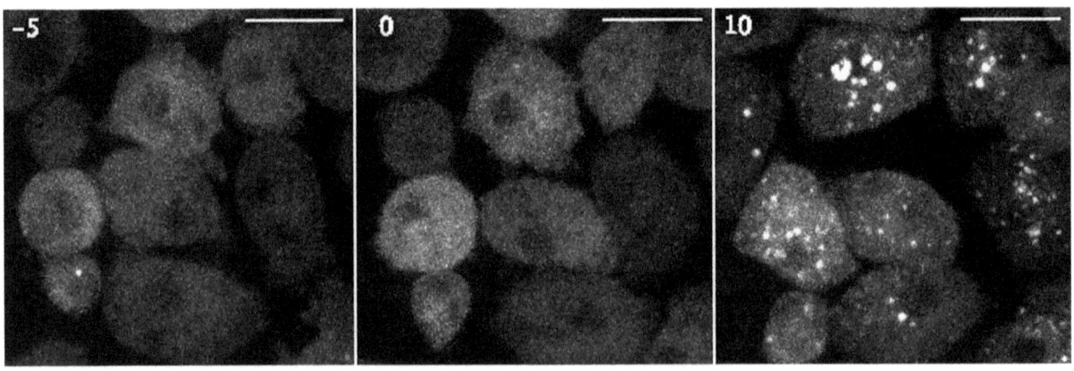

Fig. 1 Spinning disk confocal images show relocalization of the cytosolic E3-ubiquitin ligase GFP-TrafE to damaged endolysosomal membranes 10 min after addition of 4.5 mM LLOMe. Live microscopy time-course initiated 5 min before (−5) addition of LLOMe which was added at time (0). Scale bars correspond to 10 μm

the desired time-point, add gently 50 μL of 50 mM LLOMe solution to 10E6 cells in 450 μL medium to obtain a final concentration of 5 mM LLOMe.

3.2 Glycyl-L-Phenylalanine 2-Naphthylamide (GPN)-Induced Sterile Damage

Sterile damage to endolysosomes in *D. discoideum* is induced using 100–200 μM of GPN from a tenfold concentrated solution added 30 s to 5 min after the start of imaging (Fig. 2). Microscopy is performed for periods of 15–180 min with imaging interval ranging from 30 s to 5 min. Set the microscopy slide on the microscope, remove the cover, adjust the focus and the desired positions, and at the desired time-point, add gently 50 μL from 2 mM GPN to 10E6 cells in 450 μL medium to induce damage at 200 μM GPN.

3.3 Digitonin-Induced Sterile Damage

Plasma membrane sterile damage is induced using 100 μM digitonin added from fivefold concentrated solution 30 s to 5 min after the start of imaging (Fig. 3). Microscopy is performed for periods

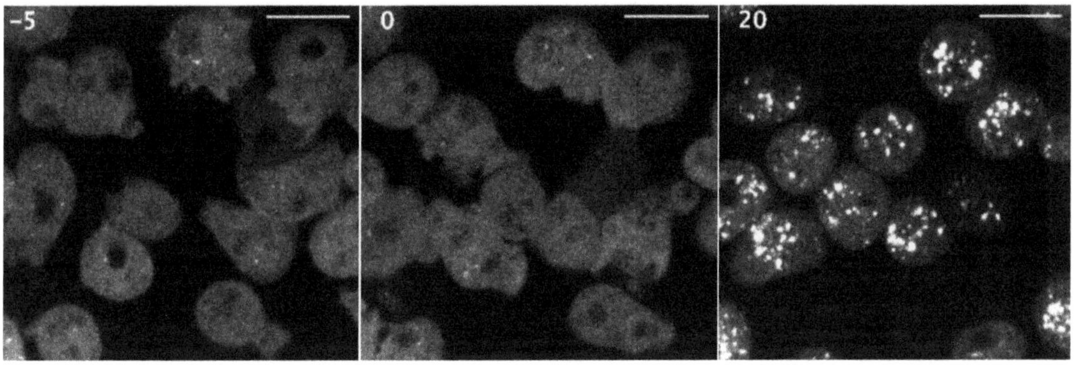

Fig. 2 Spinning disk confocal images show relocalization of cytosolic ESCRT-III subunit GFP-Vps32 to damaged endolysosomal membranes 20 min after addition of 100 μM GPN. Live microscopy time-course initiated 5 min before (−5) addition of GPN which was added at time (0). Scale bars correspond to 10 μm

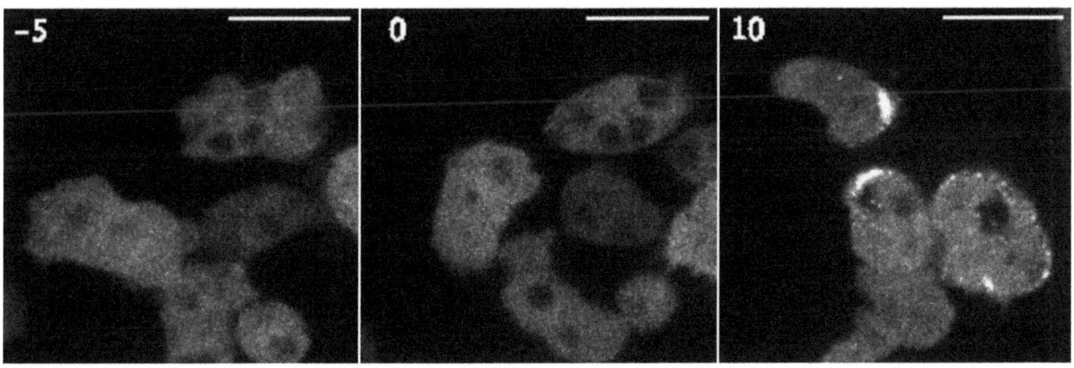

Fig. 3 Spinning disk confocal images show relocalization of cytosolic ESCRT-III subunit GFP-Vps32 to damaged plasma membranes 10 min after addition of 100 μM digitonin. Live microscopy time-course initiated 5 min before (−5) addition of digitonin which was added at time (0). Scale bars correspond to 10 μm

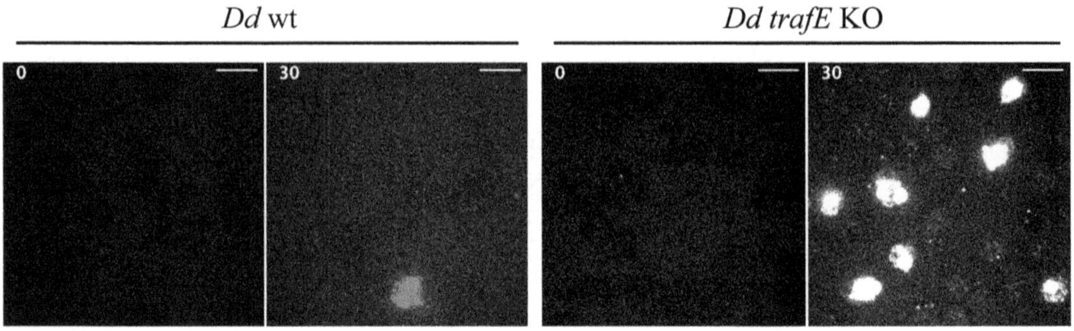

Fig. 4 Spinning disk confocal images show dead cells stained with propidium iodide (PI) 30 min after LLOMe treatment. WT or trafE KO cells were incubated in the presence of PI and monitored by live confocal time-lapse microscopy every 1 min, after addition of 5 mM LLOMe (time = 0) for 30 min (time = 30). Scale bars correspond to 10 μm

of 15–90 min with imaging interval ranging from 30 s to 1 min. Set the microscopy slide on the microscope, remove the cover, adjust the focus and the desired positions, and at the desired time-point, add gently 50 μL–10E6 cells in 450 μL medium to obtain a final concentration of 100 μM digitonin.

3.4 Hyperosmotic Shock

To induce a hyperosmotic shock, set the microscopy slide on the microscope, remove the cover, adjust the focus and the desired positions, and at the desired time-point, add gently 100 μL of 1 M sorbitol 30 s to 5 min after the start of imaging to 10E6 cells in 400 μL medium to a final concentration of 200 mM. Microscopy is performed for periods of 15–180 min with imaging interval of 30 s to 1 min.

3.5 Propidium Iodide Staining

To distinguish live from dead cells (Fig. 4), 1 μg/mL of PI ($\lambda = 536/617$) was added directly to the cell culture immediately prior to time-lapse microscopy. PI can cross compromised membranes; therefore, do not use it as a proxy for cell death while using plasma membrane damaging agents, such as digitonin.

3.6 Lysosensor pH Tracer

Lysosensor is used to monitor the pH of acidic compartments, such as endolysosomes (Fig. 5). For best results in *D. discoideum* cells, 1 μM Lysosensor Green DND-189 is added to the cell culture 3 h prior to microscopy and washed away (*see* **Note 5**) 10–20 min before visualization of the sample.

3.7 Dual-Size Dextran Leakage Assay

The labelled dextrans are hydrophilic polysaccharides that are taken up by the cells and can be used not only as endolysosomal compartments markers but also as a control for endolysosomal membrane damage (Fig. 6). Due to the small-sized membrane damage induced by LLOMe, the 3 kDa Dextran Cascade Blue leaks from

Fig. 5 Spinning disk confocal images show a decrease in lysosensor fluorescence intensity upon addition of LLOMe indicating proton leakage and pH increase within the acidic compartments which is gradually restored demonstrating an extent of repair and recovery in *D. discoideum* cells. Maximum projection of confocal sections of live WT cells incubated with Lysosensor Green DND-189, imaged every 1 min, 5 min before, and 30 min after addition of 4.5 mM LLOMe. Scale bars correspond to 10 μm

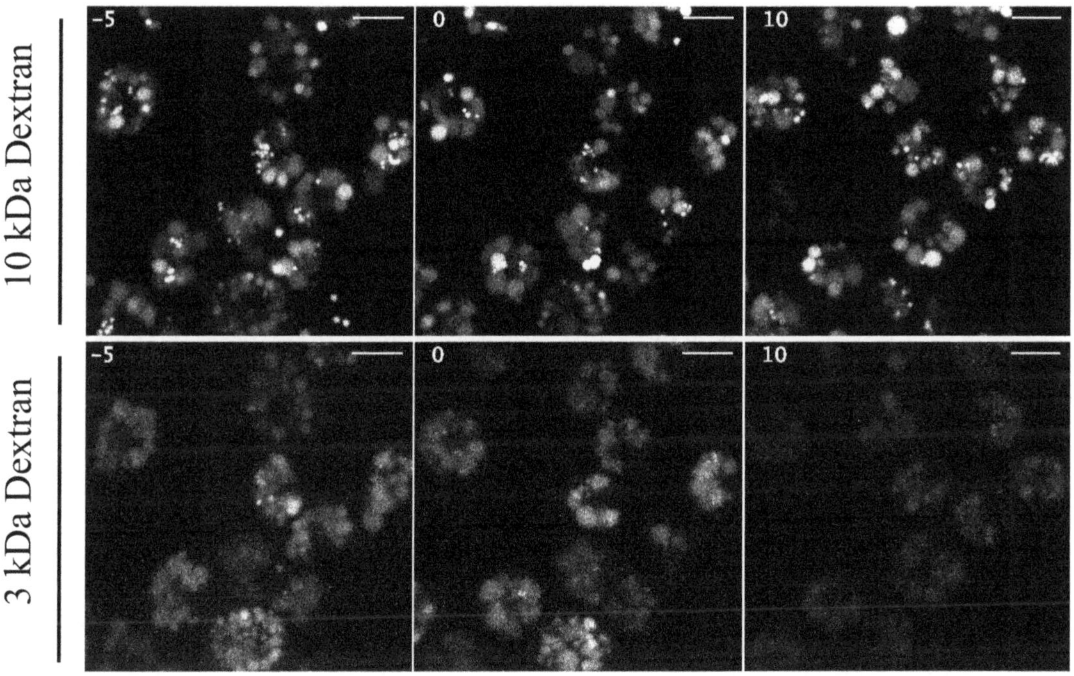

Fig. 6 Live confocal time-lapse microscopy of cells loaded with 3 kDa Cascade Blue Dextran which leaks from the LLOMe-damaged endolysosomal compartments, whereas the 10 kDa Alexa 647 Dextran remains within these compartments. Microscopy time-course initiated 5 min before (−5) addition of LLOMe which was added at time (0). Scale bars correspond to 10 μm

the endolysosomal compartments, whereas the 10 kDa Alexa 647 Dextran remains within not severely damaged compartments. Microscopy is performed for periods of 15–180 min with imaging interval ranging from 30 s to 5 min.

1. LLOMe or GPN damage or hyperosmotic conditions are induced as described in Subsections 3.1, 3.2, and 3.4.
2. *D. discoideum* cells are incubated for at least 3 h with 3 mg/mL Alexa 647 Dextran 10 kDa and/or Dextran Cascade Blue 3 kDa.
3. Dextran-containing medium is washed (*see* **Note 5**) away promptly prior to microscopy.

3.8 Microscopy

Live microscopy was performed using inverted 3i Marianas or Nikon Ti CSU-W1 spinning disk microscopes with a 63× glycerol/air or 100× oil objectives. 7–9 slices 0.5–1 μm apart were taken for z-stacks. Image analysis was performed using Fiji (ImageJ).

4 Notes

1. Cells are grown on 10 cm petri dishes from which cells are resuspended and counted in order to prepare a culture of 10E5 cells per mL in microscopy ibidi slides which overnight will reach approximately 10E6 cells per mL. The 2-well ibidi can accommodate up to 2 mL, the 4-well ibidi can accommodate up to 1 mL, and the 8-well can accommodate up to 500 μL. Prior to microscopy and addition of membranolytic components, the volume can be adjusted by gently pipetting out the desired amount of medium. Before the adjustment of the microscope focus and imaging position, the ibidi covers must be removed to allow the addition of the desired compounds.
2. Live fluorescence microscopy should be performed with the minimal laser power that allows proper visualization in the absence of a strong fluorescent background. To avoid extra phototoxicity, minimize the number of z-axis sections to 3 or 5 with 0.5 μm to 1 μm distance between them and increase the interval time between images.
3. Cells should be cultured in filtered HL5c medium at least 24 h prior to microscopy to avoid the autofluorescence of the autoclaved medium.
4. 20 min prior to microscopy, the media of cells precultured for at least 24 h in filtered HL5c is replaced with LoFlo medium.
5. To avoid the detachment of cells from the microscopy dishes and slides, perform washes by gentle pipetting the media out and adding it on the dish wall.

References

1. Dacks JB, Peden AA, Field MC (2009) Evolution of specificity in the eukaryotic endomembrane system. Int J Biochem Cell Biol 41:330–340
2. Cooper ST, McNeil PL (2015) Membrane repair: mechanisms and pathophysiology. Physiol Rev 95:1205–1240
3. Papadopoulos C, Meyer H (2017) Detection and clearance of damaged lysosomes by the endo-lysosomal damage response and lysophagy. Curr Biol 27:R1330–R1341
4. Togo T (2019) Signaling pathways involved in adaptive responses to cell membrane disruption. Curr Top Membr 84:99–127
5. Daussy CF, Wodrich H (2020) "Repair me if you can": membrane damage, response, and control from the viral perspective. Cells 9:2042
6. Vietri M, Radulovic M, Stenmark H (2020) The many functions of ESCRTs. Nat Rev Mol Cell Biol 21:25–42
7. Schuck S (2020) Microautophagy—distinct molecular mechanisms handle cargoes of many sizes. J Cell Sci 133:jcs246322
8. López-Jiménez AT, Cardenal-Muñoz E, Leuba F, Gerstenmaier L, Barisch C, Hagedorn M, King JS, Soldati T (2019) The ESCRT and autophagy machineries cooperate to repair ESX-1-dependent damage at the Mycobacterium-containing vacuole but have opposite impact on containing the infection. PLoS Pathog 14:e1007501
9. Raykov L, Mottet M, Nitschke J, Soldati T (2023) A TRAF-like E3 ubiquitin ligase TrafE coordinates ESCRT and autophagy in endolysosomal damage response and cell-autonomous immunity to Mycobacterium marinum. elife 12:e85727
10. Benhar M (2017) Application of a thioredoxin-trapping mutant for analysis of the cellular nitrosoproteome. Methods Enzymol 585:285–294
11. Repnik U, Borg Distefano M, Speth MT, Ng MYW, Progida C, Hoflack B, Gruenberg J, Griffiths G (2017) L-leucyl-L-leucine methyl ester does not release cysteine cathepsins to the cytosol but inactivates them in transiently permeabilized lysosomes. J Cell Sci 130:3124–3140
12. Yuan Y, Kilpatrick BS, Gerndt S, Bracher F, Grimm C, Schapira AH, Patel S (2021) The lysosomotrope GPN mobilises Ca(2+) from acidic organelles. J Cell Sci 134:jcs256578
13. Mercier V, Larios J, Molinard G, Goujon A, Matile S, Gruenberg J, Roux A (2020) Endosomal membrane tension regulates ESCRT-III-dependent intra-lumenal vesicle formation. Nat Cell Biol 22:947–959

Chapter 4

Assaying Lysosomal Enzyme Activity in *Dictyostelium discoideum*

William D. Kim, Adam F. DiGiacinto, and Robert J. Huber

Abstract

Lysosomes are membrane-enclosed organelles that digest intracellular material. They contain more than 50 different enzymes that can degrade a variety of macromolecules including nucleic acids, proteins, polysaccharides, and lipids. In addition to functioning within lysosomes, lysosomal enzymes are also secreted. Alterations in the levels and activities of lysosomal enzymes dysregulates lysosomes, which can lead to the intralysosomal accumulation of biological material and the development of lysosomal storage diseases (LSDs) in humans. *Dictyostelium discoideum* has a long history of being used to study the trafficking and functions of lysosomal enzymes. More recently, it has been used as a model system to study several LSDs. In this chapter, we outline the methods for assessing the activity of several lysosomal enzymes in *D. discoideum* (α-galactosidase, β-galactosidase, α-glucosidase, β-glucosidase, β-N-acetylglucosaminidase, α-mannosidase, cathepsin B, cathepsin D, cathepsin F, palmitoyl protein thioesterase 1, and tripeptidyl peptidase 1).

Key words Cathepsin, *Dictyostelium discoideum*, Enzyme, Hydrolase, Lysosome, Model system, Thioesterase

Abbreviations

DI	De-ionized
LIMP-2	Lysosomal integral membrane protein 2
LSD	Lysosomal storage disease
M6P	Mannose-6-phosphate
MES	2-(N-morpholino)ethanesulfonic acid
SM	Sussman's medium

1 Introduction

Lysosomes are membrane-bound organelles that contain enzymes capable of digesting intracellular material [1–3]. There are

Alan R. Kimmel (ed.), *Dictyostelium discoideum: Methods and Protocols*, Methods in Molecular Biology, vol. 2814, https://doi.org/10.1007/978-1-0716-3894-1_4,

hundreds of lysosomes within cells and each one maintains an acidic environment (pH 4.5–5.0) that is required for lysosomal enzyme activity [1, 4–6]. There are more than 50 different lysosomal enzymes including various nucleases, lipases, glycoside hydrolases, proteases, and phosphatases [1, 7]. Lysosomal enzymes are synthesized in the endoplasmic reticulum and trafficked to the lysosome via the mannose-6-phosphate (M6P)-dependent pathway [5, 8–10]. In this pathway, lysosomal enzymes are tagged with M6P in the trans Golgi complex and then bind to M6P receptors. Through this interaction, the enzymes are targeted to a pre-lysosomal vesicle, which buds off from the Golgi complex and fuses with an endosome that contains internalized material. The fused vesicle then undergoes intraluminal acidification to form a lysosome [5, 8–10]. There are also M6P receptor-independent pathways that target enzymes to lysosomes such as those that use clathrin, sortilin, or lysosomal integral membrane protein 2 (LIMP-2) [5, 9, 11, 12]. Additionally, lysosomal enzymes can be released from cells through various processes such as lysosomal exocytosis and autophagy induction [13–16]. Outside cells, lysosomal enzymes stimulate the endocytosis of damaged regions of the plasma membrane, modulate adhesion, and reshape the extracellular matrix [2, 16–18]. There are also pathways that re-internalize secreted lysosomal enzymes and transport them to lysosomes [8–10, 19].

In humans, alterations in the amounts and activities of lysosomal enzymes can compromise the function of lysosomes, ultimately leading to the intralysosomal accumulation of biological material and the development of lysosomal storage diseases (LSDs) [20]. A few examples of LSDs caused by the absence or reduced activity of a single lysosomal enzyme include Fabry disease (α-galactosidase), Morquio B disease (β-galactosidase), Pompe disease (α-glucosidase), Gaucher disease (β-glucosidase), Tay-Sachs disease (β-hexosaminidase A), α-mannosidosis (α-mannosidase), and several subtypes of neuronal ceroid lipofuscinosis (CLN1 disease, palmitoyl protein thioesterase 1; CLN2 disease, tripeptidyl peptidase 1; CLN10 disease, cathepsin D; CLN13 disease, cathepsin F) [21–26].

Historically, *Dictyostelium discoideum* has been used as a model system to study the trafficking and functions of lysosomal enzymes. Previous work showed that proteolytic processing of enzyme precursors is required for optimal trafficking of lysosomal enzymes in *D. discoideum* [27]. In addition, lysosomal enzymes in *D. discoideum* appear to have a common modification and are recognized by a receptor in the endoplasmic reticulum [28, 29]. *D. discoideum* cells also secrete precursor and mature forms of lysosomal enzymes during growth and the early stages of multicellular development (Table 1) [30–33]. The sorting of secreted and intracellular pools of lysosomal enzymes occurs in

Table 1
List of lysosomal enzymes detected in conditioned buffer harvested from developing wild type cells (AX2 and AX3)

Lysosomal enzyme	dictyBase protein ID	AX2[a]	AX3[b]
α-Galactosidase			
MelA	DDB0233870	No	Yes
β-Galactosidase			
Glb1	DDB0266380	Yes	Yes
Glb2	DDB0266381	No	No
α-Glucosidase			
Gaa	DDB0237578	Yes	Yes
ModA	DDB0191113	No	Yes
β-Glucosidase			
GluA	DDB0215373	Yes	Yes
β-N-Acetylglucosaminidase			
NagA	DDB0191256	No	Yes
NagB	DDB0304517	Yes	Yes
NagC	DDB0304520	No	No
NagD	DDB0304516	Yes	Yes
NagE	DDB0304521	No	No
α-Mannosidase			
ManA	DDB0201569	Yes	Yes
ManB	DDB0231612	No	Yes
ManC	DDB0231613	No	No
ManD	DDB0231614	No	No
ManE	DDB0231615	No	No
ManF	DDB0231616	No	No
ManG	DDB0231611	No	No
Cathepsin B			
CtsB	DDB0233997	Yes	Yes
Cathepsin D			
CtsD	DDB0215012	Yes	Yes
Cathepsin F			
CprA	DDB0201647	No	Yes
CprB	DDB0214998	No	Yes

(continued)

Table 1
(continued)

Lysosomal enzyme	dictyBase protein ID	AX2[a]	AX3[b]
CprE	DDB0185092	Yes	Yes
DDB0252831	DDB0252831	Yes	Yes
Palmitoyl protein thioesterase 1			
Ppt1	DDB0233890	Yes	Yes
Tripeptidyl peptidase 1			
Tpp1A	DDB0234303	No	No
Tpp1B	DDB0306176	Yes	Yes
Tpp1C	DDB0185020	Yes	Yes
Tpp1D	DDB0309140	No	No
Tpp1E	DDB0304592	No	No
Tpp1F	DDB0214912	Yes	Yes

[a]Detected in conditioned buffer during multicellular development on membranes
[b]Detected in conditioned buffer after 6 h of starvation in KK2 buffer on Petri dishes

the Golgi complex, and the release of these enzymes is due to an active secretory process, not a passive leakage of enzymes from cells or their incidental release when digested material is exocytosed [31, 32]. In *D. discoideum*, secreted lysosomal enzymes play an important role in regulating cell-substrate adhesion during the early stages of multicellular development [31, 34].

Recently, *D. discoideum* has been used to study several human diseases, including those caused by deficiencies in lysosomal enzymes [35]. The *D. discoideum* genome encodes an estimated 12,500 proteins, which includes homologs of several human lysosomal enzymes such as α-galactosidase, β-galactosidase, α-glucosidase, β-glucosidase, β-N-acetylglucosaminidase, α-mannosidase, cathepsin B, cathepsin D, palmitoyl protein thioesterase 1, and several proteins that share sequence similarity with human cathepsin F and tripeptidyl peptidase 1 [36]. Previous research in *D. discoideum* showed that knockout or suppression of genes encoding lysosomal enzymes affects various cellular and developmental processes during the life cycle (e.g., [37–39]). In this chapter, we outline methods (Figs. 1 and 2) for assessing the activity of several lysosomal enzymes including α-galactosidase, β-galactosidase, α-glucosidase, β-glucosidase, β-N-acetylglucosaminidase, α-mannosidase, cathepsin B, cathepsin D, cathepsin F, palmitoyl protein thioesterase 1, and tripeptidyl peptidase 1.

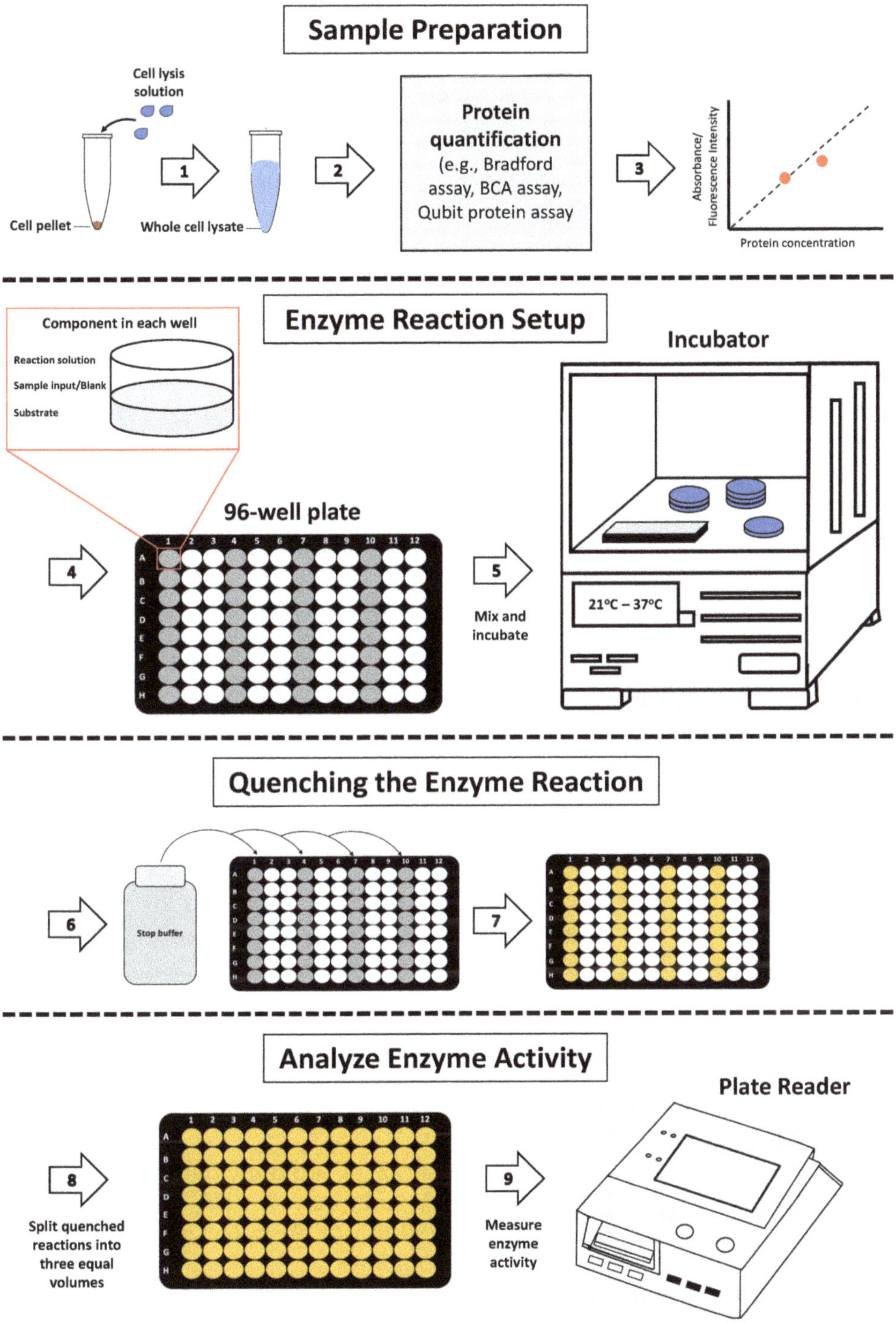

Fig. 1 *General outline of an enzyme assay.* Cell pellets during growth or starvation are lysed using cell lysis solution, and protein concentrations in whole cell lysates and conditioned buffer are quantified using biochemical techniques (*see* **Note 1**). Protein samples are diluted to equal protein concentrations with the solution the proteins are suspended in (e.g., cell lysis solution for whole cell lysate samples, KK2 buffer for conditioned buffer). In general, an enzyme reaction mixture includes the enzyme reaction solution, an equal protein amount from the whole cell lysate or conditioned buffer, and the enzyme substrate within a single well

2 Materials

2.1 Cell Culturing and Buffers

1. All *D. discoideum* cell lines are maintained on Sussman's medium (SM)/2 agar plates: 5 g/L glucose, 1 g/L $MgSO_4 \cdot 7H_2O$, 5 g/L bacto peptone, 0.5 g/L yeast extract, 1 g/L Na_2HPO_4, 2.2 g/L KH_2PO_4, 15 g/L agar, and pH 6.5 (*see* **Note 2**).
2. All cell lines are cultured in HL5 growth medium including glucose (supplier: Formedium, Hunstanton, Norfolk, United Kingdom, product code: HLG0102) in a 250 mL Erlenmeyer flask (150RPM/21 °C) (*see* **Note 3**). The axenic medium is supplemented with 100 μg/mL ampicillin and 300 μg/mL streptomycin sulfate to prevent bacterial growth. For knockout cell lines, 10 μg/mL of blasticidin S hydrochloride is added for selection.
3. Cells are starved by incubating them in KK2 buffer: 0.7 g/L K_2HPO_4, 2.2 g/L KH_2PO_4, and pH 6.5 (*see* **Note 4**).

2.2 Enzyme Assays

Deionized (DI) water is used to prepare all solutions, which are prepared at room temperature and disposed of appropriately when the assays are completed (refer to manufacturer's and MSDS recommendations for disposal). All solutions (e.g., cell lysis solution) are filtered using a syringe and 0.45 μm filter to remove potential contaminants and any free-floating particulates. HCl and NaOH are used to adjust the pH of all solutions.

1. *Cell lysis solution:* 0.1% (v/v) NP-40 in 0.05M 2-(N-morpholino)ethanesulfonic acid (MES) (pH 6.5) (100 mL). Add 0.976 g MES and 100 μL of NP-40 to 90 mL of DI water. Adjust the pH to 6.5, top up the volume to 100 mL with DI water, and adjust the pH to 6.5.

2.2.1 Lysosomal Glycoside Hydrolases

1. α-**Galactosidase**

   Substrate: p-Nitrophenyl α-D-galactopyranoside (supplier: Sigma-Aldrich, product code: N0877). Dissolve in methanol.

   Reaction solution: 0.15 M sodium citrate-phosphate solution (pH 4.5). Make solutions of 0.15 M citric acid (4.32 g citric acid in 150 mL of DI water) and 0.15 M disodium

Fig. 1 (continued) of a 96-well plate. A blank reaction mixture contains all components except proteins (e.g., cell lysis solution or KK2 buffer) and is included in the same 96-well plate to account for background signal generated by the solution. Reaction mixtures are mixed and incubated at a specific temperature and time. Subsequently, the reactions are quenched using an enzyme stop solution and mixed, and the total quenched enzyme volumes are split equally into three individual wells adjacent to each other. Finally, absorbance or fluorescence intensity (determined by substrate) is measured in each well using a plate reader

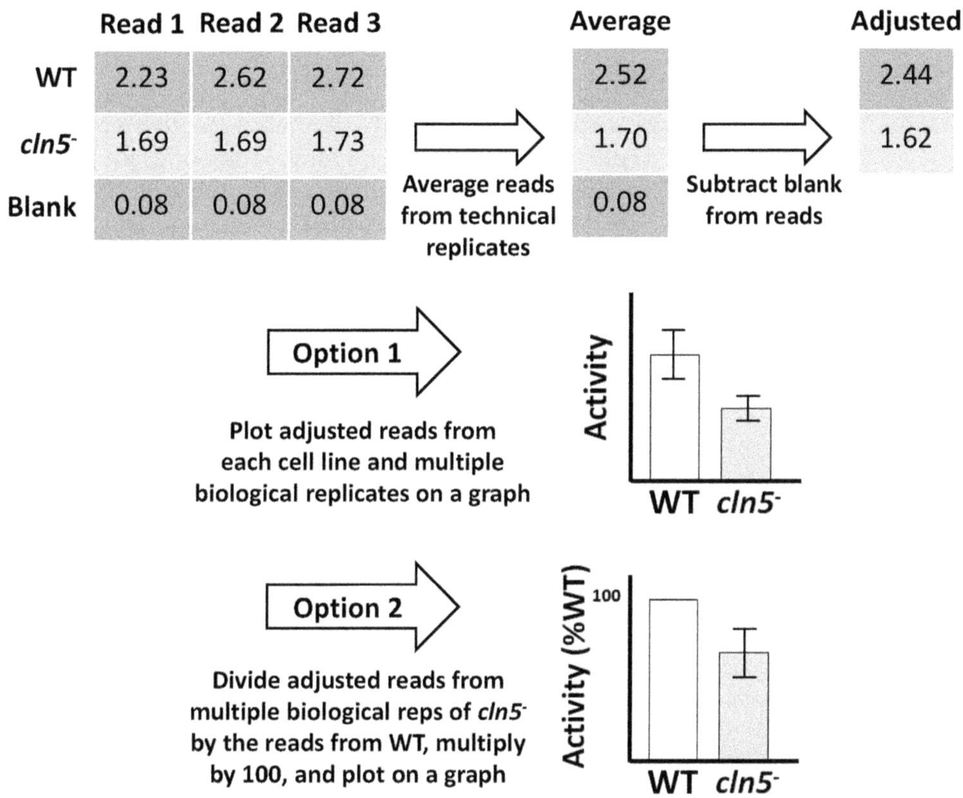

Fig. 2 Flowchart for analyzing and presenting enzyme activity data. β-N-Acetylglucosaminidase activity in AX3 (wild type, WT) and *cln5*⁻ knockout whole cell lysates. Triplicate technical readings from one biological replicate and the blank are shown. Averaged reads from samples are subtracted from the averaged blank value and can be analyzed in two different ways. Option 1 includes plotting the averaged values for all biological replicates of each cell line onto a single graph. In Option 2, the averaged read values from the knockout cell line or cells exposed to a treatment are divided by values from the WT cell line or control sample, respectively. These normalized values are plotted onto a graph as "Activity (% WT/control)." In both Options, at least three biological replicates are required for statistical analysis

phosphate (1.92 g disodium phosphate in 150 mL of DI water). Mix the two solutions until a pH of 4.5 is achieved.

Stop solution: 0.5 M sodium glycinate (pH 10.4). Make solutions of 0.5 M glycine solution (11.26 g of glycine in 400 mL of DI water) and 0.5 M NaOH (15 mL of 10 N NaOH with 285 mL of DI water). Mix the two solutions until a pH of 10.4 is achieved.

2. **β-Galactosidase**

Substrate: o-Nitrophenyl-β-D-galactopyranoside (supplier: Sigma-Aldrich, product code: 48712-M). Dissolve in reaction solution (see below).

Reaction solution: 0.1 M citrate solution (pH 4.15) (400 mL). Combine 6.17 g of citric acid with 2.32 g of trisodium citrate dihydrate in 350 mL of DI water. Adjust the pH with HCl/NaOH to 4.15. Top up volume to 400 mL with DI and adjust the pH to 4.15.

Stop solution: 0.75 M 2-amino-2methyl-propanol/HCl (pH 10.25) (400 mL). Dilute stock of 2-amino-2methyl-propanol to 0.75 M and adjust the pH to 10.25 with HCl/NaOH. Top up with DI water and adjust the pH to 10.25.

3. α-**Glucosidase**

Substrate: p-Nitrophenyl-α-D-glucopyranoside (supplier: Sigma-Aldrich, product code: 487506). Dissolve in reaction solution (see below).

Reaction solution: 0.1 M sodium succinate (pH 6.0). Add 32.41 g sodium succinate dibasic to 200 mL of DI water. Adjust the pH to 6.0 and top up volume to 250 mL with DI water.

Stop solution: 1 M sodium carbonate (400 mL). Add 42.4 g of sodium carbonate to 300 mL of DI water. Once fully dissolved, top up volume to 400 mL with DI water.

4. β-**Glucosidase**

Substrate: p-Nitrophenyl-β-D-glucopyranoside (supplier: Sigma-Aldrich, product code: 487507). Dissolve in reaction solution (see below).

Reaction solution: 50 mM sodium acetate solution (pH 5.0) (400 mL). Combine 1.042 g of sodium acetate with 420 μL of 17.4 M acetic acid in 350 mL of DI water. Adjust the pH to 5. Top up volume to 400 mL with DI and adjust the pH to 5.

Stop solution: 1 M sodium carbonate (400 mL). Add 42.4 g of sodium carbonate to 300 mL of DI water. Once fully dissolved, top up volume to 400 mL with DI water.

5. α-**Mannosidase**

Substrate: p-Nitrophenyl-α-D-mannopyranoside (supplier: Sigma-Aldrich, product code: N2127). Dissolve in DMSO.

Reaction solution: 5 mM sodium acetate solution (pH 5.0) (400 mL). Combine 0.1042 g of sodium acetate with 42 μL of 17.4 M acetic acid in 350 mL of DI water. Adjust the pH to 5. Top up volume to 400 mL with DI and adjust the pH to 5.

Stop solution: 1 M sodium carbonate (400 mL). Add 42.4 g of sodium carbonate to 300 mL of DI water. Once fully dissolved, top up volume to 400 mL with DI water.

6. **β-N-Acetylglucosaminidase**

   Substrate: p-Nitrophenyl N-acetyl-β-D-glucosaminide (supplier: Sigma-Aldrich, product code: N9376). Dissolve in reaction solution (see below).

   Reaction solution: 100 mM sodium acetate solution (pH 5.0) (250 mL). Combine 20.435 g of sodium acetate with 52.4 μL of 17.4 M acetic acid in 175 mL of DI water. Adjust the pH to 5. Top up volume to 250 mL with DI and adjust the pH to 5.

   Stop solution: 1 M sodium carbonate (400 mL). Add 42.4 g of sodium carbonate to 300 mL of DI water. Once fully dissolved, top up volume to 400 mL with DI water.

2.2.2 Lysosomal Proteases

1. **Cathepsin B**

   Substrate: Cathepsin B substrate III, Fluorogenic (supplier: Sigma-Aldrich, product code: 219392). Dissolve in DMSO.

   Reaction solution: Combine 2.73 g of disodium phosphate, 19.16 g monopotassium phosphate, and 0.596 g disodium EDTA in 325 mL of DI water. Adjust the pH to 6.0, top up volume to 400 mL with DI water, and adjust the pH to 6.0. Make a 200 mM cysteine solution using DI water. Dilute the cysteine stock solution down to 8 mM with the cathepsin B reaction solution and prepare enough solution for all reactions (determined by researcher).

   Stop solution: Combine 4.66 g of sodium chloroacetate, 1.63 g of sodium acetate, and 1.609 mL of 17.4 M acetic acid in 300 mL of DI water. Adjust the pH to 4.3, top up volume to 400 mL with DI water, and adjust the pH to 4.3.

2. **Cathepsin D**

   Kit: Cathepsin D Activity Fluorometric Assay Kit (supplier: BioVision, product code: K143–100).

3. **Cathepsin F**

   Substrate: Fluorogenic Cathepsin F Substrate (supplier: BPS Bioscience, product code: 80350).

   Reaction solution: 0.1 M potassium phosphate solution (pH 6.5) (250 mL). Combine 2.848 g of monopotassium phosphate, 0.71 g of dipotassium phosphate, 0.073 g of disodium EDTA, and 416.3 μL of 60% PEG 600 in 150 mL of DI water. Adjust the pH to 6.5, top up volume to 250 mL with DI water, and adjust the pH to 6.5.

   Other reagents: Pepsin (supplier: Sigma Aldrich, product code: 10108057001).

4. **Tripeptidyl peptidase 1**

   Substrate: Ala-Ala-Phe-7-amido-4-methylcoumarin (supplier: Sigma-Aldrich, product code: A3401). Dissolve in ethanol.

   Reaction solution: Combine 3.506 g of sodium chloride, 3.28 g of sodium acetate, and 400 μL of 100% Triton X-100 in 300 mL of DI water. Adjust the pH to 4.5, top up volume to 400 mL with DI water, and adjust the pH to 4.5.

   Stop solution: Combine 3.506 g of sodium chloride and 3.28 g of sodium acetate in 300 mL of DI water. Adjust the pH to 4.3, top up volume to 400 mL with DI water, and adjust the pH to 4.3.

2.2.3 Lysosomal Thioester Hydrolase

1. **Palmitoyl Protein Thioesterase 1**

   Substrate: 4-Methylumbelliferyl 6-thio-palmitate-β-D-glucopyranoside (supplier: Cayman Chemical, product code: 19524). Dissolve in DMSO.

   Reaction solution: McIlvain's phosphate/citrate solution supplemented with 0.375% (v/v) Triton X-100 and 15 mM dithiothreitol (pH 4.0) (500 mL). Combine 192.75 mL of 0.2 M disodium phosphate, 307.25 mL of 0.1 M citric acid, 1.157 g of dithiothreitol, and 1.875 mL of Triton X-100. Adjust the pH to 4.0.

   Stop solution: Combine 21.2 g of sodium carbonate, 0.8 mg of sodium bicarbonate, and 100 μL of Triton X-100 (final concentration: 0.025% (v/v)) in 350 mL of DI water. Adjust the pH to 10.68, top up volume to 400 mL with DI water, and adjust the pH to 10.68.

   Other reagents: Lyophilized powder of β-glucosidase (supplier: Sigma-Aldrich, product code: 49290).

2.3 Other Materials

1. *Consumables:* 96-well optical (clear) bottom black plate (for absorbance-based readings) (supplier: Fisher Scientific, product code: 165305) (*see* **Note 5**); 96-well opaque bottom black plate (for fluorescence-based readings) (supplier: VWR, product code: 76221–764) (*see* **Note 6**); Amicon Ultra-4 Centrifugal Filter protein concentration columns (supplier: Sigma-Aldrich, product code: UFC801024) (*see* **Note 7**).
2. *Equipment:* Dry block, centrifuge, and microplate reader.

3 Methods

3.1 Assessing Enzyme Activity During Growth

1. Deposit 2.25 × 10^6 cells in the mid-log phase of growth (1–5 × 10^6 cells/mL) into a 60 mm × 15 mm Petri dish containing 4 mL of HL5 medium supplemented with the appropriate antibiotics (*see* Subheading 2.1 and **Note 8**).

2. Allows the cells to adhere to the dish for at least 45 min.
3. Replace the medium with 4 mL of fresh HL5 medium supplemented with the appropriate antibiotics (*see* Subheading 2.1 and **Notes 3** and **4**).
4. Incubate the cells overnight at 21 °C (18–24 h).
5. After incubation, collect all the conditioned medium into a 15 mL tube and place the tube on ice. Then proceed to **Step 6**. Once **Step 6** is complete, return to **Step 5.1**.
   - 5.1. Spin the conditioned medium in a centrifuge (2000RPM/4 °C) for 5 min (*see* **Note 9**).
   - 5.2. Carefully aliquot 3 mL of the clarified conditioned medium into three 1.5 mL tubes (1 mL each).
   - 5.3. Store the 1.5 mL tubes at −80 °C. This sample can be used to determine the extracellular activity of a secreted lysosomal enzyme. Do not allow the aliquots to undergo more than one freeze-thaw cycle.
6. Resuspend the cells adhered to the dish in 1 mL of fresh ice-cold HL5 medium. Transfer the cell suspension to a 1.5 mL tube.
   - 6.1. Spin the cell suspension in a centrifuge (2000RPM/4 °C) for 5 min. Remove the supernatant and store the cell pellet at −80 °C. This sample can be used to determine the intracellular activity of a lysosomal enzyme.
     - 6.1.1. For the cathepsin D assay, collect 2×10^6 cells into a cell pellet and store at −80 °C.

3.2 Assessing Enzyme Activity During Starvation

1. Carry out **steps 1–4** in Subheading 3.1.
2. After incubating the cells overnight (18–24 h), check the confluency between cell lines and ensure the cell density is the same (*see* **Note 8**).
3. Remove the conditioned medium and perform two washes with 4 mL of KK2 buffer.
4. Submerge the cells in 4 mL of KK2 buffer and incubate at 21 °C for the desired time (*see* **Note 10**).
5. After starving the cells, collect all the conditioned buffer into a 15 mL tube and place the tube on ice. Then proceed to **Step 6**. Once **Step 6** is complete, return to **Step 5.1**.
   - 5.1. Spin the conditioned buffer in a centrifuge (2000RPM/4 °C) for 5 min (*see* **Note 9**).
   - 5.2. Carefully aliquot 3 mL of the clarified conditioned buffer into three 1.5 mL tubes (1 mL each) (*see* **Note 11**).

5.3. Store the 1.5 mL tubes at −80 °C. This sample can be used to determine the extracellular activity of a secreted lysosomal enzyme since previous work has detected several lysosomal enzymes in conditioned buffer from developing cells (Table 1). Do not allow the aliquots to undergo more than one freeze-thaw cycle.

6. Resuspend the cells adhered to the dish in 1 mL of fresh ice-cold KK2 buffer. Transfer the cell suspension to a 1.5 mL tube.

6.1. Spin the cell suspension in a centrifuge (2000RPM/4 °C) for 5 min. Remove the supernatant and store the cell pellet at −80 °C. This sample can be used to determine the intracellular activity of a lysosomal enzyme.

6.1.1. For the cathepsin D assay, collect 2×10^6 cells into a cell pellet and store at −80 °C.

3.3 Sample Preparation

1. Cell pellets and conditioned media/buffer are thawed on ice.
2. Lyse thawed cell pellets with ice-cold cell lysis solution and place on ice (*see* **Note 12**).
3. Determine the protein concentration in each sample (*see* **Note 1**).
4. Adjust the protein concentration by mixing the cell lysate with the cell lysis solution in a new 1.5 mL tube that results in the same protein concentration across all samples.
5. In all assays, equal amounts of protein, ranging from 100 μg or more, should be used (*see* **Note 13**).

3.4 Assessing Lysosomal Hydrolase Activity

3.4.1 α-Galactosidase

The method for assessing α-galactosidase activity is adapted from Kilpatrick and Stirling (1976) [40].

1. Designate three wells in a 96-well optical (clear) bottom black plate for the blank and three wells for the experimental sample (*see* **Note 14**).
2. Add 18 μL of 20 mM α-galactosidase substrate stock to one of the three designated wells for the blank and one of the three designated wells for the experimental sample.
3. Add the blank or the experimental sample to the designated wells containing α-galactosidase substrate.
4. Add the α-galactosidase reaction buffer to the wells containing both α-galactosidase substrate and the blank or the experimental sample to a total volume of 180 μL per well. This is referred to as the reaction solution.
5. Gently mix all reaction solutions using a pipettor.
6. Incubate the 96-well plate at 37 °C for 1 h.

7. Stop the reactions by adding an equal volume (180 μL) of α-galactosidase stop solution to the wells containing the reaction solutions. Gently mix the solutions using a pipettor.
8. Once the solutions are mixed, aliquot the quenched reaction solutions two times (120 μL each) into the two remaining wells designated for the blank and the two remaining wells designated for the experimental sample. These will serve as technical replicates for the blank and the experimental sample.
9. Using a microplate reader, read the 96-well plate at an absorbance wavelength of 405 nm.

3.4.2 β-Galactosidase

The method for assessing β-galactosidase activity is adapted from Maruhn (1976) [41] (*see* **Note 15**).

1. Designate three wells in a 96-well optical (clear) bottom black plate for the blank and three wells for the experimental sample (*see* **Note 14**).
2. Add 36 μL of 25 mM β-galactosidase substrate stock to one of the three designated wells for the blank and one of the three designated wells for the experimental sample.
3. Add the blank or the experimental sample to the designated wells containing β-galactosidase substrate.
4. Add the β-galactosidase reaction buffer to the wells containing both β-galactosidase substrate and the blank or the experimental sample to a total volume of 180 μL per well. This is referred to as the reaction solution.
5. Gently mix all reaction solutions using a pipettor.
6. Incubate the 96-well plate at 37 °C for 1 h.
7. Stop the reactions by adding an equal volume (180 μL) of β-galactosidase stop solution to the reaction solutions. Gently mix the solutions using a pipettor.
8. Once the solutions are mixed, aliquot the quenched reaction solutions two times (120 μL each) into the two remaining wells designated for the blank and the two remaining wells designated for the experimental sample. These will serve as technical replicates for the blank and the experimental sample.
9. Using a microplate reader, read the 96-well plate at an absorbance wavelength of 405 nm.

3.4.3 α-Glucosidase

The method for assessing α-glucosidase activity is adapted from Wimmer et al. (1997) [42] (*see* **Note 16**).

1. Add 15 μL of 20 mM α-glucosidase substrate stock into two 1.5 mL tubes: one for the blank and one for the experimental sample.

2. Add the blank or the experimental sample to each designated 1.5 mL tube containing the α-glucosidase substrate (*see* **Note 14**).
3. Add α-glucosidase reaction buffer to both 1.5 mL tubes to achieve a total volume of 150 μL. This is referred to as the reaction solution.
4. Gently mix all reaction solutions using a pipettor.
5. Incubate the 1.5 mL tubes at 65 °C overnight.
6. Stop the reactions by adding an equal volume (150 μL) of α-glucosidase stop solution to the reaction solutions. Gently mix the solutions using a pipettor.
7. Once the solutions are mixed, aliquot the quenched reaction solutions three times (100 μL each) into three wells designated for the blank and the three wells designated for the experimental sample. These will serve as technical replicates for the blank and the experimental sample.
8. Using a microplate reader, read the 96-well plate at an absorbance wavelength of 395 nm.

3.4.4 β-Glucosidase

The method for assessing β-glucosidase activity is adapted from Coston and Loomis (1969) [43].

1. Designate three wells in a 96-well optical (clear) bottom black plate for the blank and three wells for the experimental sample (*see* **Note 14**).
2. Add 18 μL of 100 mM β-glucosidase substrate stock to one of the three designated wells for the blank and one of the three designated wells for the experimental sample.
3. Add the blank or the experimental sample to the designated wells containing β-glucosidase substrate.
4. Add the β-glucosidase reaction buffer to the wells containing both β-glucosidase substrate and the blank or the experimental sample to a total volume of 180 μL per well. This is referred to as the reaction solution.
5. Gently mix all reaction solutions using a pipettor.
6. Incubate the 96-well plate at 35 °C for 1 h.
7. Stop the reactions by adding an equal volume (180 μL) of β-glucosidase stop solution to the reaction solutions. Gently mix the solutions using a pipettor.
8. Once the solutions are mixed, aliquot the quenched reaction solutions two times (120 μL each) into the two remaining wells designated for the blank and the two remaining wells designated for the experimental sample. These will serve as technical replicates for the blank and the experimental sample.
9. Using a microplate reader, read the 96-well plate at an absorbance wavelength of 405 nm.

3.4.5 α-Mannosidase

The method for assessing α-mannosidase activity is adapted from Loomis (1970) [44] (*see* **Note 17**).

1. Designate three wells in a 96-well optical (clear) bottom black plate for the blank and three wells for the experimental sample (*see* **Note 14**).
2. Add 18 μL of 50 mM α-mannosidase substrate stock to one of the three designated wells for the blank and one of the three designated wells for the experimental sample.
3. Add the blank or the experimental sample to the designated wells containing α-mannosidase substrate.
4. Add the α-mannosidase reaction buffer to the wells containing both α-mannosidase substrate and the blank or the experimental sample to a total volume of 180 μL per well. This is referred to as the reaction solution.
5. Gently mix all reaction solutions using a pipettor.
6. Incubate the 96-well plate at 35 °C for 1 h.
7. Stop the reactions by adding an equal volume (180 μL) of α-mannosidase stop solution to the reaction solutions. Gently mix the solutions using a pipettor.
8. Once the solutions are mixed, aliquot the quenched reaction solutions two times (120 μL each) into the two remaining wells designated for the blank and the two remaining wells designated for the experimental sample. These will serve as technical replicates for the blank and the experimental sample.
9. Using a microplate reader, read the 96-well plate at an absorbance wavelength of 405 nm.

3.4.6 β-N-Acetylglucosaminidase

The method for assessing β-N-acetylglucosaminidase activity is adapted from Loomis (1969) and Huber and Mathavarajah (2018a) [45, 46] (*see* **Note 18**).

1. Designate three wells in a 96-well optical (clear) bottom black plate for the blank and three wells for the experimental sample (*see* **Note 14**).
2. Add 75 μL of 7 mM β-N-acetylglucosaminidase substrate stock to one of the three designated wells for the blank and one of the three designated wells for the experimental sample.
3. Add the blank or the experimental sample to the designated wells containing β-N-acetylglucosaminidase substrate.
4. Add β-N-acetylglucosaminidase reaction buffer to the wells containing both β-N-acetylglucosaminidase substrate and the blank or the experimental sample to a total volume of 125 μL per well. This is referred to as the reaction solution.
5. Gently mix all reaction solutions using a pipettor.

6. Incubate the 96-well plate at 37 °C for 15 min.
7. Stop the reactions by adding 175 μL of β-N-acetylglucosaminidase stop solution to the reaction solutions. Gently mix the solutions using a pipettor.
8. Once the solutions are mixed, aliquot the quenched reaction solutions two times (100 μL each) into the two remaining wells designated for the blank and the two remaining wells designated for the experimental sample. These will serve as technical replicates for the blank and the experimental sample.
9. Using a microplate reader, read the 96-well plate at an absorbance wavelength of 405 nm.

3.5 Assessing Lysosomal Protease Activity

3.5.1 Cathepsin B

The method for assessing cathepsin B activity is adapted from Barrett (2017) [47].

1. Add powdered cysteine to the cathepsin B reaction solution to achieve a final concentration of 8 mM cysteine. Ensure this solution is made fresh for each assay.
2. Obtain two 1.5 mL tubes. Add the blank into one tube and the experimental sample into the other tube (*see* **Note 14**).
3. Add the cathepsin B reaction solution containing 8 mM cysteine to each 1.5 mL tube containing either the blank or the experimental sample to a total volume of 150 μL per well.
4. Gently mix all solutions using a pipettor.
5. Incubate the 1.5 mL tubes at 40 °C for 5 min.
6. Add 27 μL of 10 mM cathepsin B substrate stock to each 1.5 mL tube. This is referred to as the reaction solution.
7. Gently mix all reaction solutions using a pipettor.
8. Incubate the 1.5 mL tubes at 40 °C for 1 h.
9. Designate three wells in a 96-well opaque bottom black plate for the blank and three wells for the experimental sample.
10. Stop the reactions by adding two equal volumes (354 μL) of cathepsin B stop solution to the reaction solutions. Gently mix the solutions using a pipettor.
11. Once the solutions are mixed, aliquot the quenched reaction solutions three times (177 μL each) into the three wells designated for the blank and the three wells designated for the experimental sample. These will serve as technical replicates for the blank and the experimental sample.
12. Using a microplate reader, read the 96-well plate at a fluorescence excitation/emission wavelength pair of 320–360 nm/ 460 nm.

3.5.2 Cathepsin D

The method for assessing cathepsin D activity follows the manufacturer's instructions (BioVision K143–100) with minor modifications.

1. Designate three wells in a 96-well opaque bottom black plate for the blank and three wells for the experimental sample (*see* **Note 14**).
2. Lyse cells with 100 μL of cathepsin D cell lysis solution provided in the kit and chill on ice for 10 min.
3. Spin down lysate (10,000RPM/4 °C) for 5 minutes and collect the clarified whole cell lysate in a new 1.5 mL tube.
4. Make enough cathepsin D substrate-reaction solution for all reactions (e.g., a blank/experimental sample will require 52uL of cathepsin D substrate-reaction solution, which is 50 μL of cathepsin D reaction buffer + 2 μL of cathepsin D substrate).
5. Add the blank or experimental sample into one of the three designated blank or experimental sample wells, respectively.
6. Add 52 μL of cathepsin D substrate-reaction solution to the wells containing either the blank or the experimental sample. This is referred to as the reaction solution.
7. Incubate the 96-well plate at 37 °C for 1 h.
8. Aliquot the reaction solutions two times (34 μL each) into the two remaining wells designated for the blank and the two wells designated for the experimental sample. These will serve as technical replicates for the blank and the experimental sample.
9. Using a microplate reader, read the 96-well plate at a fluorescence excitation/emission wavelength pair of 320–360 nm/ 460 nm.

3.5.3 Cathepsin F

The method for assessing cathepsin F activity was adapted from Fonovič et al. (2004) [48] (*see* **Note 19**).

1. Designate three wells in a 96-well opaque bottom black plate for the blank and three wells for the experimental sample.
2. Add the blank or experimental sample into one of the three designated blank or experimental sample wells, respectively (*see* **Note 14**).
3. Add 0.5 μL of 25% (v/v) HCl, along with a final concentration of 5 μg/mL pepsin, into each well that contains either the blank or the experimental sample.
4. Add the cell lysis solution to the wells containing pepsin, HCl, and either the blank or the experimental sample to a total volume of 50 μL per well.
5. Incubate the plate at 37 °C for 1 h.

6. Add 0.24 μL of 0.5 mM cathepsin F substrate stock to the wells containing pepsin, HCl, and either the blank or the experimental sample.
7. Add 64 μL of cathepsin F reaction solution to the wells containing pepsin, HCl, and either the blank or the experimental sample.
8. Add 5.76 μL of DMSO to adjust the final concentration of DMSO to 5% (v/v) in each well containing pepsin, HCl, cathepsin F reaction buffer, and either the blank or the experimental sample. This is referred to as the reaction solution.
9. Incubate the 96-well plate at 27 °C for 1 h.
10. Aliquot the reaction solutions two times (40 μL each) into the two remaining wells designated for the blank and the two remaining wells designated for the experimental sample. These will serve as technical replicates for the blank and the experimental sample.
11. Read the 96-well plate at a fluorescence excitation/emission wavelength pair of 320–360 nm/460 nm.

3.5.4 Tripeptidyl Peptidase 1

The method for assessing tripeptidyl peptidase 1 activity was adapted from Stumpf et al. (2017) [49] (*see* **Note 20**).

1. Designate three wells in a 96-well opaque bottom black plate for the blank and three wells for the experimental sample (*see* **Note 14**).
2. Add 80 μL of 200 μM tripeptidyl peptidase 1 substrate stock to one of the three designated wells for the blank and one of the three designated wells for the experimental sample.
3. Add the blank or experimental sample to the designated wells containing tripeptidyl peptidase 1 substrate.
4. Add the tripeptidyl peptidase 1 reaction buffer to the wells containing both tripeptidyl peptidase 1 substrate and the blank or experimental sample to a total volume of 130 μL per well. This is referred to as the reaction solution.
5. Gently mix all reaction solutions using a pipettor.
6. Wrap the 96-well plate in aluminum foil (to protect from light) and incubate at 37 °C for 1 h.
7. Stop the reactions by adding an equal volume (130 μL) of tripeptidyl peptidase 1 stop solution to the reaction solutions. Gently mix the solutions using a pipettor.
8. Aliquot the reaction solutions two times (86 μL each) into the two remaining wells designated for the blank and the two remaining wells designated for the experimental sample. These will serve as technical replicates for the blank and the experimental sample.

9. Using a microplate reader, read the 96-well plate at a fluorescence excitation/emission wavelength pair of 360 nm/460 nm.

3.6 Assessing Lysosomal Thioesterase Hydrolase Activity

The method for assessing palmitoyl protein thioesterase 1 activity was adapted from van Diggelen et al. (1999) [50].

3.6.1 Palmitoyl Protein Thioesterase 1

1. Add 2 μL of 9 mM palmitoyl protein thioesterase 1 substrate stock to 1.5 mL two tubes: one for the blank and one for the experimental sample.
2. Add the blank or experimental sample to each designated 1.5 mL tube containing the palmitoyl protein thioesterase 1 substrate (*see* **Note 14**).
3. Add palmitoyl protein thioesterase 1 reaction buffer to each 1.5 mL tube to a total volume of 90 μL. This is referred to as the reaction solution.
4. Incubate the 1.5 mL tubes at 37 °C for 1 h.
5. Place the 1.5 mL tubes on a dry block set at 95 °C for 3 min.
6. Leave the 1.5 mL tubes at 21 °C until the solutions cool and briefly pulse in a centrifuge to collect condensates at the bottom of the tubes.
7. Add 2.75 μL of 2.5 M NaOH to each 1.5 mL tube.
8. Add 4 μL of 0.25 U/μL β-glucosidase enzyme (dissolved in 0.2% (w/v) BSA in DI water) to each 1.5 mL tube.
9. Gently mix the solutions using a pipettor.
10. Incubate the 1.5 mL tubes at 37 °C for 1 h.
11. Stop the reactions by adding 90 μL of palmitoyl protein thioesterase 1 stop solution.
12. Gently mix the solutions using a pipettor.
13. Designate three adjacent wells in a 96-well opaque bottom black plate for the blank and three adjacent wells for the experimental sample.
14. Aliquot the reaction solutions three times (60 μL each) into the three wells designated for the blank and the three wells designated for the experimental sample. These will serve as technical replicates for the blank and the experimental sample.
15. Using a microplate reader, read the 96-well plate at a fluorescence excitation/emission wavelength pair of 320–360 nm/460 nm.

3.7 Data Analysis

1. Average the signals from the three technical replicates for each sample, as well as the blank solution.
2. To determine the signal derived from the enzymatic activity, subtract the averaged sample signal value from the averaged blank signal value.

3. If a researcher is interested in comparing enzyme activity in a wild type and knockout cell line, the enzyme signal in the knockout line can be expressed relative to the enzyme signal in the wild type cell line.
4. Use the Student's t-test to determine significance. If data is expressed relative to a control, then the one-sample t-test should be used (*see* **Note 21**).

4 Notes

1. A variety of protein quantification assays can be used: Qubit Protein Assay Kit, Bradford Assay, or BCA Protein Assay Kit (Supplier: Fisher Scientific).
2. Cell lines can be maintained as fruiting bodies on SM/2 agar for 2 weeks. After this time, new plates should be prepared by replating spores from the original dish and spreading with *Klebsiella aerogenes* on fresh SM/2 agar. Cells and spores can also be placed in an ultralow temperature freezer for long-term storage [51].
3. Low fluorescence (LoFlo) medium (supplier: Formedium, Hunstanton, Norfolk, United Kingdom, product code: LF0501) should be used if the signal from the enzyme assay is florescence-based.
4. This article describes the methods used for conducting enzyme assays on whole cell lysates generated from cells adhered to Petri dishes. If cells are known to display reduced cell-substrate adhesion, they can be incubated in suspension in HL5 medium, LoFlo medium, or KK2 buffer (150RPM/21 °C).
5. The 96-well optical (clear) bottom black plate (for absorbance-based readings) referenced in this article has a maximum volume of 400 μL per well. If the number of technical replicates is increased, ensure that the total volume of the quenched reaction solution does not exceed 400 μL per well.
6. The 96-well opaque bottom black plate (for fluorescence-based readings) referenced in this article has a maximum volume of 350 μL per well. If the number of technical replicates is increased, ensure that the total volume of the quenched reaction solution does not exceed 350 μL per well.
7. This article describes the use of a concentration column that retain proteins with molecular weights greater than 10 kDa. Alternative molecular weight cut-offs are also available from suppliers (e.g., 3 kDa).

8. Petri dishes (60 mm × 15 mm) allow 2.25×10^6 AX3 cells to grow to 90–100% confluency in 18–24 h. Alternate Petri dish sizes can be used (e.g., 100 mm × 15 mm) if the number of total cells deposited is adjusted. In that case, the amount of medium or buffer added to the dish would need to be increased to ensure complete coverage of the deposited cells. Also, some mutant cell lines may proliferate slower or faster than wild type cells. Consequently, this will require seeding either less or more cells to achieve similar confluency after 18–24 h.
9. Spinning down conditioned medium or buffer (1500RPM/4 °C) for 5 min will pellet free-floating cells in the solution to ensure the clarified solution contains only secreted material.
10. The secretion of most lysosomal enzymes peaks after 2–4 h of starvation [31, 34].
11. Some enzymes are not heavily secreted, which could result in the signal from the enzyme assay being weak and close to the signal generated by the blank solution. If the amount of a secreted enzyme is known to be low, an Amicon Ultra-4 Centrifugal Filter protein concentration column can be used to concentrate the conditioned buffer. To do this, collect the clarified conditioned buffer from the starved cells and place it in a concentration column (*see* above). Spin the column in a centrifuge (14000RPM/4 °C) for 25 min. After this time, the volume of conditioned buffer should be 150–200 μL. This concentrated conditioned buffer can then be aliquoted and stored at −80 °C. Concentrating clarified conditioned HL5 medium clogs the Amicon Ultra-4 Centrifugal Filter columns. Thus, concentrating conditioned HL5 medium with this column is not recommended.
12. The cell lysis solution does not contain protease inhibitors. Chilling the cell lysis solution ensures that enzyme breakdown via proteolytic enzymes is slowed. If the whole cell lysate is not clarified, ensure that the whole cell lysate is gently mixed using a pipettor or the tube is flicked. If cell debris within the lysate is an issue in the enzyme assay (e.g., inconsistent signal between replicates), the crude lysate can be spun down (10,000RPM/4 °C) for 5 min, and the clarified whole cell lysate can then be placed into a new 1.5 mL tube. If signal detection is low in the enzyme assay, one could also create a cell lysis solution of 0.1% NP-40 in the reaction solution (i.e., to reduce possible effect of detergent) and lyse the cells in accordance with the enzyme assay protocol of interest. Pipetting the solution is strongly recommended to lyse cells as vortexing cells in cell lysis solution may affect enzyme activity. Finally, performing the enzyme assays on fresh samples is recommended but not required.

13. If a low protein amount (e.g., 0.5 μg) is added to the reactions and/or the enzyme activity is low, the incubation time can be extended to 2 h or even overnight. For all reactions, it is recommended to limit the sample input to no more than 25 μL.
14. The blank sample is the cell lysis solution. The blank volume will be equivalent across all experimental sample volumes (i.e., if 25 μL is chosen as the input volume for all experimental samples, then the blank volume should be 25 μL of cell lysis solution).
15. There are two β-galactosidase genes in the *D. discoideum* genome (*glb1* and *glb2*). *glb1* is more highly expressed during growth and after 8–20 h of development [52]. *glb2* is more highly expressed after 4 and 24 h of development. Since there are two β-galactosidases, it is not possible to determine the exact amount of activity that each protein contributes within a given sample.
16. There are two α-glucosidase genes in the *D. discoideum* genome (*gaa* and *modA*). In general, *modA* expression is higher than *gaa* throughout the life cycle [52]. However, gene expression is nearly equal during growth and after 16–20 h of development. Since there are two α-glucosidases, it is not possible to determine the exact amount of activity that each protein contributes within a given sample.
17. There are seven α-mannosidase genes in the *D. discoideum* genome (*manA*, *manB*, *manC*, *manD*, *manE*, *manF*, *manG*). *manA* is the most highly expressed gene during growth, after 4–12 h of development, and after 24 h of development [52]. *manF* is the most highly expressed gene after 16–20 h of development. The remaining genes, *manB*, *manC*, *manD*, *manE*, and *manG*, are all expressed at relatively lower amounts. Since there are several α-mannosidases, it is not possible to determine the exact amount of activity that each protein contributes within a given sample.
18. There are five β-N-acetylglucosaminidase genes in the *D. discoideum* genome (*nagA*, *nagB*, *nagC*, *nagD*, *nagE*). *nagA* is the most highly expressed gene during growth [52]. *nagD* is the most highly expressed gene after 4–20 h of development. Expression of *nagA* and *nagD* is nearly equal after 24 h of development. The remaining genes, *nagB*, *nagC*, and *nagE*, are all expressed at relatively lower amounts. Since there are several β-N-acetylglucosaminidases, it is not possible to determine the exact amount of activity that each protein contributes within a given sample.

19. There are four proteins encoded by the *D. discoideum* genome that share sequence similarity with human cathepsin F (CprA, CprB, CprE, uncharacterized protein DDB0252831). The gene encoding uncharacterized protein DDB0252831 (*DDB_G0291191*) is the most highly expressed gene during growth [52]. The expression of *cprA* and *cprB* is nearly equal after 4 h of development. *cprB* is the most highly expressed gene after 8–24 h of development. Since there are four cathepsin F-like proteins, it is not possible to determine the exact amount of activity that each protein contributes within a given sample.
20. There are six proteins encoded by the *D. discoideum* genome that share sequence similarity with human tripeptidyl peptidase 1 (Tpp1A, Tpp1B, Tpp1C, Tpp1D, Tpp1E, Tpp1F). *tpp1B* is the most highly expressed gene during growth and after 4 h and 24 h of development [52]. *tpp1A* is the most highly expressed gene after 8–20 h of development. The remaining genes, *tpp1C*, *tpp1D*, *tpp1E*, and *tpp1F*, are all expressed at relatively lower amounts. Since there are six tripeptidyl peptidase 1-like proteins, it is not possible to determine the exact amount of activity that each protein contributes within a given sample.
21. As noted above, some enzyme activities can be attributed to multiple proteins recognizing the same substrate. Thus, the interpretation of the data may need to be softened.

References

1. Bouhamdani N, Comeau D, Turcotte S (2021) A compendium of information on the lysosome. Front Cell Dev Biol 9:796262
2. Trivedi PC, Bartlett JJ, Pulinilkunnil T (2020) Lysosomal biology and function: modern view of cellular debris bin. Cells 9:1–35
3. Yang C, Wang X (2021) Lysosome biogenesis: regulation and functions. J Cell Biol 220: e202102001
4. Cocchiaro P, De Pasquale V, Della Morte R, Tafuri S, Avallone L, Pizard A et al (2017) The multifaceted role of the lysosomal protease cathepsins in kidney disease. Front Cell Dev Biol 5:114
5. Saftig P, Klumperman J (2009) Lysosome biogenesis and lysosomal membrane proteins: trafficking meets function. Nat Rev Mol Cell Biol 10:623–635
6. Verkruyse LA, Hofmann SL (1996) Lysosomal targeting of palmitoyl-protein thioesterase. J Biol Chem 271:15831–15836
7. Schwake M, Schröder B, Saftig P (2013) Lysosomal membrane proteins and their central role. Traffic 14:739–748
8. Kornfeld S (1987) Trafficking of lysosomal enzymes. FASEB J 1:262–468
9. Staudt C, Puissant E, Boonen M (2017) Subcellular trafficking of mammalian lysosomal proteins: an extended view. Int J Mol Sci 18:47
10. Dahms NM, Lobel P, Kornfeld S (1989) Mannose 6-phosphate receptors and lysosomal enzyme targeting. J Biol Chem 264:12115–12118
11. Reczek D, Schwake M, Schröder J, Hughes H, Blanz J, Jin X et al (2007) LIMP-2 is a receptor for lysosomal mannose-6-phosphate-independent targeting of β-glucocerebrosidase. Cell 131:770–783
12. Ruscetti T, Cardelli JA, Niswonger ML, O'Halloran TJ (1975) Clathrin heavy chain functions in sorting and secretion of lysosomal enzymes in *Dictyostelium discoideum*. Nature 258:273–273

13. New J, Thomas SM (2019) Autophagy-dependent secretion: mechanism, factors secreted, and disease implications. Autophagy 15:1682–1693
14. Buratta S, Tancini B, Sagini K, Delo F, Chiaradia E, Urbanelli L et al (2020) Lysosomal exocytosis, exosome release and secretory autophagy: the autophagic- and endo-lysosomal systems go extracellular. Int J Mol Med 21:2576
15. Cavalli G, Cenci S (2020) Autophagy and protein secretion. J Mol Biol 432:2525–2545
16. Tancini B, Buratta S, Delo F, Sagini K, Chiaradia E, Pellegrino RM et al (2020) Lysosomal exocytosis: the extracellular role of an intracellular organelle. Membranes 10:406
17. Castro-Gomes T, Corrotte M, Tam C, Andrews NW (2016) Plasma membrane repair is regulated extracellularly by proteases released from lysosomes. PLoS One 11:e0152583
18. Fujita K, Teramura N, Hattori S, Irie S, Mitsunaga-Nakatsubo K, Akimoto Y et al (2010) Mammalian arylsulfatase a functions as a novel component of the extracellular matrix. Connect Tissue Res 51:388–396
19. Ni X, Canuel M, Morales CR (2006) The sorting and trafficking of lysosomal proteins. Histol Histopathol 21:899–913
20. Platt FM, Azzo A, Davidson BL, Neufeld EF, Tifft CJ (2018) Lysosomal storage diseases. Nat Rev Dis Primers 4:27
21. Stirnemann JÔ, Belmatoug N, Camou F, Serratrice C, Froissart R, Caillaud C et al (2017) A review of Gaucher disease pathophysiology, clinical presentation and treatments. Int J Mol Sci 18:441
22. Ceccarini MR, Codini M, Conte C, Patria F, Cataldi S, Bertelli M et al (2018) Alpha-mannosidosis: therapeutic strategies. Int J Mol Sci 19:1500
23. Solovyeva VV, Shaimardanova AA, Chulpanova DS, Kitaeva KV, Chakrabarti L, Rizvanov AA (2018) New approaches to Tay-Sachs disease therapy. Front Physiol 9:1663
24. Meena NK, Raben N (2020) Pompe disease: New developments in an old lysosomal storage disorder. Biomol Ther 10:1339
25. Yuskiv N, Higaki K, Stockler-Ipsiroglu S (2020) Morquio B disease. Disease characteristics and treatment options of a distinct GLB1-related dysostosis multiplex. Int J Mol Sci 21:9121
26. Li X, Ren X, Zhang Y, Ding L, Huo M, Li Q (2022) Fabry disease: mechanism and therapeutics strategies. Front Pharmacol 13:1025740
27. Richardson JM, Woychik NA, Ebert DL, Dimond RL, Cardelli JA (1988) Inhibition of early but not late proteolytic processing events leads to the missorting and oversecretion of precursor forms of lysosomal enzymes in *Dictyostelium discoideum.* J Cell Biol 107:2097–2107
28. Knecht DA, Dimond RL (1981) Lysosomal enzymes possess a common antigenic determinant in the cellular slime mold, *Dictyostelium discoideum.* J Biol Chem 256:3564–3575
29. Cardelli JA, Golumbeski GS, Dimond RL (1986) Lysosomal enzymes in *Dictyostelium discoideum* are transported to lysosomes at distinctly different rates. J Cell Biol 102:1264–1270
30. Burns RA, Livi GP, Dimond RL (1981) Regulation and secretion of early developmentally controlled enzymes during axenic growth in *Dictyostelium discoideum.* Dev Biol 84:407–416
31. Dimond RL, Burns RA, Jordan KB (1981) Secretion of lysosomal enzymes in the cellular slime mold, *Dictyostelium discoideum.* J Biol Chem 256:6565–6572
32. Mierendorf RC Jr, Cardelli JA, Dimond RL (1985) Pathways involved in targeting and secretion of a lysosomal enzyme in *Dictyostelium discoideum.* J Cell Biol 100:1777–1787
33. Cardelli JA, Schatzle J, Bush JM, Richardson J, Ebert D, Freeze H (1990) Biochemical and genetic analysis of the biosynthesis, sorting, and secretion of *Dictyostelium* lysosomal enzymes. Dev Genet 11:454–462
34. Rossomando EF, Maldonado B, Crean EV, Kollar EJ (1978) Protease secretion during onset of development in *Dictyostelium discoideum.* J Cell Sci 30:305–318
35. Huber RJ, Williams RSB, Müller-Taubenberger A (2022) Editorial: *Dictyostelium*: a tractable cell and developmental model in biomedical research. Front Cell Dev Biol 10:909619
36. Eichinger L, Pachebat JA, Glöckner G, Rajandream M-A, Sucgang R, Berriman M et al (2005) The genome of the social amoeba *Dictyostelium discoideum.* Nature 435:43–57
37. Dimond RL, Brenner M, Loomis WF (1973) Mutations affecting N-acetylglucosaminidase in *Dictyostelium discoideum.* Proc Natl Acad Sci USA 70:3356–3360
38. Phillips JE, Gomer RH (2015) Partial genetic suppression of a loss-of-function mutant of the neuronal ceroid lipofuscinosis-associated protease TPP1 in *Dictyostelium discoideum.* Dis Model Mech 8:147–156

39. Kim WD, Huber RJ (2022) An altered transcriptome underlies *cln5*-deficiency phenotypes in *Dictyostelium discoideum*. Front Genet 13: 1045738
40. Kilpatrick DC, Stirling JL (1976) Properties and developmental regulation of an α-D-galactosidase from *Dictyostelium discoideum*. Biochem J 158:409–417
41. Maruhn D (1976) Rapid colorimetric assay of β-galactosidase and N-acetyl–βglucosaminidase in human urine. Clin Chim Acta 73:453–461
42. Wimmer B, Lottspeich F, Ritter J, Bronnenmeier K (1997) A novel type of thermostable α-D-glucosidase from *Thermoanaerobacter thermohydrosulfuricus* exhibiting maltodextrinohydrolase activity. Biochem J 328:581–586
43. Coston MB, Loomis WF (1969) Isozymes of beta-glucosidase in *Dictyostelium discoideum*. J Bacteriol 100:1208–1217
44. Loomis WF (1970) Developmental regulation of α-mannosidase in *Dictyostelium discoideum*. J Bacteriol 103:375–381
45. Loomis WF (1969) Acetylglucosaminidase, an early enzyme in the development of *Dictyostelium discoideum*. J Bacteriol 97:1149–1154
46. Huber RJ, Mathavarajah S (2018) Cln5 is secreted and functions as a glycoside hydrolase in *Dictyostelium*. Cell Signal 42:236–248
47. Barrett A (1980) Fluorimetric assays for cathepsin B and cathepsin H with methylcoumarylamide substrates. Biochem J 187:909–912
48. Fonovič M, Brömme D, Turk V, Turk B (2004) Human cathepsin F: expression in baculovirus system, characterization and inhibition by protein inhibitors. Biol Chem 385:505–509
49. Stumpf M, Müller R, Gaßen B, Wehrstedt R, Fey P, Karow MA et al (2017) A tripeptidyl peptidase 1 is a binding partner of the Golgi pH regulator (GPHR) in *Dictyostelium*. Dis Model Mech 10:897–907
50. van Diggelen OP, Keulemans JLM, Winchester B, Hofman IL, Vanhanen SL, Santavuori P et al (1999) A rapid fluorogenic palmitoyl-protein thioesterase assay: pre- and postnatal diagnosis of INCL. Mol Genet Metab 66:240–244
51. Fey P, Kowal AS, Gaudet P, Pilcher KE, Chisholm RL (2007) Protocols for growth and development of *Dictyostelium discoideum*. Nat Protoc 2:1307–1316
52. Stajdohar M, Rosengarten RD, Kokosar J, Jeran L, Blenkus D, Shaulsky G, Zupan B (2017) dictyExpress: a web-based platform for sequence data management and analytics in *Dictyostelium* and beyond. BMC Bioinformatics 18:291

Chapter 5

Procedures to Measure *Dictyostelium* Phagocytosis and Macropinocytosis

Lou W. Kim and Cong Gu

Abstract

Uptaking particulate objects and bulk liquid by eucaryotic cells is critical for their growth, survival, and defense. *Dictyostelium* is a model organism spearheaded to uncover mechanisms behind various types of uptaking activities. Here, we describe assays measuring phagocytosis and macropinocytosis using *Dictyostelium discoideum*.

Key words *Dictyostelium*, Phagocytosis, Latex-Bead, Heat-killed yeast, Macropinocytosis, Dextran

1 Introduction

Vegetative *Dictyostelium* amoeba forage nutrient through phagocytosis and macropinocytosis. Particulate objects bigger than 0.5 μm of diameter can be internalized through phagocytic receptors that can recognize different types of targets [1]. *Dictyostelium* cells uptake various types of particulate objects which include but not limited to latex bead, yeast, and bacteria. Phagocytosis consists of the following four stages: particle recognition, signal pathway activation for internalization, phagosome formation, and phagosome maturation. These four steps are generally conserved in varieties of phagocytic cells [2], but there are multiple types of phagocytic receptors, signaling pathways, regulatory circuits [1, 3], and distinct maturation mechanisms exist for distinct types of phagocytosis [4].

In contrast, *Dictyostelium* cells internalize liquid by generating large protrusion of the plasma membrane which remodels through multiple steps and eventually form an internalized vesicle called the macropinosome. The widely used *Dictyostelium* strains, such as *Ax2*, *Ax3*, and *JH10*, exhibit a prominent constitutive rate of macropinocytosis, but the rate of macropinocytosis is influenced

Alan R. Kimmel (ed.), *Dictyostelium discoideum: Methods and Protocols*, Methods in Molecular Biology, vol. 2814, https://doi.org/10.1007/978-1-0716-3894-1_5,

by a number of factors. These factors include nutrient source, incubation temperature, pH, adhesion, and cell cycle and developmental stages [5–11]. It is noteworthy that the presence of bacteria has been observed to suppress macropinocytosis, which aligns with the inhibitory effect of folate on this process [12]. Additionally, starvation is a well-known condition that inhibits macropinocytosis, and it has been reported to be a cAMP/PKA-dependent event [12]. To accurately measure macropinocytosis, researchers commonly track the uptake of fluorescent dextran, a widely used fluid-phase marker [5, 13, 14].

2 Materials

2.1 Dictyostelium Development and Growth

Grow *Dictyostelium discoideum* cells in D3T media (dissolve 14.3 g of Peptone #3, 7.15 g of yeast extract, 15.4 g of glucose, 0.48 g of KH_2PO_4, and 0.525 g of $Na_2HPO_4.7H_2O$ in deionized water and top off to 1 L). Autoclave for 40 min; let it cool down (*see* **Note 1**). In addition to D3T medium, JH10 cells will be grown with thymidine (0.5 mg/mL). Grow cells in shaking cultures at 19 °C at 150 rpm or in culture flasks at 19 °C for about 24 h until they are in log phase. Monitor cell numbers using a hemocytometer. Harvest and use cells when the cell numbers are between 2×10^6 cells/mL and 2.5×10^6 cells/mL.

2.2 Buffers and Reagents

1. *Sorensen Buffer*
   Sorensen buffer contains 2 mM Na_2HPO_4 and 14.6 mM KH_2PO_4, pH 6.0
2. *Development Buffer*
   Development buffer (5 mM Na_2HPO_4, 5 mM KH_2PO_4, 1 mM $CaCl_2$, 2 mM $MgCl_2$, pH 6.5)
3. *TTG Cell Lysis Buffer*
   TTG cell lysis buffer (50 mM Na_2HPO_4, pH 9.4, 0.2% triton X-100)
4. *FITC-dextran*
   Fluorescein isothiocyanate—dextran and relative molecular weight 70,000 (Sigma-Aldrich)
5. *Tryphan blue*
   Tryphan blue is from Sigma-Aldrich
6. *BCA Assay*
   Pierce BCA protein Assay (Thermo Scientific)
7. *Latex Bead*
   1 μm Latex beads is from Sigma-Aldrich
8. *Yeast Cells*
   Yeast cell is from Sigma-Aldrich and inactivated by heating and labelled with TRITC.

9. *TRITC Labelling Buffer*
   Labeling buffer contains 0.1 mg/mL TRITC (T5646, Sigma), 50 mM Na2HPO4, and pH 9.2.

2.3 Equipment

1. *Spectrophotometer*
   GENESYS/louv
2. *Spectrofluorometer*
   Shimazu RF-1501, excitation 544 nm; emission 574 nm
3. *Fluorescent Microscope*
   Leica DM IRB inverted epifluorescence microscope
4. *Sonicator*
   Fisher 100

3 Methods

3.1 Latex Bead Uptake Assay

1. Prepare four separate 15 mL centrifuge tubes with a total of 4×10^6 log-phase *Dictyostelium discoideum* cells.
2. Centrifuge the tubes at 2000 rpm for 5 min at 4 °C and discard the supernatant.
3. Resuspend each cell pellet in 1 mL of cold D3T media, and add 50 fluorescent FITC-latex beads per cell to each tube.
4. Shake the mixtures at 150 rpm and room temperature for several time points: start, 10, 20, and 30 min (*see* **Note 2**). For analyses that do not require a time course, refer to **Note 3**.
5. After shaking, wash 1×10^6 *Dictyostelium discoideum* cells four times with 10 mL of cold DB buffer.
6. Resuspend the final cell pellet in 1 mL of DB media at room temperature, and place 500 μL of the final suspension into a four-chamber slide. Allow the cells to settle on the surface of the slide for 10 min at 4 °C.
7. Randomly select 20 *Dictyostelium* cells and capture photographs of them at 100× magnification at various layers along the z-axis to obtain a three-dimensional view of the cells (*see* **Note 4**).
8. A representative image of internalized FITC-latex beads is shown in Fig. 1.

3.2 Phagocytosis of Heat-Killed Yeast

3.2.1 Heat-Killed Yeast Labelling with TRITC

1. Pellet yeast cells by centrifugation and wash twice with cold water.
2. Resuspend the yeast cells in water at a concentration of 1×10^8 cells/mL.
3. Heat the cells at 65 °C for 30 min with occasional shaking.

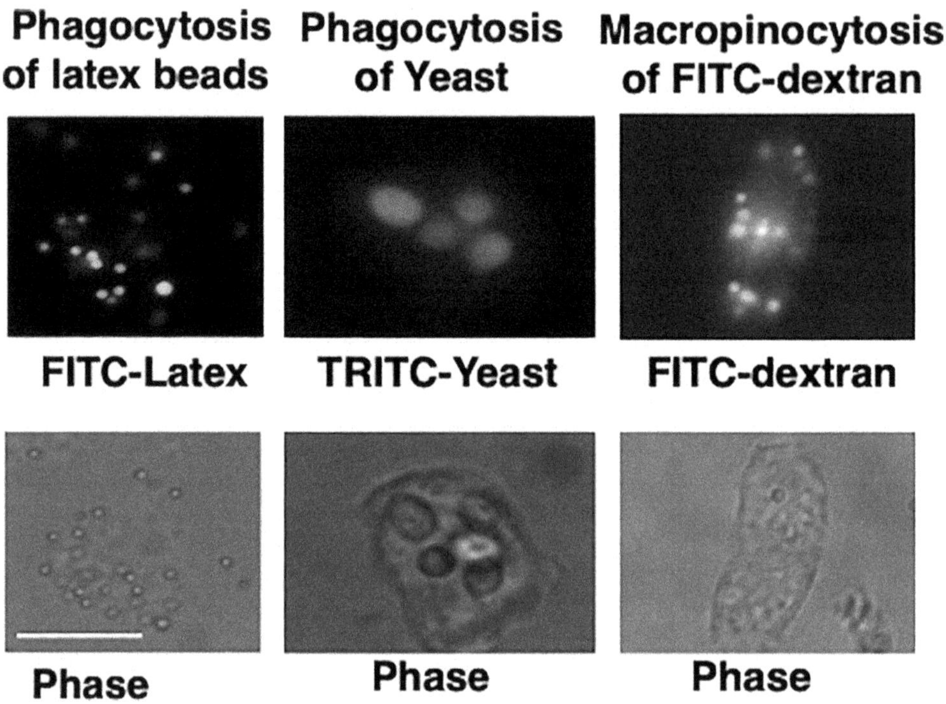

Fig. 1 Images of *JH10* (wild-type) *Dictyostelium* cells uptaking latex beads, heat-killed yeasts, and bulk liquid containing FITC-dextran. Cells and target materials were incubated for 30 min and both phase contrast and fluorescent images are shown. Scale bar represents 10 μm

4. Pellet the cells by centrifugation and wash them twice with cold water.
5. Resuspend the heat-inactivated yeasts in Sorensen buffer at a concentration of 1×10^9 yeast particles per milliliter (*see* **Note 5**).
6. Incubate 20 mL of yeast particles with labeling buffer containing TRITC for 30 min at 30 °C, and subsequently wash them twice with labeling buffer that does not contain TRITC.

3.2.2 Analysis of Phagocytosis Using TRITC-Labeled Yeast

1. Pellet log-phase *Dictyostelium* cells in a tabletop centrifuge and resuspend them in D3T media at a concentration of 2×10^6 cells per milliliter.
2. Allow the cells to recover on a shaker at 19 °C and 150 rpm for 30 min.
3. Take a 0.5 mL sample of the cell suspension for a BCA assay (*see* **Note 6**).
4. Centrifuge the samples, wash them with 1 mL of Sorensen buffer, and lyse them with 400 μL of TTG lysis buffer. Determine the protein concentration by performing the Pierce BCA protein assay (Thermo Scientific) using 50 μL of the lysates.

5. Add TRITC-yeasts to the *Dictyostelium* cell suspension at a concentration six times higher than that of the *Dictyostelium* cells (*see* **Note** 7). To differentiate acidic and neutral phagosomes containing heat-killed yeasts, FITC yeasts were useful [15] (*see* **Note 8**).
6. Incubate the mixtures in shaking cultures at 19 °C and 150 rpm for 30, 60, and 120 min.
7. Withdraw 1 mL of the sample at different time intervals and add 100 μL of trypan blue to each sample (*see* **Note 9**).
8. Incubate the samples for 3 min.
9. Centrifuge the samples, resuspend them in 3 mL of Sorensen buffer, and measure the fluorescence using a spectrofluorometer with excitation at 544 nm and emission at 574 nm.
10. To obtain a percentile value, divide the fluorescence value by the protein concentration, and express the result as a percentage with the data from the wild-type cells at the 2 h time point set as 100%.

3.2.3 Phagocytosis Imaging

1. Take 0.2 million log-phase *Dictyostelium discoideum* cells and add them to a #1.5 German borosilicate cover glass (Nalgen Nunc International). Allow the cells to settle for 5 min.
2. Discard the media on the top and resuspend the cells with 0.5 mL of Development Buffer. Allow the cells to settle for another 5 min.
3. Add one million TRITC-yeasts to the cover glass and mix thoroughly.
4. Capture live images of the phagocytosed TRITC-yeasts using an inverted epifluorescence microscope with a 60× or 100× lens.
5. A representative image of yeast uptaking wild-type cell is shown in Fig. 1.

3.3 Macropinocytosis

3.3.1 Macropinocytosis Assay Using FITC-Labeled Dextran

1. Pellet log-phase *Dictyostelium* cells in a tabletop centrifuge and resuspend them with 5 mL of D3T media at a concentration of 2.5×10^6 cells/mL.
2. Let the cells recover on a shaker at 19 °C at 150 rpm for 30 min.
3. Add FITC-dextran (Sigma Aldrich) to cell suspensions at a concentration of 1 mg/mL (*see* **Note 10**).
4. Incubate the mixtures of *Dictyostelium* cells and FITC-dextran in shaking cultures at 19 °C and 150 rpm for 0, 30, 60, and 120 min.
5. At each time intervals, withdraw 1 mL of the sample; pellet and wash them twice with 1 mL of ice-cold Sorensen buffer.

6. Resuspend the pellets with 1 mL of TTG lysis buffer, take 50 μL of the lysates, and determine protein concentration with the BCA assay.
7. Resuspend the remaining lysates with 2 mL of TTG lysis buffer, and measure the fluorescence using a spectrofluorometer (Shimadzu RF-1501, excitation 470 nm, emission 515 nm).
8. To obtain a percentile value, divide the fluorescence value by the protein concentration, and express the result as a percentage with the data from the wild-type cells at the 2 h time point set as 100%.
9. Repeat the assay at least three times.
10. Capture images of internalized FITC-dextran using an inverted epifluorescence microscope equipped with a 100× oil lens. A representative image of FITC-dextran up-taking Dictyostelium cell is shown in Fig. 1.
11. Quantify the mean number of vacuoles and the mean size of vacuoles from 90 cells and perform unpaired Students' *t* test.

4 Notes

1. To prevent contamination, add an antibiotics mixture to the D3T media to achieve final concentrations of 100 μg/mL ampicillin, 25 μg/mL tetracycline, and 25 μg/mL chloramphenicol.
2. A starting period consists of 0–2 min to account for the time the cells are in contact with the beads before and during spinning.
3. For analyses that do not require a time course analysis, two different procedures were employed. In the first procedure, *Dictyostelium* cells were settled down on a chamber and allowed to uptake latex beads for 30 min or longer [16]. In the second procedure, *Dictyostelium* cells and latex beads were incubated on a rotary shaker for 1 h [17]. These procedures were designed to investigate images of internalized latex beads.
4. To identify and quantify internalized latex beads, examine each cell along the z-axis, and count the number of beads present. Move through the z-axis by approximately 5 μm between each picture to ensure comprehensive coverage.
5. Dilute heat-inactivated yeasts 1000 times and count using a hemocytometer. Resuspend the heat-inactivated yeasts in Sorensen buffer at a concentration of 1×10^9 yeast particles per milliliter.
6. The sample for BCA assay can be kept on an ice bucket till the phagocytosis assay is completed.

7. To prepare for mixing TRITC-yeast with *Dictyostelium* cells, sonicate the TRITC-yeast in a 50 mL Falcon tube using a microtip for 15 s at medium strength.
8. While TRITC is pH insensitive, the FITC signal is known to be sensitive to acidic environments. Consequently, when phagosomes undergo acidification, the FITC signals are significantly diminished. In order to distinguish between acidic and neutral phagosomes that contain heat-killed yeasts, FITC-labeled yeasts were utilized [15].
9. The trypan blue stock solution should be kept at room temperature.
10. TRITC-dextran is a pH-insensitive liquid phase marker that is suitable for determining whole liquid phase uptake. However, to determine the internalized dextran portions in vesicles with neutral pH, the acid-sensitive FITC-dextran should be used [18].

References

1. Uribe-Querol E, Rosales C (2020) Phagocytosis: our current understanding of a universal biological process. Front Immunol 11:1066
2. Dunn JD, Bosmani C, Barisch C, Raykov L, Lefrançois LH, Cardenal-Muñoz E, López-Jiménez AT, Soldati T (2018) Eat prey, live: *Dictyostelium discoideum* as a model for cell-autonomous defenses. Front Immunol 8:1906
3. Rosales C, Uribe-Querol E (2017) Phagocytosis: a fundamental process in immunity. Biomed Res Int 2017:9042851
4. Pauwels AM, Trost M, Beyaert R, Hoffmann E (2017) Patterns, receptors, and signals: regulation of phagosome maturation. Trends Immunol 38(6):407–422
5. Kayman SC, Clarke M (1983) Relationship between axenic growth of *Dictyostelium discoideum* strains and their track morphology on substrates coated with gold particles. J Cell Biol 97:1001–1010
6. Aguado-Velasco C, Bretscher MS (1999) Circulation of the plasma membrane in *Dictyostelium*. Mol Biol Cell 10:4419–4427
7. Katoh M, Chen G, Roberge E, Shaulsky G, Kuspa A (2007) Developmental commitment in *Dictyostelium discoideum*. Eukaryot Cell 6:2038–2045
8. Maeda Y (1983) Axenic growth of *Dictyostelium discoideum* wild-type NC-4 cells and its relation to endocytotic ability. J Gen Microbiol 129:2467–2473
9. Maeda Y, Kawamoto T (1986) Pinocytosis in *Dictyostelium discoideum* cells: a possible implication of cytoskeletal actin for pinocytotic activity. Exp Cell Res 164:516–526
10. Maeda Y (1988) Changes of endocytic activities during the cell cycle of *Dictyostelium* cells. Develop Growth Differ 30:15–24
11. Novak KD, Peterson MD, Reedy MC, Titus MA (1995) *Dictyostelium* myosin I double mutants exhibit conditional defects in pinocytosis. J Cell Biol 131:1205–1221
12. Williams TD, Kay RR (2018) The physiological regulation of macropinocytosis during *Dictyostelium* growth and development. J Cell Sci 131:jcs213736
13. Thilo L, Vogel G (1980) Kinetics of membrane internalization and recycling during pinocytosis in *Dictyostelium discoideum*. Proc Natl Acad Sci USA 77:1015–1019
14. Hacker U, Albrecht R, Maniak M (1997) Fluid-phase uptake by macropinocytosis in *Dictyostelium*. J Cell Sci 110:105–112
15. Clarke M, Maddera L, Engel U, Gerisch G (2010) Retrieval of the vacuolar H^+-ATPase from phagosomes revealed by live cell imaging. PLoS One 5:e8585

16. Giorgione J, Clarke M (2008) Heterogeneous modes of uptake for latex beads revealed through live cell imaging of phagocytes expressing a probe for phosphatidylinositol-(3,4,5)-trisphosphate and phosphatidylinositol-(3,4)-bisphosphate. Cell Motil Cytoskeleton 65: 721–733
17. Fey P, Stephens S, Titus MA, Chisholm RL (2002) SadA, a novel adhesion receptor in *Dictyostelium.* J Cell Biol 159:1109–1119
18. Padh H, Ha J, Lavasa M, Steck TL (1993) A post-lysosomal compartment in *Dictyostelium discoideum.* J Biol Chem 288:6742–6747

Chapter 6

Quantitative Analysis of the Inactivation Process of Internalized Bacteria in *Dictyostelium* Cells

Lou W. Kim and Victor Osorio-Castillo

Abstract

The understanding of the inactivation process of ingested bacteria by phagocytes is a key focus in the field of host-pathogen interactions. *Dictyostelium* is a model organism that has been at the forefront of uncovering the mechanisms underlying this type of interaction. In this study, we describe an assay designed to measure the inactivation of *Klebsiella aerogenes* in the phagosomes of *Dictyostelium discoideum*.

Key words *Dictyostelium discoideum*, *Klebsiella aerogenes*, Phagosome maturation, Bactericidal activity

1 Introduction

Phagosome maturation is a highly ordered process that enables the phagosome to effectively kill and digest bacteria [1]. During this process, the nascent phagosome undergoes multiple fusion and fission events with early and late endosomes and with lysosomes [2, 3], ultimately resulting in the formation of the phagolysosome, which has an acidic environment critical for pathogen killing and waste disposal [3, 4]. Acidification is essential for bacterial digestion because most lysosomal enzymes have optimal activity at acidic pH, ranging from 3 to 6, and the production of reactive oxygen species requires the presence of protons [4, 5]. Phagosome maturation involves several components that are essential for digestion and bacterial killing. Some of these components include the following:

1. Vacuolar ATPase (V-ATPase): This multi-subunit ATP proton pump is acquired early in the maturation process [1]. It acidifies the lumen by hydrolyzing ATP and quickly decreases the luminal pH [6, 7].

Alan R. Kimmel (ed.), *Dictyostelium discoideum: Methods and Protocols*, Methods in Molecular Biology, vol. 2814, https://doi.org/10.1007/978-1-0716-3894-1_6,

2. Lysosomal enzymes: These enzymes belong to different classes and hydrolyze sugar groups like mannosidases and peptide bonds. They are responsible for eliminating several microbes [1, 8].
3. NADPH oxidase (NOX) complex: The NOX complex catalyzes the production of reactive oxygen species (ROS) from oxygen and NADPH [9]. ROS species participate in phagosomal pH regulation [1].
4. Nramps: *Dictyostelium discoideum* has two Nramp proteins (Nramp1 and Nramp2), which mediate the transport of divalent metals through cellular membranes [10]. NRAMP1 is a metal ion transporter present in the phagosome membrane. It produces a metal-ion-depleted environment that affects the survival and replication of bacteria by transporting iron and manganese outside the phagosome [8]. Therefore, Nramp1 makes the phagosome resistant to pathogenic bacteria. Nramp2 synergistically regulates iron homeostasis with Nramp1 but is located only in the contractile vacuole membrane [10].
5. fAR1, LrrkA, and Kil2: Bodinier and others [11] uncovered that the stimulation of intracellular killing by folate involves Far1 (the cell surface receptor for folate), LrrkA, and Kil2. This study identified a signaling pathway regulating intraphagosomal bacterial killing in *Dictyostelium* cells. However, in 2011, Walk and others [12] reported that treatment of *D. discoideum* cells with the bacterial cell wall PAMP, lipopolysaccharide (LPS), resulted only in a modest enhancement of clearance of the mycobacterial species *M. smegmatis*. In addition, no significant difference was observed in the clearance of Gram-negative bacteria such as *K. pneumoniae*, *E. coli*, and *S. typhimurium* or Gram-positive bacteria such as *B. subtilis* and *E. faecalis* following LPS treatment. This study suggests the presence of an interaction between *D. discoideum* cells and ingested bacteria, the nature of which requires further investigation. It remains unclear whether fAR1/LrrkA/Kil2 pathway is essential for killing the abovementioned diverse groups of bacteria.
6. Kil1: It is a sulfotransferase found in the Golgi apparatus of *Dictyostelium*. Kil1 participates in the maturation of lysosomal enzymes. When Kil1 is genetically deactivated, lysosomal enzymes are defectively sulfated. This deficiency results in a decreased ability of phagocytes to effectively eliminate ingested *Klebsiella pneumoniae* [13].
7. Kil2: It is a type of P-type ATPase located in the phagosome, which is believed to be responsible for transporting magnesium ions from the cytosol to the phagosomal lumen. If Kil2 is genetically deactivated, it reduces proteolytic activity in the phagosomes and decreases the ability to eliminate *Klebsiella pneumoniae* intracellularly [13].

8. Ctr-type copper permease and putative copper-translocating P-type ATPases: Numerous antimicrobial qualities are found in copper. By switching between two oxidation states, copper can generate hydroxyl radicals through the Fenton and Haber-Weiss reactions. Moreover, copper can damage protein structure, and Cu^{2+} may also affect Fe-S clusters. A Ctr-type (copper transporter) copper permease [1, 14] and three putative copper-translocating P-type ATPases (atpl, atp2, and atp3) have been found in *Dictyostelium*. The CTR family of copper transporters is responsible for the obtention of copper from the environment. These kinds of transporters do not require energy from ATP hydrolysis to function; they use a concentration gradient since intracellular copper is rapidly sequestered [15, 16]. In *Dictyostelium*, the copper transporter is p80, which could facilitate the transport of copper into the cytosol. P-type ATPases are transporters that use energy from the hydrolysis of ATP to pump ions across the cell membrane against their concentration gradient [17]. In *Dictyostelium*, copper tolerance is associated with Atpl. Atp2 and Atp3 are believed to be associated with copper trafficking during phagocytosis and the killing of bacteria [1, 14].
9. *Dictyostelium* LvsB is involved in the lysosome maturation, lysosome biogenesis regulation, and phagosome acidification and secretion [8].

2 Materials

2.1 Dictyostelium Cell Culture

Grow *Dictyostelium discoideum* cells in D3T media (dissolve 14.3 g of Peptone #3, 7.15 g of yeast extract, 15.4 g of glucose, 0.48 g of KH_2PO_4, and 0.525 g of $Na_2HPO_4.7H_2O$ in deionized water and top off to 1 L). Autoclave for 40 min; let it cool down (*see* **Note 1**). In addition to D3T medium, JH10 cells will be grown with thymidine (0.5 mg/mL). Grow cells in shaking cultures at 19 °C at 150 rpm or in culture flasks at 19 °C for about 24 h until they are in log phase. Monitor cell numbers using a hemocytometer. Harvest and use cells when the cell number are between 2×10^6 cells/mL and 2.5×10^6 cells/mL.

2.2 Bacterial Cell Culture

10 mL LB medium was inoculated with *Klebsiella aerogenes* from a frozen stock. These bacteria were grown overnight (16 h) at 37 °C with shaking (200 rpm) [18, 19]. After that, a single colony of *Klebsiella aerogenes* was isolated by streaking a plate, and a single colony was picked and overnight incubated until the late-exponential phase of growth [19, 20] (*see* **Note 2**).

To determine the number of CFU per OD per ml, *Klebsiella aerogenes* overnight culture was titrated ($1:10^6$). Then, a spectrophotometer (Genesis 10 UV, Thermo Scientific, MA, USA) was used to determine *Klebsiella aerogenes* optical density (OD) at 600 nm. After that, 10 μL of the titrated ($1:10^6$) solution was plated in an LB agar plate. CFU per OD per mL was established by multiplying the colony count for the dilution multiple and then dividing it for the OD at 600 nm. The OD and the CFU were determined by the mean of three replicas [20–22].

2.3 Buffers and Reagents

1. **Development buffer** (5 mM Na_2HPO_4, 5 mM KH_2PO_4, 1 mM $CaCl_2$, 2 mM $MgCl_2$, pH 6.5).
2. **1 × PBS** (137 mM NaCl, 2.7 mM KCl, 10 mM Na_2HPO_4, and 1.8 mM KH_2PO_4)
3. **LB agar plate** (10 g of tryptone, 5 g of yeast extract, 10 g of NaCl 10 g, and 1.5% of agar in 1 liter of water).
4. **Triton X-100** (Sigma-Aldrich).

2.4 Equipment

1. **Spectrophotometer** GENESYS/10 UV.
2. **Tabletop Microcentrifuge** AccuSpin Micro 17

3 Methods

3.1 Bactericidal Assay

1. Dilute the overnight *Klebsiella aerogenes* culture 10,000 times with 1 × PBS (total volume of 1 mL), and incubate for 30 min at 140 RPM shaking.
2. Spin down 2×10^6 *Dictyostelium* cells for 5 min at 300 × g (AccuSpin Micro 17) and resuspend with 1 mL of 1 × PBS.
3. Add 2×10^4 bacteria (around 10 μL) to *Dictyostelium* cells to make a 1:10 mixture and incubate on a shaker (140 RPM) for 0 and 120 min at 21 °C.
4. Add 10 μL of the reaction cell mixture (which includes 200 bacteria) to 140 μL of 1 × PBS or 140 μL of 1X PBS + 0.1% Triton solution.
5. Plate the mixture to 10 cm diameter LB agar plates and incubate overnight at 37 °C and count the bacterial colonies the next day (*see* **Note 3**). Make two to three replicas per experiment.
6. Before the experiments, determine the total *Klebsiella aerogenes* colony-forming units (CFU) by diluting a bacterial stock ($1: 10^6$) on the agar LB plates. Use spectrophotometer (Genesis 10 UV, Thermo Scientific, MA, USA) to determine the optical density (OD) at 600 nm (Thomas et al., 2015). Determine the total number of bacteria using the average CFU at time zero.
7. Determine the percentage of the inactivated *Klebsiella aerogenes* inside the phagosome by subtracting the average CFUs

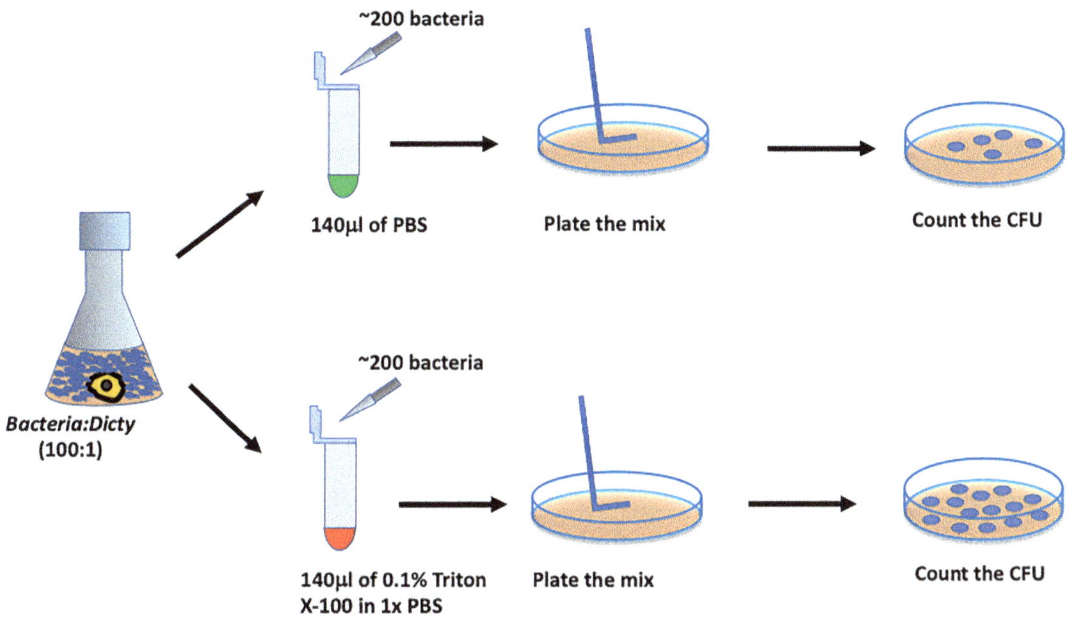

Fig. 1 The bactericidal assay involves several steps as outlined in the diagram
Initially, an excess of bacteria is incubated with *Dictyostelium* cells. Bacteria that are not engulfed by the *Dictyostelium* cells are quantified, as indicated in the upper portion of the diagram. The bacteria that do get trapped within the phagosome are subsequently released by the addition of 0.1% Triton X-100, as shown in the lower portion of the figure. To determine the percentage of live bacteria inside the phagosome, the colony-forming units (CFUs) found after 2 h of phagocytosis with 1X PBS treatment are subtracted from the CFUs obtained after 2 h of the 0.1% Triton in 1X PBS treatment. This calculation allows for the quantification of the surviving bacteria within the phagosome

from samples treated with 0.1% Triton in 1X PBS after 2 h of phagocytosis from the average CFUs at time zero.

8. Determine the percentage of live bacteria inside the phagosome. The process of determining the efficiency of ingested bacterial killing is described in *see* **Note 4**.
9. The whole procedure is summarized in Fig. 1.
10. For analyzing the process of a single bacterium, *see* **Note 5**.

4 Notes

1. To prevent contamination, add an antibiotics mixture to the D3T media to achieve final concentrations of 100 µg/mL ampicillin, 25 µg/mL tetracycline, and 25 µg/mL chloramphenicol.
2. When conducting experiments that involve the addition of compounds dissolved in dimethyl sulfoxide (DMSO), such as LY294002, it is crucial to consider that DMSO has been

observed to impede the growth of different bacteria when used in concentrations exceeding 5% [23]. Our observations have also shown that even at lower concentrations, such as 0.3% DMSO, the viability of *Klebsiella aerogenes* decreases. To overcome this issue, we recommend cultivating *Klebsiella aerogenes* in a growth medium containing a concentration of 0.3% DMSO. Furthermore, it is advisable to isolate *Klebsiella aerogenes* colonies from an LB agar plate that also contains 0.3% DMSO. This approach will help ensure the growth and viability of *Klebsiella aerogenes* while working with DMSO-containing compounds in the experiments.

3. In case 5 cm diameter plates are preferred, plate 40 μL of reaction mixture onto the agar plate.
4. To quantitatively analyze the inactivation process of internalized bacteria in *Dictyostelium* cells, the following steps were performed. First, determine the mean number of ingested bacteria (CFUi) by subtracting CFU count at the 2 h time point (CFU_{2h}, for example, 150) from the CFU count at time zero (CFU_{0h}, for example, 200). Second, determine the mean number of live bacteria ingested by *Dictyostelium* (CFU_{il}) by subtracting the CFU count after 2 h of phagocytosis with 1X PBS treatment (CFU_{2h}, 150) from the CFU count after 2 h of phagocytosis with the 0.1% Triton in 1X PBS-treated sample ($CFU_{2h,Triton}$, for example, 170). The percentile of ingested but live bacteria is $(CFU_{2h,Triton} - CFU_{2h})/(CFU_{0h} - CFU_{2h})$ *100, which is (170–150)/(200–150)*100 = 20/50 = 60%. Therefore, the killing rate of ingested bacteria is 40%.
5. The above-described method is an excellent procedure to quantitatively analyze the bacterial inactivation in the phagosome. However, when mechanistic analysis of the inactivation of a single bacterium during the phagosome maturation is the goal, following procedure is the recommended. Feed *Dictyostelium* cells with bacteria that express fluorescent protein tag such as GFP and monitor the intensity of the fluorescence of the bacterium in the phagosome. This allows for monitoring the intensity of fluorescence emitted by the bacterium within the phagosome. The decrease in fluorescence intensity reflects the digestion of the bacterium expressing the fluorescent protein within the phagosome. It is important to note that the expression level of fluorescent proteins in bacteria is typically not uniform. Therefore, this procedure is specifically designed to quantitatively measure the inactivation of a single bacterium within a phagosome. To obtain reliable quantitative data, a number of repetitions will be necessary [11].

References

1. Dunn JD, Bosmani C, Barisch C, Raykov L, Lefrançois LH, Cardenal-Muñoz E, López-Jiménez A, Soldati T (2018) Eat prey, live: *Dictyostelium discoideum* as a model for cell-autonomous defenses. Front Immunol 8:1906
2. Levin R, Grinstein S, Canton J (2016) The life cycle of phagosomes: formation, maturation, and resolution. Immunol Rev 273:156–179
3. Yeo JC, Wall AA, Luo L, Stow JL (2016) Sequential recruitment of Rab GTPases during early stages of phagocytosis. Cell Logist 6: e1140615
4. Sun-Wada GH, Tabata H, Kawamura N, Aoyama M, Wada Y (2009) Direct recruitment of H^+-ATPase from lysosomes for phagosomal acidification. J Cell Sci 122:2504–2513
5. Marchetti A, Lelong E, Cosson PA (2009) Measure of endosomal pH by flow cytometry in *Dictyostelium*. BMC Res Notes 2:7
6. Clarke M, Köhler J, Arana Q, Liu T, Heuser J, Gerisch G (2002) Dynamics of the vacuolar H^+-ATPase in the contractile vacuole complex and the endosomal pathway of *Dictyostelium* cells. J Cell Sci 115:2893–2905
7. Pamarthy S, Kulshrestha A, Katara GK, Beaman KD (2018) The curious case of vacuolar ATPase: regulation of signaling pathways. Mol Cancer 17:41
8. Cosson P, Lima WC (2014) Intracellular killing of bacteria: is *Dictyostelium* a model macrophage or an alien? Cell Microbiol 16:816–823
9. Lardy B, Bof M, Aubry L, Paclet MH, Morel F, Satre M, Klein G (2005) NADPH oxidase homologs are required for normal cell differentiation and morphogenesis in *Dictyostelium* discoideum. Biochim Biophys Acta 1744:199–212
10. Buracco S, Peracino B, Cinquetti R, Signoretto E, Vollero A, Imperiali F, Castagna M, Bossi E, Bozzaro S (2015) *Dictyostelium* Nramp1, which is structurally and functionally similar to mammalian DMT1 transporter, mediates phagosomal iron efflux. J Cell Sci 128:3304–3316
11. Bodinier R, Leiba J, Sabra A, Jauslin TN, Lamrabet O, Guilhen C, Marchetti A, Iwade Y, Kawata T, Lima WC, Cosson P (2020) LrrkA, a kinase with leucine-rich repeats, links folate sensing with Kil2 activity and intracellular killing. LrrkA, a kinase with leucine-rich repeats, links folate sensing with Kil2 activity and intracellular killing. Cell Microbiol 22(1):e13129
12. Walk A, Callahan J, Srisawangvong P, Leuschner J, Samaroo D, Cassilly D, Snyder ML (2011) Lipopolysaccharide enhances bactericidal activity in Dictyostelium discoideum cells. Dev Comp Immunol 35:850–856
13. Jauslin T, Lamrabet O, Crespo-Yañez X, Marchetti A, Ayadi I, Ifrid E, Guilhen C, Leippe M, Cosson P (2021) How phagocytic cells kill different bacteria: a quantitative analysis using Dictyostelium discoideum. MBio 12: e03169-20
14. Hao X, Lüthje F, Rønn R, German NA, Li X, Huang F, Kisaka J, Huffman D, Alwathnani HA, Zhu YG, Rensing C (2016) A role for copper in protozoan grazing—two billion years selecting for bacterial copper resistance. Mol Microbiol 102:628–641
15. Lee J, Peña MM, Nose Y, Thiele DJ (2002) Biochemical characterization of the human copper transporter Ctr1. J Biol Chem 277(6): 4380–4387
16. Page MD, Kropat J, Hamel PP, Merchant SS (2009) Two Chlamydomonas CTR copper transporters with a novel cys-met motif are localized to the plasma membrane and function in copper assimilation. Plant Cell 21:928–943
17. Kühlbrandt W (2004) Biology, structure and mechanism of P-type ATPases. Nat Rev Mol Cell Biol 5(4):282–295
18. Denkovskienė E, Paškevičius Š, Misiūnas A, Stočkūnaitė B, Starkevič U, Vitkauskienė A, Hahn-Löbmann S, Schulz S, Giritch A, Gleba Y, Ražanskienė A (2019) Broad and efficient control of Klebsiella pathogens by peptidoglycan-degrading and pore-forming Bacteriocins Klebicins. Sci Rep 9:15422
19. Marcoleta AE, Varas MA, Ortiz-Severín J, Vásquez L, Berríos-Pastén C, Sabag AV, Chávez FP, Allende ML, Santiviago CA, Monasterio O, Lagos R (2018) Evaluating different virulence traits of *Klebsiella pneumoniae* using *Dictyostelium discoideum* and zebrafish larvae as host models. Front Cell Infect Microbiol 8:30
20. Elbing KL, Brent R (2019) Growth of *E. Coli* on solid media. Curr Protoc Mol Biol 125:e82
21. Thomas P, Sekhar AC, Upreti R, Mujawar MM, Pasha SS (2015) Optimization of single plate-serial dilution spotting (SP-SDS) with sample anchoring as an assured method for bacterial and yeast cfu enumeration and single colony isolation from diverse samples. Biotechnol Rep (Amst) 8:45–55

22. Beal J, Farny NG, Haddock-Angelli T, Selvarajah V, Baldwin GS, Buckley-Taylor R, Gershater M, Kiga D, Marken J, Sanchania V, Sison A, Workman CT, iGEM Interlab Study Contributors (2020) Robust estimation of bacterial cell count from optical density. Commun Biol 3(1):512
23. Basch H, Gadebusch HH (1968) *In vitro* antimicrobial activity of Dimethylsulfoxide. Appl Microbiol 16:1953–1954

Chapter 7

Methods to Assess Autophagic Activity in *Dictyostelium*

Laura Antón-Esteban, Alba Tornero-Écija, Olivier Vincent, and Ricardo Escalante

Abstract

Autophagy is an intracellular clearance and recycling pathway that delivers different types of cargos to lysosomes for degradation. In recent years, autophagy has attracted considerable medical interest, and many different techniques are being developed to study this process in experimental models such as *Dictyostelium*. Here we describe the use of different autophagic markers in confocal microscopy, in vivo and also in fixed cells. In particular, we describe the use of the GFP-Atg8-RFP-Atg8ΔG marker and the optimization of the GFP-PgkA cleavage assay to detect small differences in autophagy flux.

Key words *Dictyostelium*, Autophagy, Autophagosome, Autophagic flux

1 Introduction

Macroautophagy (hereafter called autophagy) is a highly conserved process in all eukaryotic organisms by which cells degrade cytoplasmic components for cellular cleaning and recycling. These components can range from proteins, lipids, or organelles to aberrant cellular structures (such as protein aggregates or damaged mitochondria) or pathogens. It consists of the formation of intracellular double-membrane vesicles called autophagosomes, which eventually fuse with lysosomes where degradation takes place [1–3].

The process of autophagy occurs constitutively at low levels, known as basal autophagy. But under certain stimuli, such as starvation or cellular stress, autophagy is induced to a greater extent to overcome a lack of nutrients and energy or to eliminate harmful material. Autophagy is a cytoprotective process essential for cell survival and the maintenance of cellular homeostasis, differentiation,

Authors Laura Antón-Esteban and Alba Tornero-Écija have equally contributed to this chapter.

Alan R. Kimmel (ed.), *Dictyostelium discoideum: Methods and Protocols*, Methods in Molecular Biology, vol. 2814, https://doi.org/10.1007/978-1-0716-3894-1_7,

Table 1
Markers and antibodies for autophagic structures

Protein	Marker	Antibody
Atg8	GFP-Atg8 [10] GFP-Atg8-RFP-Atg8ΔG [11]	Anti-Atg8 [12]
Atg18	GFP-Atg18 [13]	
Wdr45l	FLAG-Wdr45l [11]	Anti-FLAG
Autophagy flux markers	GFP-PgkA [10, 14] RFP-GFP-Atg8 [10]	

and development and is thus involved in many aspects of human health and disease [4].

Given its importance, the autophagy mechanism is tightly regulated by a series of complexes formed by the Atg proteins, first described in *Saccharomyces cerevisiae*. These control the initiation, elongation, and maturation of autophagosomes [5].

The social amoeba *Dictyostelium discoideum* is a great model for the study of autophagy due to the strong conservation of the pathway [6–8]. Several markers can be used to study the autophagy process [9, 10]. Most of them are based on Atg proteins fused to different tags that allow visualization of autophagic structures (Table 1).

In this chapter, we describe in detail the use of some of these markers in confocal microscopy, in vivo and also in fixed cells. The GFP-Atg8-RFP-Atg8ΔG marker, described previously in mammalian cells [15], enables the visualization of autophagosomes by specifically distinguishing them from protein aggregates. In addition, other techniques can be used to determine the autophagy flux as a measure of the degradation of specific markers such as GFP-PgkA. Here we describe an update of this technique that allows the detection of subtle defects in autophagic degradation.

2 Materials

2.1 Chemicals, Buffers, and Solutions

1. H50: 20 mM HEPES, 50 mM KCL, 10 mM NaCl, 1 mM $MgSO_4$, 5 mM $NaHCO_3$, and 1 mM NaH_2PO_4, pH 7.
2. PDF buffer: 20 mM KCl, 9 mM K_2HPO_4, 13 mM KH_2PO_4, 1 mM $CaCl_2$, 2.5 mM $MgSO_4$, and pH 6.4.
3. Phosphate-buffered saline (PBS) 1X: 133 mM NaCl, 8 mM Na_2HPO_4, 2 mM KH_2PO_4, and pH 7.4.
4. 10% paraformaldehyde (stock solution) (Merck, 104005.1000) in water. To dissolve the paraformaldehyde, heat the solution at 60–65 °C while stirring and add a few drops of 1 M NaOH until dissolved. This stock solution can be stored at -20 °C.

5. 2% paraformaldehyde in PBS 2X. Use 10% paraformaldehyde (stock solution) and 10X PBS to make this solution (*see* **Note 1**). Keep on ice until use.
6. Blocking solution: PBS 1X with 2% BSA (Sigma-Aldrich, A3059) and 0.5% NP-40 (Nonidet® P 40 Substitute, Sigma-Aldrich, 74,385).
7. ProLong Diamond Antifade Mountant (Molecular Probes, P36970).
8. NH_4Cl (Merck, 413-A844845).
9. Pierce BCA Protein Assay Kit (Thermo Fisher Scientific, 23,225).
10. RIPA buffer: 50 mM Tris-HCl, pH 8, 1% NP-40, 0.1% SDS, 0.5% sodium deoxycholate, 150 mM NaCl, and 2 mM EDTA.
11. Protease inhibitors (Sigma-Aldrich, P8340).
12. 5X Loading buffer: 250 mM Tris–HCl, pH 6.8, 10% SDS, 25% glycerol, 0.04% bromophenol blue, and 100 mM dithiothreitol (DTT).
13. TBST buffer: 20 mM Tris–HCl, pH 7.5, 150 mM NaCl, and 0.05% Tween 20.
14. Skim Milk Powder (Sigma-Aldrich, 70,166).
15. ECL SuperSignal West Pico PLUS, Chemiluminescent Substrate (Thermo Fisher Scientific, 34,580).
16. Antibodies. Primary antibodies: anti-GFP (Sigma-Aldrich, G1544), anti-Atg8 (kindly provided by Jason King, Department of Biomedical Sciences, University of Sheffield, Sheffield, United Kingdom), and anti-FLAG (Cell Signaling Technology, #8146). Secondary antibodies: mouse anti-rabbit IgG-HRP (Santa Cruz Biotechnology, sc-2357) and immunofluorescence secondary antibodies conjugated to Alexa Fluorophores (Molecular Probes).

2.2 Culture, Cells, and Markers

1. HL5 medium (Formedium, HLB0102).
2. *Dictyostelium discoideum* AX4 and mutants *atg1*$^-$, *atg18*$^-$ [11, 16].
3. Autophagic markers shown in Table 1.

2.3 Equipment

1. Confocal images were acquired with an inverted Zeiss spectral LSM710 microscope.
2. μ-Slide 8-well chambered for in vivo visualization and for cell fixation (Ibidi, 80,826).
3. Glass coverslips 9 mm (VWR, 631–0169).
4. Six-well multiwell plates (Falcon).
5. 15 mL conical tubes (Falcon).
6. Eppendorf tubes.

7. Gel electrophoresis (Bio-Rad Mini-Protean) and transfer (Trans-Blot) systems.
8. Polyvinylidene fluoride (PVDF) transfer membranes (Millipore, IPVH00010).
9. Electroporation cuvette 0.1 cm (Bio-Rad, 165–2089).
10. Electroporation system: Gene Pulser II (Bio-Rad).

3 Methods

3.1 Detection of Autophagic Markers by Confocal Microscopy

Dictyostelium cells expressing autophagic markers fused to fluorescent proteins can be easily visualized in vivo as they do not require special environmental conditions.

3.1.1 GFP-Atg8-RFP-Atg8ΔG Marker

1. Grow *Dictyostelium discoideum* cells at a density of approximately 1–3 × 10^6 cells/mL in HL5 medium.
2. Place 5 × 10^6 cells in a 15 mL Falcon® conical tube and chill the cells on ice for 15 min.
3. Collect cells by centrifugation at 1000 rpm for 4 min and wash twice with 1 mL of ice-cold H50 buffer.
4. Pellet by centrifugation at 1000 rpm for 4 min and resuspend the cells in 100 μL of ice-cold H50 buffer.
5. Mix the cells with 1 μg of GFP-Atg8-RFP-Atg8ΔG plasmid, and put the cell suspension in a cold 0.1 cm electroporation cuvette.
6. For electroporation, apply two consecutive 0.85 kV pulses at 25 μF of capacitance. Wait 5 seconds between pulses.
7. Plate cells from the transformation in a 100 mm culture dish containing 10 mL of HL5 medium.
8. The next day, add the selection drug, G418 at 25 μg/mL.
9. Allow the drug-resistance colonies to grow for at least 5 days.
10. Deposit 2 × 10^5 cells on an Ibidi μ-Slide 8-well chambered coverslip (*see* **Note 2**).
11. Allow the cells to adhere for 15 min.
12. For induction of autophagy by starvation, replace HL5 medium with PDF and incubate the cells for 2 h.
13. Visualize the cells in vivo with a confocal microscope. Wild-type cells (AX4 strain) show the presence of green-only puncta. Since only GFP-Atg8 can be lipidated, green puncta marks autophagic structures specifically. On the other hand, green and red aggregates can be detected in autophagic mutants of *Dictyostelium* such as *atg1*$^-$, a strain that shows a block of autophagy at an early stage. An example is shown in Fig. 1.

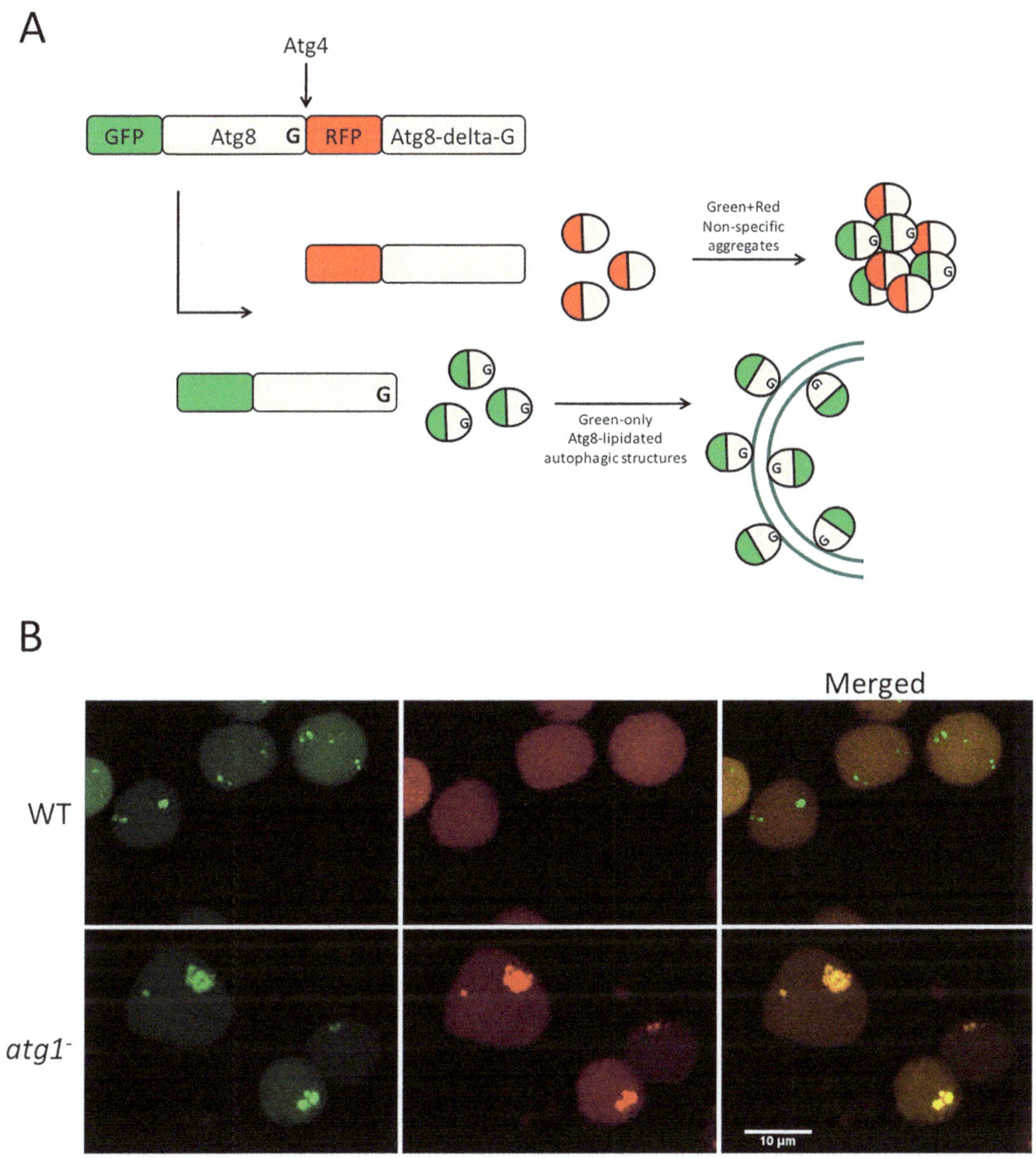

Fig. 1 GFP-Atg8-RFP-Atg8ΔG assay. (**a**) Schematic representation of the function of the GFP-Atg8-RFP-Atg8ΔG marker. The Atg4 protease cleaves the recombinant protein, so that equimolar amounts of GFP-Atg8 and RFP-Atg8ΔG are released. GFP-Atg8 can be conjugated to autophagic membranes, whereas the fusion to RFP lacks the essential glycine (G) that allows covalent bonding to phosphatidylethanolamine. (**b**) Wild-type (WT) and *atg1*$^{-}$ cells expressing the GFP-Atg8-RFP-Atg8ΔG marker were incubated in PDF, and fluorescence was observed in vivo by confocal microscopy. Green/not-red dots indicate the presence of autophagic structures (present only in WT). Green/red puncta marks protein aggregates (present only in the Atg1 mutant). The lack of green/not-red dots in a mutant strain indicates a complete block of autophagy. Bar, 10 μm

3.1.2 Analysis of Autophagic Markers in Fixed Cells

Here we describe a general protocol for the fixation and immunodetection of autophagic markers in *Dictyostelium* cells. This protocol can be used for the detection of endogenous Atg8 and expressed FLAG-Wdr45l.

1. Grow axenically *Dictyostelium discoideum* cells expressing the marker protein in HL5 medium to a density of 1–3 × 10^6 cells/mL.
2. Place 3–4 × 10^5 cells in one well of an Ibidi μ-Slide 8-well chambered coverslips. Wait for at least 1 h for the cells to adhere (*see* **Note 2**).
3. If necessary, replace the medium with PDF buffer for 2 h for nutrient-starved conditions.
4. Remove the medium and fix the cells with 200 μL of 2% paraformaldehyde in 2X PBS for 15 min at room temperature.
5. Replace the paraformaldehyde solution with 1X PBS. Incubate for 5 min.
6. Repeat **step 5** two more times (*see* **Note 3**).
7. Add 200 μL of blocking solution and incubate for 30 min at room temperature.
8. Replace the buffer with 100 μL of primary antibody prepared at the appropriate dilution in blocking solution and incubate the cells overnight at 4 °C (*see* **Note 4**).
9. Remove the primary antibody and wash three times with 1X PBS as in **steps 5** and **6**.
10. Add 100 μL of secondary antibody prepared at the appropriate dilution in blocking solution, and incubate for 1 h in the dark at room temperature (*see* **Note 5**).
11. Remove the secondary antibody and wash three times with 1X PBS as in **steps 5** and **6**.
12. Put 5 μL of ProLong anti-fade reagent on a coverslip and lay it down over cells, avoiding trapping air bubbles. Keep the plate at room temperature overnight to dry, and then store at 4 °C until visualization with a confocal microscope.

3.2 Protein Cleavage Assay to Monitor Autophagic Flux

In this assay, free-GFP is generated as a result of degradation of the cytosolic marker GFP-PgkA in autolysosomes [14]. The level of free-GFP follows a curve that reveals the optimum concentration of NH_4Cl required for maximal accumulation of free GFP. An example of this protocol is shown in Fig. 2. The use of NH_4Cl concentrations ranging between 75 and 200 mM allows the detection of a slight shift in the degradation curve in the *atg18*⁻ mutant, suggesting a slight defect in autophagy flux. The amount of free GFP is related to the degradative capacity of the autophagolysosomes in AX4 and *atg18*⁻ cells under NH_4Cl treatment (*see* **Note 6**).

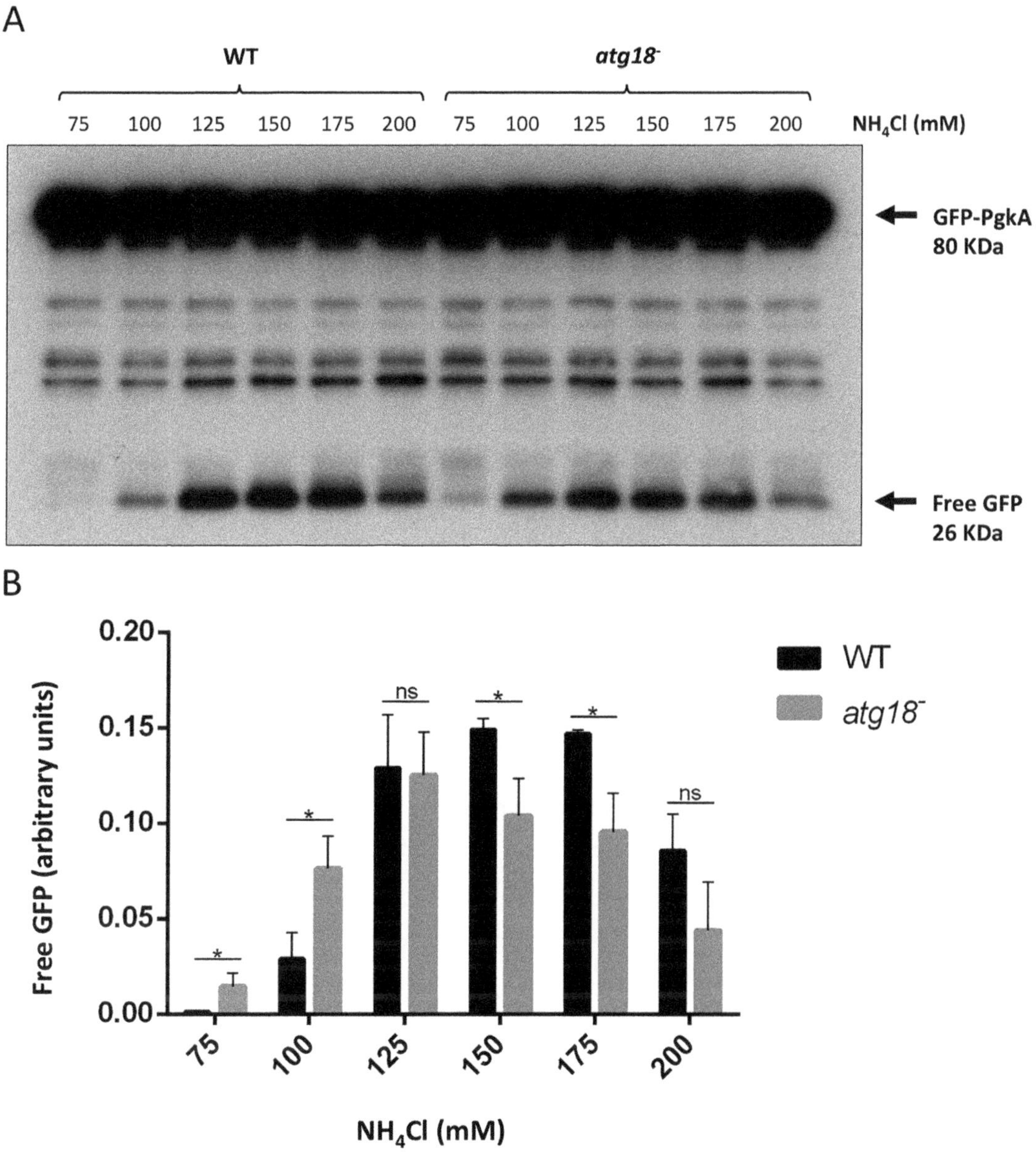

Fig. 2 Autophagic flux assay. (**a**) Western blot showing GFP-PgkA and free-GFP signals from WT and *atg18*⁻ mutant cells. (**b**) Quantification of free-GFP signal at different NH_4Cl concentrations. Mean and standard deviation of three independent experiments are shown. Significance was calculated by student t-test for planned comparisons within each treatment (* $p < 0.05$)

1. Grow *Dictyostelium* cells expressing the GFP-PgkA construct in HL5 medium to a density of 1–3 × 10^6 cells/mL.
2. For each condition, place 1.3 × 10^6 cells in a 15 mL Falcon® conical tube and pellet the cells by centrifugation at 1000 rpm for 5 min.

3. Resuspend the cells with 1 mL of HL5 and place them in one well of a 6-well plate. Allow the cells to adhere for 30 min.
4. Add NH_4Cl from a 2 M stock solution to achieve the following concentrations: 75, 100, 125, 150, 175 and 200 mM. Using this specific range of concentrations allows detecting slight differences in GFP-PgkA degradation. Incubate the cells for 2 h at 22 °C.
5. Repeat the NH_4Cl treatment, as in **step 4**.
6. Resuspend the cells in the same medium and place in Eppendorf® tubes. Pellet the cells by centrifugation at 2000 rpm for 5 min.
7. Wash the cells with 500 μL of PDF buffer.
8. Pellet by centrifugation at 2000 rpm for 5 min, and resuspend cells in 50 μL of 1X RIPA buffer supplemented with protease inhibitors.
9. Keep the cells on ice for 1 h and vortex every 10 min to ensure complete lysis.
10. Centrifuge at 12000 rpm, 4 °C, for 10 min. Transfer the supernatant containing the protein extract to clean Eppendorf® tubes.
11. Determine the amount of protein by BCA assay.
12. Use 4 μg of protein for Western blot analysis. Add loading buffer and boil samples at 95 °C for 10 min.
13. Run samples on a standard 12% SDS-PAGE gel.
14. Transfer the samples from the gel to a PVDF membrane.
15. Block the membrane with TBST with 5% skim milk for 1 h at room temperature.
16. Incubate the membrane with anti-GFP antibody diluted 1: 4000 in TBST with 5% skim milk, overnight at 4 °C.
17. Wash the membrane with TBST during 1 h. Change the buffer every 10 min.
18. Incubate the membrane with the secondary antibody diluted 1: 5000 in TBST with 2% skim milk for 1 h at room temperature.
19. Wash the membrane, as in **step 17**.
20. Incubate the membrane with ECL to detect the protein signal. Use appropriate exposure to avoid saturation.
21. Perform densitometric analysis of the Western blot images using ImageJ software.

4 Notes

1. Heat the paraformaldehyde stock solution (10% in water) at 50 °C until completely redissolved. The first time the stock is used, check that the pH of the final fixative solution (2% paraformaldehyde in 2X PBS) is around 7.5 using a pH test strip.
2. Make sure that the cells are homogeneously distributed in the well.
3. If running multiple wells, prevent cells from drying out at each step by replacing the buffer in each sample immediately after removing the previous buffer.
4. Seal the plate to prevent evaporation of the antibody solution.
5. Keep cells expressing fluorescent markers and immunofluorescence secondary antibodies in the dark in order to avoid loss of fluorescent signal.
6. In this technique, the use of NH_4Cl is necessary to slow down the degradative capacity of autolysosomes, allowing the accumulation of free GFP. A shift in the concentration of NH_4Cl that is required for maximum accumulation of the degradative product may indicate a change in the degradative capacity and/or pH of the autolysosomes. The use of only one or two NH_4Cl concentrations may not allow detection of these minor defects and/or lead to erroneous conclusions.

Acknowledgments

This work has been supported by the "Ministerio de Ciencia e Innovación," grant number PGC2018-093604-B-I00 (MCIU/AEI/FEDER,UE).

References

1. Lahiri V, Hawkins WD, Klionsky DJ (2019) Watch what you (self-) eat: Autophagic mechanisms that modulate metabolism. Cell Metab 29:803–826
2. Melia TJ, Lystad AH, Simonsen A (2020) Autophagosome biogenesis: from membrane growth to closure. J Cell Biol 219: e202002085
3. Gubas A, Dikic I (2022) A guide to the regulation of selective autophagy receptors. FEBS J 289:75–89
4. Mizushima N, Levine B (2020) Autophagy in human diseases. N Engl J Med 383:1564–1576
5. Al-Bari MAA (2020) A current view of molecular dissection in autophagy machinery. J Physiol Biochem 76:357–372
6. Mesquita A, Cardenal-Muñoz E, Dominguez E, Muñoz-Braceras S, Nuñez-Corcuera B, Phillips BA, Tábara LC, Xiong Q, Coria R, Eichinger L, Golstein P, King JS, Soldati T, Vincent O, Escalante R (2017) Autophagy in *Dictyostelium*: mechanisms, regulation and disease in a simple biomedical model. Autophagy 13:24–40
7. Calvo-Garrido J, Carilla-Latorre S, Kubohara Y, Santos-Rodrigo N, Mesquita A, Soldati T, Golstein P, Escalante R (2010)

Autophagy in Dictyostelium: genes and pathways, cell death and infection. Autophagy 6: 686–701

8. Fischer S, Eichinger L (2019) Dictyostelium discoideum and autophagy – a perfect pair. Int J Dev Biol 63:485–495
9. Domínguez-Martín E, Cardenal-Muñoz E, King JS, Soldati T, Coria R, Escalante R (2017) Methods to monitor and quantify autophagy in the social amoeba *Dictyostelium discoideum*. Cells 6:18
10. Mesquita A, Calvo-Garrido J, Carilla-Latorre S, Escalante R (2013) Monitoring autophagy in Dictyostelium. Methods Mol Biol 983:461–470
11. Tornero-Écija A, Tábara LC, Bueno-Arribas M, Antón-Esteban L, Navarro-Gómez C, Sánchez I, Vincent O, Escalante R (2022) A Dictyostelium model for BPAN disease reveals a functional relationship between the WDR45/WIPI4 homolog Wdr45l and Vmp1 in the regulation of autophagy-associated PtdIns3P and ER stress. Autophagy 18:661–677
12. Cardenal-Muñoz E, Arafah S, López-Jiménez AT, Kicka S, Falaise A, Bach F, Schaad O, King JS, Hagedorn M, Soldati T (2017) Mycobacterium marinum antagonistically induces an autophagic response while repressing the autophagic flux in a TORC1- and ESX-1-dependent manner. PLoS Pathog 13: e1006344
13. Gerstenmaier L, Pilla R, Herrmann L, Herrmann H, Prado M, Villafano GJ, Kolonko M, Reimer R, Soldati T, King JS, Hagedorn M (2015) The autophagic machinery ensures nonlytic transmission of mycobacteria. Proc Natl Acad Sci USA 112:E687–E692
14. Calvo-Garrido J, Carilla-Latorre S, Mesquita A, Escalante R (2011) A proteolytic cleavage assay to monitor autophagy in Dictyostelium discoideum. Autophagy 7:1063–1068
15. Morishita H, Kaizuka T, Hama Y, Mizushima N (2017) A new probe to measure autophagic flux in vitro and in vivo. Autophagy 13:757–758
16. Mesquita A, Tábara LC, Martinez-Costa O, Santos-Rodrigo N, Vincent O, Escalante R (2015) Dissecting the function of Atg1 complex in Dictyostelium autophagy reveals a connection with the pentose phosphate pathway enzyme transketolase. Open Biol 5:150088

Chapter 8

Controlling the Subcellular Localization of Signaling Proteins Using Chemically Induced Dimerization and Optogenetics

Mariam Beshay, Yu Deng, and Chris Janetopoulos

Abstract

A given protein can perform numerous roles in a cell with its participation in protein complexes and distinct localization within the cell playing a critical role in its diverse functions. Thus, the ability to artificially dimerize proteins and recruit proteins to specific locations in a cell has become a powerful tool for the investigation of protein function and the understanding of cell biology. Here, we discuss two systems that have been used to activate signal transduction pathways, a chemically inducible dimerization (CID) and a light-inducible (LI) system to control signaling and cytoskeletal regulation in a spatial and temporal manner.

Key words Optogenetics, Chemically induced dimerization, Rapamycin, FKBP, FRB, Lov domain

1 Introduction

Having the capacity to localize or activate molecules in a spatially and temporally defined manner has revolutionized our ability to unmask the function of many components in cell biology. One method, the chemically inducible dimerization (CID) system, has been used in a variety of systems and has been successful in elucidating and studying various proteins and signaling pathways (**reviewed in** [1–4]). The dimerization field has been steered by a system that takes advantage of the mechanism of TORC1 inhibition by rapamycin, an immunosuppressive drug, with its use first demonstrated in the control of gene expression [5]. Rapamycin controls the dimerization of two components; it binds first to FKBP, and then the complex can bind to FRB, which is the binding domain from the TOR protein (Fig. 1). One protein (either FKBP or FRB) can be anchored at a specific cellular localization, while the other protein component is fused to a protein of interest (POI) and

Alan R. Kimmel (ed.), *Dictyostelium discoideum: Methods and Protocols*, Methods in Molecular Biology, vol. 2814,
https://doi.org/10.1007/978-1-0716-3894-1_8,

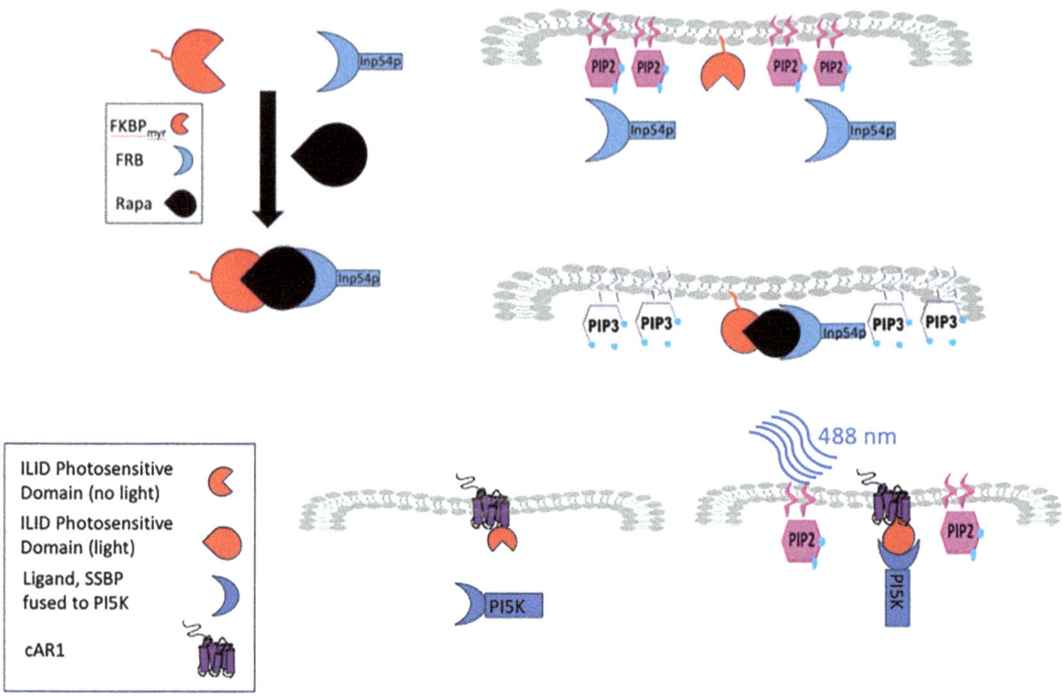

Fig. 1 Two methods for triggering signal transduction using synthetic biology. (Top) The FKBP/FRB system lowers plasma membrane PI(4,5)P2 levels. Diagram of FKBP/FRB system. (Top, Left) FKBP and FRB form a heterodimer in the presence of rapamycin (rapa). (Top, Right) Illustration of an experimental setup for recruiting a protein to the membrane with the FKBP/FRB system in *Dictyostelium*. FKBP domains are anchored to membrane by myristoylation (myr), and the actuator attached to an FRB domain is diffusing in the cytosol. When rapamycin is added, the actuator is recruited to membrane by binding to FKBP. The actuator is fused to the lipid phosphatase Inp54p which lowers PI(4,5)P2 locally and results in PI 3 kinase activity and the synthesis of PI(3,4,5)P3. (Bottom) Optogenetic system for controlling protein localization in *Dictyostelium*. (Bottom, Left) Diagram of blue light dimerization systems. In the dark, a photosensitive protein (iLID) has low affinity for its ligand (SSBP). When exposed to 488 nm light (optimally, it would be 440 nm light), a conformational change in the photosensitive domain allows it to bind to the ligand with affinity. (Bottom, right) Diagram of iLID-based systems for recruiting proteins to the membrane in *Dictyostelium*. The actuator (blue), fused to PI5 Kinase, is recruited to the uniformly distributed membrane anchor fused to cAR1 (red) only at the area exposed to blue light. Levels of PI(4,5)P2 should be elevated

is free in the cytoplasm. Both components can be color coded by genetically fused proteins, and the readout is typically the translocation of the cytoplasmic molecule to the anchored protein. The addition of rapamycin, which is membrane permeable, forms this irreversible ternary complex in a rapid manner, taking place on the order of seconds to minutes.

The second technique (Fig. 1) discussed here uses optogenetics, which has several advantages [6]. Unlike CID, optogenetics can be reversible, and because specific wavelengths of light are used to activate the dimerization, it is rapid (msec to second timescales) and can be localized to specific subdomains of a cell. With optogenetics, light-induced conformational changes can expose binding domains or separate protein domains so that one can perform dimerization experiments similar to those seen with CID [7, 8]. In *Dictyostelium*, protein-protein interactions have been studied using the Crytochrome CRY2-mediated approaches [7, 9], but more recently, studies have also taken advantage of improvement in the light-oxygen voltage (LOV) domains used for optogenetic control, which will be discussed here [8, 10]. The newer system relies on an improved light-induced dimer (iLID) which has an added benefit of a quicker turnover of the light-induced state (half-life of 30 seconds vs. about 6 minutes for Cry2). This means that a response can be recorded over and over again on the same sample more readily. In addition, for systems where neither component is anchored, the cellular response is more limited to the area that was targeted, whereas in the Cry2 system, the activated molecules could diffuse away and provide a response in areas of the cell outside of your targeted zone. The main disadvantage for the optogenetics systems is that for targeted activation, one needs to use a scanning confocal system or a digital micromirror device (DMD), which adds a significant cost to the technique. Experiments using optogenetics investigating the small Rho GTPases have already made important contributions to our understanding of cellular mechanics during cell migration [11–13]. This approach is now being rapidly incorporated for use in the model system *Dictyostelium discoideum*. Here we demonstrate that CID can be used to get robust signaling responses critical for control of cell morphology and also show that iLID can be used to spatially control the recruitment of an enzyme that will be used for future cell migration studies. The ability to use these techniques in *D. discoideum* will permit use of this system to continue to play a critical role in furthering our understanding of signal transduction, cell polarity, and cell migration.

2 Materials

2.1 Cell Lines Used

1. Wild-type *Dictyostelium discoideum* cells of the AX2 strain were obtained from the R. Kay laboratory (MRC Laboratory of Molecular Biology, UK). Cells were cultured at 22 °C in HL5 medium supplemented with 300 μg/mL streptomycin. Cells were passaged on cell culture dishes every 3–4 days, and experiments were performed within 7–8 weeks of thawing cells from frozen stocks. Stable cell lines were grown similarly with selection antibiotics.
2. Heat-killed *Klebsiella aerogenes* (KA), from *Dictybase* (http://dictybase.org).

2.2 Media and Solutions

1. HL5 axenic medium for CID experiments: Formedium liquid media; 22 g of Formedium HL-5 powder and 10 g of dextrose were added to 1 L of distilled water and then autoclaved. HL5 media for iLID experiments: Recipe was obtained from DictyBase and includes 5 g proteose peptone, 5 g thiotone E peptone, 10 g glucose, 5 g yeast extract, 0.35 g Na2HPO4*7H2O, 0.35 g KH2PO4, and 0.05 g dihydrostreptomycin-sulfate (GoldBio, #S-150-100) were added to distilled water, and the volume was brought to 1 liter. pH was adjusted with HCl to pH 6.4–6.7. Medium was autoclaved for 20 mins to sterilize.
2. Development buffer (DB): Recipe was obtained from DictyBase. 5 mM Na2HPO4, 5 mM KH2PO4, 1 mM CaCl2, and 2 mM MgCl2 were added to distilled water. pH was adjusted to 6.5and filter sterilized.
3. H-50 buffer: Recipe was obtained from DictyBase. Briefly, 20 mM HEPES, 1 mM MgSO4, 50 mMKCl, 5 mM NaHCO3, 10 mM NaCl, and 1 mM NaH2PO4 were added to distilled water. pH was adjusted to 7 and filter sterilized.
4. SM agar plate (peptone 10 g/L; yeast extract 1 g/L; glucose 10 g/L; KH_2PO_4 1.9 g/L; K_2HPO_4 x 3 H_2O, 1.3 g/L; $MgSO_4$ anhydrous 0.49 g/L; 1.7% agar) for KA growth.
5. LB-medium (lysogeny broth) for bacterial cell growth and plasmid preparation.

2.3 Recombinant DNA Used

1. Both the sequences encoding the first 150 amino acids at the N-terminus of PKBR1 (denoted as "myr") for the membrane-anchored FKBP constructs, as well as tandem FKBPs from the ARGENT™ Regulated Heterodimerization Kit (Version 2.0), were joined with a linker sequence of AGTGCTGGTGGT and cloned into pCV5 and also pDM358 [14].
2. Sequences encoding mCherry, FRB, and 1-331 amino acids of Inp54p (or inactivate form Inp54pD281A) were linked and cloned into pCV5 as well as pB18. The linker sequence between mCherry and FRB is GGAGCAAGTGGA, and two repeats of AGTGCTGGTGGT were used between FRB and Inp54p. PH_{crac}-YFP was also cloned in pCV5 [14].
3. pDM358 plasmid (DictyBase #534) expressing cAR1-CIBN or CAAX-deleted Venus-iLID fused to N150 gene fragment (N150-Venus-iLID; Addgene #201763) [15].
4. pCV5 plasmid (DictyBase #23) expressing CRY2PHR-mCherry (Addgene #26866) or tgRFPt-SspB R73Q fused to the N-terminal of the catalytic domain of RacGEF1 (gift from Firtel lab, UCSD).

5. PI5K Truncation 367-719aa ORF was amplified from PI5K 367-719aa-GFP/pCV5 and then subcloned into mRFPmars-SspB-R73Q/pCV5 at the NheI and NotI site to generate mRFPmars-SspB-PI5K 367-719aa/pCV5.

2.4 Cell Transformation

1. Gene Pulser Electroporation Cuvettes, 0.1 cm gap (BIO-RAD #1652089).
2. Gene Pulser Xcell Electroporation System (BIO-RAD #1652660).

2.5 Microscopy

1. CID experiments were performed on an Intelligent Imaging and Innovations (3I) Marianas system outfitted with a Yokogawa spinning disk confocal system. The microscope base was an inverted Zeiss-Axiovert 200 M epifluorescence microscope (Zeiss Inc., Oberkochen, Germany) with a Photometrix Evolve EMCCD camera. The lens used was a 40X PlanNeofluar 1.3NA oil immersion objective.
2. iLID experiments were performed on a Zeiss LSM800 GaAsP single-point, laser scanning confocal microscope with a wide-field camera using ZEN Blue software. All images were acquired with 63X/1.40 PlanApo oil DIC objective, along with digital zoom.
3. 8-well chamber (Lab-Tek, #155409 PK).

3 Methods

All procedures were carried out at 22 °C unless otherwise noted.

3.1 Stable Cell Transformations

1. *Dictyostelium* stable cell lines were generated by electroporation (*see* **Note 1**). For CID, cells were first transformed with mCherry-FRB-Inp54p in pB18 and PH_{crac}-YFP in CV5 by electroporation and selected with G418 (20 μg/mL). Then the selected cells were transformed again with myr-FKBP in pDM358 and cultured with G418 (20 μg/mL) and hygromycin (50 μg/mL) (*see* **Note 2**). For iLID, cells were electroporated simultaneously with both the N150-Venus-iLID/pDM358 and mRFPmars-SspB-PI5K 367-719aa/pCV5 constructs. To perform the transformation, 1×10^7 cells were harvested, washed twice, and resuspended in 100 μL ice-cold H-50 buffer. For each transformation, 2 μg of plasmid DNA was mixed with the cell suspension and then moved to an ice-cold 0.1 cm-gap cuvette (*see* **Note 3**). Electroporation was performed with two pulses at 0.85 kV/25 μF with a 5 s interval (*see* **Note 4**). Cuvettes were then incubated on ice for an additional 10 min. Next, cells were transferred to a 10 cm culture dish containing 12 mL HL5 culture medium

supplemented with heat-killed bacteria. On the next day, 10–20 μg/mL G418 sulphate and when appropriate 30–40 μg/mL hygromycin B were added to cells and selected over 1–3 weeks (*see* **Note 5**).

2. Once cells have reached confluency, cells are maintained in the 10 cm culture dishes with HL5 and streptomycin and are passed every few days as needed (*see* **Note 6**).

3.2 CID Microscopy

1. Vegetative cells were placed in an eight-well cover slip chamber and allowed to adhere for 10–15 min. Next, the media was replaced with 200 μL DB (*see* **Note 7**).
2. 10 μM rapamycin was prepared fresh from a 10 mM DMSO stock solution. This was diluted with nuclease-free water (NFW) to reach a final concentration of 100 μM. Then, 25 μL of 100 μM rapamycin were mixed with 25 μL of NFW in a separate tube. 50 μL of this 50 μM solution is added to a well containing 200 μL of HL5 with adhering vegetative cells (*see* **Note 8**).
3. Time-lapse microscopy was performed on an inverted Zeiss Axiovert 200 M using spinning disk confocal microscopy. Cells were imaged with a 40X lens and images were acquired every 5 s using 488 nm and 561 nm illumination to acquire GFP and RFP images depicted in Fig. 2. As can be seen, the

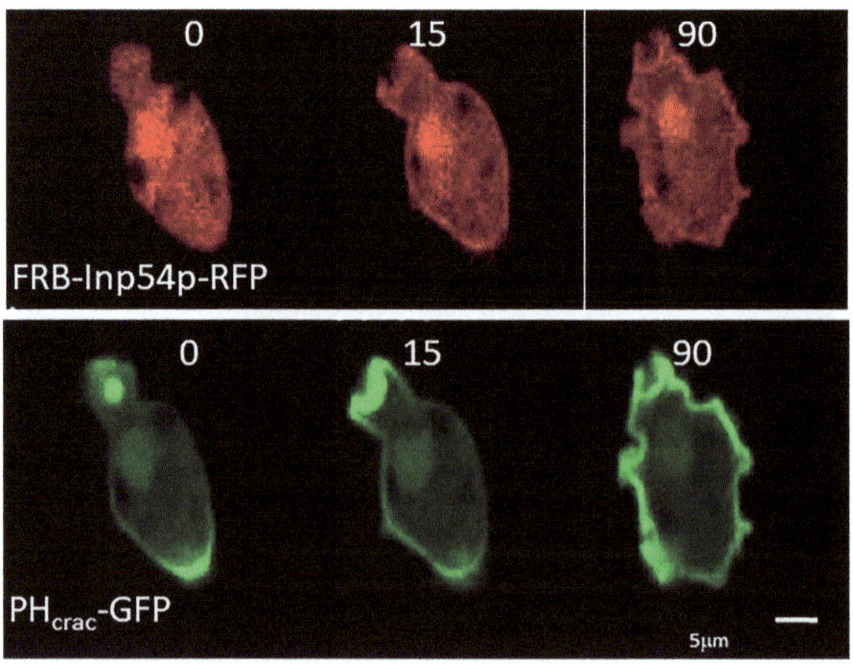

Fig. 2 Synthetic lowering of the PM PI(4,5)P2 levels triggers cell spreading and PI3 Kinase activity. (TOP) Time-lapse spinning disk confocal images showing the recruitment of FRB-Inp54p-RFP after 10 μM rapamycin addition. (Bottom) The biosensor PH_{crac}-GFP translocates to the plasma membrane after PI(4,5)P2 is depleted

Inp54P-RFP was recruited to the plasma membrane after addition of rapamycin. Interestingly, there was a simultaneous burst of PI3 Kinase activity, demonstrated by the near simultaneous recruitment of PH_{crac}-GFP to the plasma membrane, a biosensor for PI(3,4)P2 and PI(3,4,5)P3. This occurs after the PI (4,5)P2 is depleted upon recruitment of the Inp54p, a 5′ lipid phosphatase. These studies provide further evidence that a PI (4,5)P2 threshold regulates signaling responses [14, 16].

3.3 iLID Microscopy

1. Vegetative cells were seeded in an eight-well cover slip chamber and allowed to adhere for 10–15 min (*see* **Note 9**). Next, the media was replaced with 250 μL DB. Cells were then imaged. At time zero, the 488 nm laser is not yet on, so the Venus channel is blank, whereas 561 nm excitation was used for mRFPMars and the mRFPmars-SspB-PI5K 367-719aa is still in the cytosol (Fig. 3). For uniform recruitment, a 0.4% 488 nm argon laser was turned on to acquire a full-field image. The membrane anchor was uniformly distributed on the plasma membrane, as previously described [17]. Once the 488 nm laser is turned on, the mRFPmars-SspB-PI5K 367-719aa is recruited uniformly to the plasma membrane. Images were acquired ~8 s.

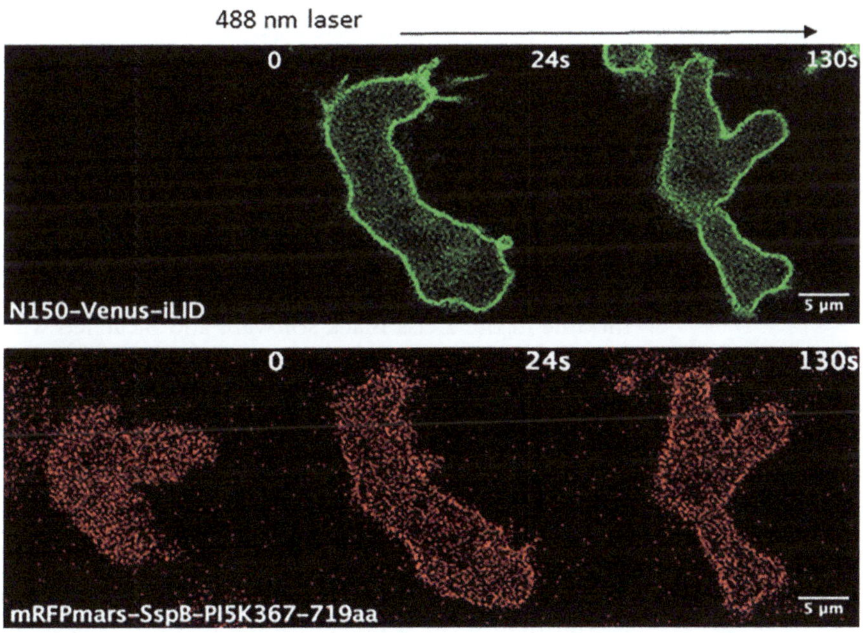

Fig. 3 Uniform recruitment of PI5 Kinase to the plasma membrane in response to illumination. At time zero, the 488 nm laser is not yet on, so the Venus channel is blank, whereas 561 nm excitation was used for mRFPMars and the mRFPmars-SspB-PI5K 367-719aa is still in the cytosol. For uniform recruitment, a 0.4% 488 nm argon laser was turned on to acquire a full-field image, and the mRFPmars-SspB-PI5K 367-719aa is recruited uniformly to the plasma membrane

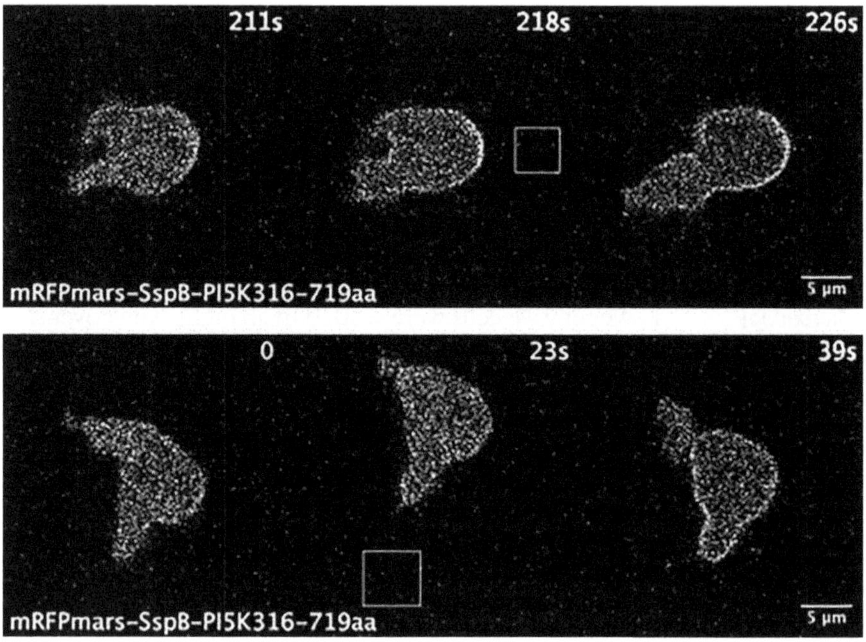

Fig. 4 Two examples of localized recruitment of PI5 kinase in response to targeted illumination. For local recruitment, the interactive photobleaching module on Zeiss LSM800 was used to perform local recruitment experiments. A small region of interest (box) was placed in front or back of migrating cells and bleached with 488 nm laser (laser power of ~0.4%) for one iteration. In the first example (top), the 488 box is placed near the right side of the cell, and a localized recruitment of the mRFPmars-SspB-PI5K 367-719aa is observed. In the second cell (bottom), a box is placed below the cell, and a similar recruitment of mRFPmars-SspB-PI5K 367-719aa takes place

2. For local recruitment, the interactive photobleaching module on Zeiss LSM800 was used to perform local recruitment experiments (Fig. 4). A 514 nm (Ar laser) was used to excite YFP and mVenus (not shown). A small region of interest was placed in front or back of migrating cells (box) using the region module in the Zeiss Black software and bleached with 488 nm laser (laser power of ~0.4%) for one iteration. Since *Dictyostelium* are light sensitive, the laser intensity is low, ~0.06–0.08 W/cm^2 at the objective. Time interval of photoactivation and image acquisition was ~8 s. Two examples of the mRFPMars are shown. In the first cell, the 488 box is placed near the right side of the cell, and a localized recruitment of the mRFPmars-SspB-PI5K 367-719aa is observed. In the second cell (bottom panel), a box is placed below the cell, and a similar recruitment of mRFPmars-SspB-PI5K 367-719aa takes place. Images were acquired ~8 s (*see* **Note 10**).

4 Notes

1. Cells should be frozen immediately once transformants reach confluency. Cells can often change their phenotype when expressing a protein of interest, so it is critical that they be frozen so that the experimenter can thaw new cells out. The selection process for extrachromosomal plasmids is typically faster due to high copy numbers present in *D. discoideum* cells. Transfectants can sometimes be there in just a few days. Check your 10 cm plates regularly under a phase contrast microscope when transfectants come up so that you can freeze them quickly [18]. To obtain consistent results, new cultures should be started every 3–4 weeks, and cultures should never be allowed to grow to densities higher than 4×10^6 cells/mL. If overgrowth occurs, a new culture should be thawed out.
2. While many motifs (e.g., myristoylation membrane localization) have been used for membrane localization of signaling regulators, it is recommended that one use a receptor that should be even more uniformly distributed across the membrane, like the g protein coupled receptor cAR1, which is used here for the iLID experiments. We have found that including two FKBP domains on the C-terminus tail increases recruitment of the POI. The POI should have any localization domains removed so that targeting is controlled by the CID or iLID systems.
3. Going higher than 2 μg of DNA can be toxic to cells and decrease transfection efficiency. Keep the DNA volume below 5 μL.
4. When expressing an optogenetic pair, use pCV5 for your actuator and pDM358 for your anchor. Use a dual expression system for imaging your biosensors. *Caution*: The antibiotics used as selectable marker for your plasmids are toxic. Wear gloves.
5. When only expressing a CID pair in a cell, the actuator can be expressed off pCV5 or PB18 and the anchor pDM358. If including a biosensor, use pCV5 vectors for both CID vectors and pDM358 for the biosensor. Make sure to include a proper control. For enzymes, a catalytically inactive protein should be used as an actuator. The best controls are done under blinded conditions.
6. If one is having difficulty expressing two different fluorescently colored probes in the same line, the creation of giant cells through cell fusion can be used to overcome this problem. This just requires two individual cell lines each expressing a given fluorescent protein. To fuse cells, start with a total of 1.5×10^8 axenically grown cells, washed two times and

resuspended in 10 mL SB (17 mM Soerensen buffer, 15 mM KH2PO4 and 2 mM Na2HPO4, pH 6.0). Next, the cells were put inside a conical tube and gently rolled for 30–40 min to induce cluster formation. Subsequently, 800 μL of rolled cells were transferred to a 0.4 cm gap Bio-Rad cuvette, using pipette tips with cutoff tips (to ensure aggregated clusters remain intact). Using a BioRad Gene Pulser, the electroporation was performed at 1 kV, 3 μF once, and then with 1 kV, 1 μF twice more to induce cell fusion. Cells were allowed to recover, settle, and adhere to substrate for ~1 hour before starting the image acquisition. One could alter the proportion of cells to increase the concentration of one fluorescent cell line over the other. This technique also allows you to fuse mutant and wild-type backgrounds and is not limited to just two cell lines.

7. *D. discoideum* can be very sensitive to the illumination laser. Incubating the cells in DB for at least 30 minutes helps with viability prior to microscopic observation. Moreover, we have found that growing cells on an SM lawn seeded with *K. aerogenes* overnight can make cells extremely resistant to photodamage [19]. The majority of bacteria should be removed prior to imaging. *K. aerogenes* bacteria can be removed by a low speed spin (1200 g). The *D. discoideum* will pellet, while most of the bacteria will remain in the supernatant. Similarly, we found that allowing the *D. discoideum* to feed on a low density of *K. aerogenes* for an hour or two, once in DB, also helps with protection from photodamage. Phototoxicity is a big problem and can lead to artifacts, but keep in mind that growing cells in the presence of bacteria will change their phenotype and may influence your results.

8. We have found that gently adding 50 μL dropwise to an 8 well chamber containing 200 μL of DB results in quick mixing of the compound with the DB, with minimal loss of cells from dispersion. The final concentration of rapamycin used for CID experiments is 5–10 μM, depending on the CID pair. Allow sufficient time (minimal, 1–2 minutes) for membrane recruitment during the imaging, but keep in mind that rapamycin is a toxin, and experiments beyond an hour may begin to show phenotypes as a result of cell death.

9. Always record the temperature in the microscopy room. Many microscope rooms are notorious for overheating from running equipment. Temperature fluctuations can influence cell physiology.

10. For statistical analysis, Fiji ImageJ and Microsoft Excel with the XLSTAT add-on can be used to analyze collected data. Fiji ImageJ is typically used to analyze videos taken from all time lapses and isolate desired frames. Microsoft Excel with

XLSTAT can be used in order to organize numerical data, run student t-tests, perform norming, derive standard deviation, and create graphs.

Acknowledgments

This work was supported in part by NIH RO1GM080370 to CJ.

References

1. Fegan A, White B, Carlson JC, Wagner CR (2010) Chemically controlled protein assembly: techniques and applications. Chem Rev 110(6):3315–3336. https://doi.org/10.1021/cr8002888
2. Putyrski M, Schultz C (2012) Protein translocation as a tool: the current rapamycin story. FEBS Lett 586(15):2097–2105. https://doi.org/10.1016/j.febslet.2012.04.061. Epub 2012 May 11
3. DeRose R, Miyamoto T, Inoue T (2013) Manipulating signaling at will: chemically-inducible dimerization (CID) techniques resolve problems in cell biology. Pflugers Arch 465(3):409–417. https://doi.org/10.1007/s00424-012-1208-6. Epub 2013 Jan 9. PMID: 23299847; PMCID: PMC3584178
4. Kuhn J, Lin Y, Devreotes PN (2021) Using live-cell imaging and synthetic biology to probe directed migration in Dictyostelium. Front Cell Dev Biol 9:740205. https://doi.org/10.3389/fcell.2021.740205. PMID: 34676215; PMCID: PMC8523838
5. Rivera VM, Clackson T, Natesan S, Pollock R, Amara JF, Keenan T, Magari SR, Phillips T, Courage NL, Cerasoli F Jr, Holt DA, Gilman M (1996) A humanized system for pharmacologic control of gene expression. Nat Med 2(9):1028–1032. https://doi.org/10.1038/nm0996-1028
6. Toettcher JE, Voigt CA, Weiner OD, Lim WA (2011) The promise of optogenetics in cell biology: interrogating molecular circuits in space and time. Nat Methods 8:35–38. https://doi.org/10.1038/nmeth.f.326
7. Kennedy MJ, Hughes RM, Peteya LA, Schwartz JW, Ehlers MD, Tucker CL (2010) Rapid blue-light-mediated induction of protein interactions in living cells. Nat Methods 7:973–975. https://doi.org/10.1038/nmeth.1524
8. Guntas G, Hallett RA, Zimmerman SP, Williams T, Yumerefendi H, Bear JE et al (2015) Engineering an improved light-induced dimer (iLID) for controlling the localization and activity of signaling proteins. Proc Natl Acad Sci USA 112:112–117. https://doi.org/10.1073/pnas.1417910112
9. Liu H, Yu X, Li K, Klejnot J, Yang H, Lisiero D, Lin C (2008) Photoexcited CRY2 interacts with CIB1 to regulate transcription and floral initiation in Arabidopsis. Science 322(5907):1535–1539. https://doi.org/10.1126/science.1163927. Epub 2008 Nov 6
10. Christie JM, Reymond P, Powell GK, Bernasconi P, Raibekas AA, Liscum E, Briggs WR (1998) Arabidopsis NPH1: a flavoprotein with the properties of a photoreceptor for phototropism. Science 282(5394):1698–1701. https://doi.org/10.1126/science.282.5394.1698
11. Wu YI, Frey D, Lungu OI, Jaehrig A, Schlichting I, Kuhlman B, Hahn KM (2009) A genetically encoded photoactivatable Rac controls the motility of living cells. Nature 461(7260):104–108. https://doi.org/10.1038/nature08241. Epub 2009 Aug 19. PMID: 19693014; PMCID: PMC2766670
12. Wagner E, Glotzer M (2016) Local RhoA activation induces cytokinetic furrows independent of spindle position and cell cycle stage. J Cell Biol 213(6):641–649. https://doi.org/10.1083/jcb.201603025. Epub 2016 Jun 13. PMID: 27298323; PMCID: PMC4915195
13. Chandrasekar S, Beach JR, Oakes PW (2023) Shining a light on RhoA: optical control of cell contractility. Int J Biochem Cell Biol 161:106442. https://doi.org/10.1016/j.biocel.2023.106442. Epub 2023 Jun 20. PMID: 37348811; PMCID: PMC10530351
14. Miao Y, Bhattacharya S, Edwards M, Cai H, Inoue T, Iglesias PA, Devreotes PN (2017) Altering the threshold of an excitable signal transduction network changes cell migratory modes. Nat Cell Biol 19(4):329–340. https://doi.org/10.1038/ncb3495. Epub

2017 Mar 27. PMID: 28346441; PMCID: PMC5394931

15. Pal DS, Lin Y, Zhan H, Banerjee T, Kuhn J, Providence S, Devreotes PN (2023) Optogenetic modulation of guanine nucleotide exchange factors of Ras superfamily proteins directly controls cell shape and movement. Front Cell Dev Biol 11:1195806. https://doi.org/10.3389/fcell.2023.1195806. PMID: 37492221; PMCID: PMC10363612
16. Janetopoulos, Christopher. Phosphoinositide (4, 5) bisphosphate as a diagnostic tool and target for cancer treatment. U.S. Patent Application No. 15/067,874
17. Banerjee T, Matsuoka S, Biswas D, Miao Y, Pal DS, Kamimura Y, Ueda M, Devreotes PN, Iglesias PA (2023) A dynamic partitioning mechanism polarizes membrane protein distribution. bioRxiv [Preprint] 2023.01.03.522496. https://doi.org/10.1101/2023.01.03.522496. PMID: 36712016; PMCID: PMC9881856
18. Gaudet P, Fey P, Chisholm R (2008) Making permanent stocks of *Dictyostelium*. CSH Protoc 2008:pdb.prot5101. https://doi.org/10.1101/pdb.prot5101. PMID: 21356747
19. Aufderheide KJ, Janetopoulos C (2016) Migration of Dictyostelium discoideum to the chemoattractant folic acid. Methods Mol Biol 1407:25–39. https://doi.org/10.1007/978-1-4939-3480-5_3. PMID: 27271892

Chapter 9

APEX2-Mediated Proximity Protein Labeling in *Dictyostelium*

Jamie A. Takashima, Helena A. Woroniecka, and Pascale G. Charest

Abstract

Largely due to its simplicity, while being more like human cells compared to other experimental models, *Dictyostelium* continues to be of great use to discover basic molecular mechanisms and signaling pathways underlying evolutionarily conserved biological processes. However, the identification of new protein interactions implicated in signaling pathways can be particularly challenging in *Dictyostelium* due to its extremely fast signaling kinetics coupled with the dynamic nature of signaling protein interactions. Recently, the proximity labeling method using engineered ascorbic acid peroxidase 2 (APEX2) in mammalian cells was shown to allow the detection of weak and/or transient protein interactions and also to obtain spatial and temporal resolution. Here, we describe a protocol for successfully using the APEX2-proximity labeling method in *Dictyostelium*. Coupled with the identification of the labeled proteins by mass spectrometry, this method expands *Dictyostelium*'s proteomics toolbox and should be widely useful for identifying interacting partners involved in a variety of biological processes in *Dictyostelium*.

Key words *Dictyostelium discoideum*, Cellular signaling, Protein-protein interaction, Proximity labeling, Engineered ascorbic peroxidase, APEX, Biotinylation

1 Introduction

The recent development of protein proximity labeling methods allows for capturing and identifying transient and/or weak interactions. In these methods, the protein of interest (POI) is fused to an enzyme that catalyzes the covalent linking of the small molecule biotin to proteins that come in its vicinity (within estimated radius of ~20 nm) [1]. The biotinylated proteins can then be purified and identified by mass spectrometry (MS). The proximity-dependent biotin identification (BioID) method, and the improved TurboID, makes use of a promiscuous mutant of the *E. coli* biotin ligase BirA that, in the presence of ATP and biotin, generates an activated biotinoyl-5'-AMP molecule, which then reacts with primary amines of exposed lysine residues on neighboring proteins,

Alan R. Kimmel (ed.), *Dictyostelium discoideum: Methods and Protocols*, Methods in Molecular Biology, vol. 2814,
https://doi.org/10.1007/978-1-0716-3894-1_9,

resulting in their covalent biotin modification [2–4]. The BioID method was successfully adapted for use in *Dictyostelium* [5], but better temporal resolution is needed to capture dynamic protein interactions such as those involved in chemotactic signaling pathways. The more recently developed protein proximity labeling method that makes use of ascorbate peroxidase (APEX), and the improved APEX2, offers this possibility [6, 7].

1.1 APEX2 Proximity Labeling

APEX2 is a 27 kDA engineered monomeric ascorbate peroxidase from pea/soybean [6, 7]. In the presence of hydrogen peroxide (H_2O_2), APEX2 catalyzes the oxidation of biotin-phenol (BP) to a short-lived biotin-phenoxyl radical, which then reacts with tyrosine residues, and to a lesser extent with tryptophan and cysteine residues, on proteins proximal to APEX2, resulting in their covalent biotinylation [6–8]. Since BP alone does not trigger biotinylation and H_2O_2 readily penetrates cells, the rapid addition of H_2O_2 to cells previously incubated with BP is used to trigger APEX2-mediated biotinylation within seconds, with optimal labeling reached after ~1 min, and the reaction can be equally rapidly quenched using excess ascorbate and reducing agents. More recently, the development of an improved APEX2 enzyme with higher stability and catalytic efficiency really makes this method ideal for mapping interactomes in live cells when used in combination with appropriate spatial references and data analyses [9–11]. Indeed, because APEX2 labeling is so efficient, APEX2-fused POIs were shown to label not only proteins in the POI's interaction network but also neighboring proteins that diffuse through the reactive biotin cloud and/or reside in the same compartment (bystander proteins), and spatial references coupled with quantitative proteomics are then necessary to distinguish those from the POI's interactome [11]. Using the APEX2 protocol published by Hung et al. [10], and cAR1 as a prototypical POI, we have adapted the APEX2 method for use in *Dictyostelium*.

1.2 APEX2 Proximity Labeling for Proteomics Applications in Dictyostelium

Since APEX2 is the same size as GFP (27 kDa), its fusion to proteins that retain functionality when fused to GFP is expected to generate equally functional proteins. However, a recent study reported that fusing APEX2 to some proteins may result in reduced expression due to the presence of a cysteine on APEX2 (C32), and a C32S mutant was shown to fix this issue [12]. If expression issues are observed, it is then recommended to try the APEX2-C32S to generate POI-APEX2 fusion proteins. For proof of concept, we adapted the APEX2 proximity protein labeling method to *Dictyostelium* using a cAR1-APEX2 construct expressed in *carA* null cells. In troubleshooting experiments, we discovered that BP is not taken up by *Dictyostelium* without first permeabilizing the cells, which had also been reported to be the case in *S. cerevisiae* [13]. Further, as previously described, there are three endogenously biotinylated proteins in *Dictyostelium*, which can be observed by Western blot using streptavidin-HRP: methylcrotonyl-CoA carboxylase alpha

(77 kDa) and propionyl-CoA carboxylase alpha (80 kDa), which seem to migrate together (strong band around 75–80 kDa), and acetyl-CoA carboxylase (257 kDa) [5]. Only when cells were first permeabilized did we observe significant biotinylation of other proteins. We then performed a large-scale labeling reaction with cAR1-APEX2-expressing cells and purified the biotinylated proteins using streptavidin beads. Finally, to identify the biotinylated proteins, we performed in-gel trypsin digestion followed by liquid chromatography and tandem MS. Although we did not have a PM reference to properly distinguish bystanders from the cAR1 interactome, the cAR1-APEX2 proximity labeling allowed identification of many known cAR1-interacting proteins (e.g., $G\alpha_2$, $G\beta$, $G\gamma$, and arrestin-like proteins) as well as cell cortex proteins and transport compartment-specific proteins (e.g., actin, myosin, clathrin, secretory complex proteins), consistent with similar APEX2 experiments performed with GPCRs in mammalian cells [11]. Therefore, this experiment serves as proof of principle demonstrating the successful adaptation of the APEX2 proximity labeling method for proteomics applications in *Dictyostelium*. Here, we describe this protocol, including the purification of biotin-labeled proteins and sample preparation for subsequent MS analysis (Fig. 1), which we originally published in the Journal of Biological Methods [14].

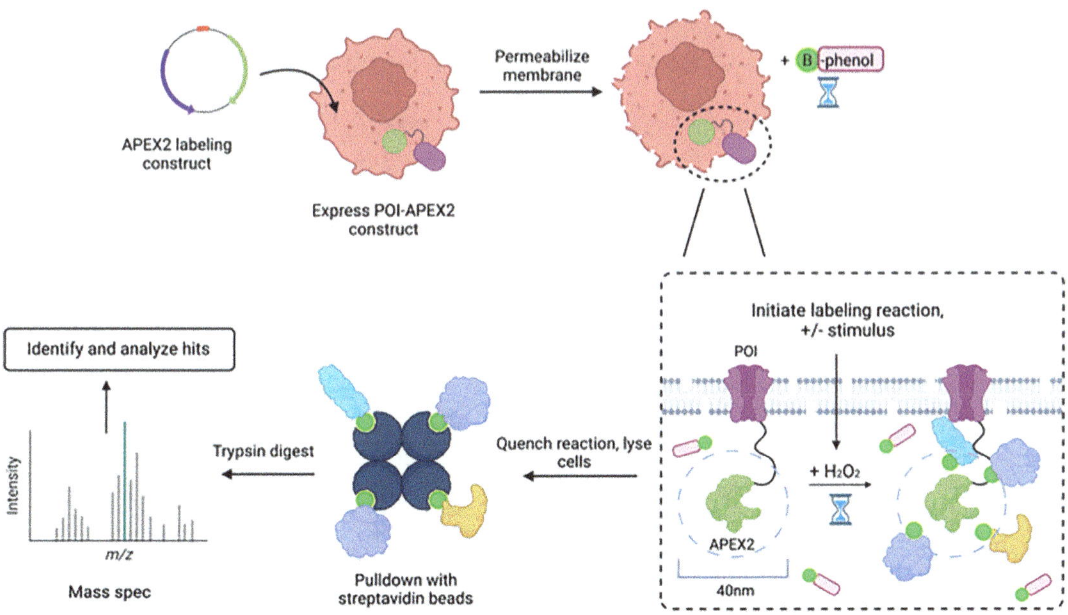

Fig. 1 Overview of the APEX2 protocol for proteomics applications in *Dictyostelium*. APEX2 constructs are prepared by fusing APEX2 in-frame to a POI, or reference control (not shown), and expressed in *Dictyostelium* cells. Cells are first permeabilized with digitonin followed by incubation with biotin-phenol (B-phenol). The reaction is started by adding H_2O_2, with or without stimulus, which will induce labeling of proteins within ~20 nm radius of APEX2. The reaction is then stopped by quenching at the desired time. Biotinylated proteins are then purified with streptavidin beads, eluted, and digested with trypsin prior LC/MS/MS analysis. Created with BioRender.com

2 Materials

Prepare all solutions with ultrapure deionized water (dH_2O) and sterilize the ones that are used with cells.

2.1 Preparing APEX2 Fusion Constructs

1. Any *Dictyostelium* expression vector (e.g., extrachromosomal pDM304): available through the Dicty Stock Center [15].
2. cDNA or gene of the POI and of control protein to be used.
3. APEX2 cDNA: Addgene ID# 72480, depositor: A. Ting, Stanford.

2.2 Validating the APEX2 Constructs

1. SDS-PAGE and Western blotting apparatus and necessary solutions.
2. Antibody recognizing the POI and control protein or the epitope tags used.
3. Fluorescence microscope.

2.3 Cell Permeabilization

1. 12 mM Na/K phosphate buffer: 2.5 mM Na_2HPO_4, 9.5 mM KH_2PO_4, and pH 6.1. Autoclave to sterilize and store at 22 °C (*see* **Note 1**).
2. 0.2 mM digitonin: Dissolve in 12 mM Na/K phosphate buffer, sonicate or heat to 95 °C to dissolve completely, and then cool on ice. Prepare immediately before use.
3. Gyratory shaker with rack for tubes.

2.4 APEX2 Labeling

1. 500 mM BP: Dissolve in DMSO. It may need to be sonicated to completely dissolve. Store as small aliquots at −80 °C. Thaw on ice before use and warm further if BP has precipitated.
2. 30 mM cAMP: Dissolve in 12 mM Na/K phosphate buffer. Store aliquots at −20 °C. Dilute in 12 mM Na/K phosphate buffer to make all working solutions.
3. 100 mM H_2O_2: Dilute the 30% (w/w) H_2O_2 reagent (~10 M H_2O_2 in water) in Na/K phosphate buffer. Prepare immediately before use.
4. 1 M sodium azide (NaN_3): Dissolve in dH_2O. Aliquot and stored at −20 °C.
5. 1 M sodium L-ascorbate: Dissolve in dH_2O. Prepare immediately before use.
6. 0.5 M Trolox: Dissolve in DMSO and sonicate to completely dissolve. Prepare immediately before use.
7. Quenching solution: 10 mM NaN_3, 10 mM sodium ascorbate, and 5 mM Trolox in Na/K phosphate buffer. Keep on ice. Prepare immediately before use.

8. RIPA lysis buffer: 50 mM Tris pH 7.5, 150 mM NaCl, 0.1% (w/v) SDS, 0.5% (w/v) sodium deoxycholate, 1% (v/v) Triton X-100, 10 mM NaN_3, 10 mM sodium ascorbate, 5 mM Trolox, 1 mM PMSF, 5 μg/mL leupeptin, and 5 μg/mL aprotinin, in dH_2O.

2.5 Purifying Biotin-Labeled Proteins

1. Streptavidin agarose beads (e.g., 6% slurry from Gold Biotechnology). Alternatively, streptavidin magnetic beads can also be used.
2. 1 M KCl: Dissolve in dH_2O. Store at room temperature.
3. 0.1 M Na_2CO_3: Dissolve in dH_2O. Store at room temperature.
4. 2 M urea: Dissolve in 10 mM Tris-HCl pH 8.0. Make the day it is to be used.
5. 100 mM biotin: Dissolve in DMSO. Aliquot and store at −20 °C.
6. 1 M DTT stock: Dissolve in dH_2O. Aliquot and store at −20 °C.

2.6 Analyzing Biotin Labeling by Western Blot

1. 6X Protein sample buffer: 0.375 mM Tris-HCl pH 6.8, 30% glycerol, 6% SDS, 600 mM DTT, and 6% bromophenol blue, in dH_2O. Aliquot and store at −20 °C.
2. Streptavidin-horseradish peroxidase (HRP).
3. Tris-buffer saline with Tween-20 (TBST): Dilute 0.1% (v/v) Tween-20 in TBS (50 mM Tris-HCl, 150 mM NaCl in dH_2O, pH 7.5).

2.7 Preparing Samples for LC-MS/MS

1. 50 mM and 100 mM ammonium bicarbonate buffers: Dissolve ammonium bicarbonate in dH_2O; pH should be ~7.8.
2. Destaining solution: 50% (v/v) methanol and 50 mM ammonium bicarbonate. Mix 1 volume methanol with 1 volume 100 mM ammonium bicarbonate buffer.
3. Reducing solution: Dissolve DTT to 25 mM (3.8 mg/mL) in 50 mM ammonium bicarbonate buffer.
4. Alkylating solution: Dissolve iodoacetamide to 55 mM (10.2 mg/mL) in 50 mM ammonium bicarbonate buffer.
5. Trypsin digestion solution: Dissolve 20 μg/mL mass spectrometry-grade Trypsin/Lysine-C mix in 50 mM ammonium bicarbonate buffer containing 0.1% ProteaseMAX™ surfactant Trypsin Enhancer.
6. Peptide extraction solution: 0.2% (v/v) formic acid, 45% acetonitrile (v/v), and 5% (v/v) isopropanol, in dH2O.

3 Methods

The protocol described is for identifying protein interactions involved in cAMP-induced signaling mechanisms, and for this, using ~5.5 h developed cells is recommended to have cells highly responsive to cAMP. Otherwise, no special preparation of the cells is necessary besides normal cell culturing.

3.1 Preparing APEX2 Fusion Constructs

1. Clone the cDNA or gene of the POI in-frame with the APEX2 cDNA in a *Dictyostelium* expression vector of choice. If possible, two APEX2-POI constructs should be made, one with APEX2 fused in N-terminus and one fused at the C-terminus to test both conformations.
2. Ideally, an APEX2 labelling control protein that localizes to the same subcellar compartment as the POI should also be prepared, which will help distinguish between specific interactions and bystander proteins (*see* **Note 2**).
3. An epitope tag or GFP can also be added to the APEX2 constructs to facilitate their analysis by Western blotting and imaging.

3.2 Validating the APEX2 Constructs

1. Culture and transform cells according to standard protocols (*see* **Note 3**).
2. Verify the integrity of APEX2 fusion proteins by Western blot of whole cell lysates using an appropriate antibody (*see* **Note 4**).
3. Verify the functionality of the APEX2-fused POI using appropriate activity assay and/or phenotypic rescue of null strains. This will vary with the POI under study.
4. Verify the localization of the APEX2-fused POI and APEX2 control by either immunofluorescence or fluorescence microscopy if using GFP (*see* **Note 5**).
5. Verify APEX2 functionality in a small-scale labelling assay after cell permeabilization, described in **steps 3.3** and **3.4**.

3.3 Cell Permeabilization

1. Transfer 5.5 h developed cells (10^8 for small-scale and 10^9 for large-scale assays) to 50 mL conical tubes, pellet by centrifugation at 500 × g for 3 min, and wash twice with cold 12 mM Na/K phosphate buffer.
2. Resuspend the cells to 2×10^7 cells/mL in ice-cold 12 mM Na/K phosphate buffer. Transfer cells to an ice-cold Erlenmeyer flask.
3. Add 1 volume of ice-cold 0.2 mM digitonin to the cells (0.1 mM final). Incubate at 4 °C or on ice with shaking for 5 min.

4. Transfer the cells to cold 50 mL conical tubes and pellet by centrifugation at 500 × g for 3 min, at 4 °C.
5. Gently wash the cells twice with ice-cold 12 mM Na/K phosphate buffer with moderate up and down pipetting.
6. Resuspend the cells to 10^8 cells/mL in 12 mM Na/K phosphate buffer at room temperature.

3.4 Small-Scale APEX2 Labeling

1. Perform **step 3.3** using APEX2-POI-expressing cells, APEX2 control-expressing cells, and wild-type cells as negative control. For each strain or condition to be tested, resuspend 10^8 5.5 h developed cells in 1 mL 12 mM Na/K phosphate buffer in a 50 mL conical tube (*see* **Note 6**).
2. Add 2 μL of 500 mM BP (1 mM final) and incubate for 30 min at 22 °C with shaking at ~140 rpm.
3. Start the labeling reaction by adding 20 μL of 100 mM H_2O_2 (2 mM final). Mix by swirling or pipetting up and down and incubate for 30 sec.
4. Stop the reaction by adding 1 mL ice-cold quenching solution and mix well by pipetting up and down. Transfer to 2 mL tubes.
5. Pellet the cells by centrifugation at 500 × g for 3 min, and quickly wash twice with 1 mL ice-cold quenching solution.
6. Resuspend the cells in 0.5 mL RIPA lysis buffer; vortex and incubate on ice for 2 min.
7. Analyze the samples by Western blot as described in Subheading 3.7. If the labeling is satisfactory, proceed with the large-scale labeling for subsequent MS/MS analysis.

3.5 Large-Scale APEX2 Labeling

1. For each cell strain, resuspend 10^9 5.5 h developed cells in 10 mL 12 mM Na/K phosphate buffer in a 50 mL conical tube.
2. Add 20 μL of 500 mM BP (1 mM final) and incubate for 30 min at 22 °C with shaking at ~140 rpm.
3. Start the labeling reaction and stimulate with cAMP at the same time by simultaneously adding 0.2 mL of 50 μM cAMP solution (1 μM final) and 0.2 mL of 100 mM H_2O_2 (2 mM final). Mix and incubate for 30 sec (*see* **Note 7**).
4. Stop the reaction by adding 40 mL ice-cold quenching solution and mix well.
5. Pellet the cells by centrifugation at 500 × g for 3 min, and quickly wash the cells twice with 40 mL ice-cold quenching solution (*see* **Note 8**).
6. Resuspend the cells in 5 mL RIPA lysis buffer; vortex and incubate at 4 °C with agitation for 4 hrs (*see* **Note 9**).

7. Vortex and transfer to tubes appropriate for high-speed centrifugation, and clarify the cell lysates by centrifugation at 15,000 × g for 10 min at 4 °C.
8. Transfer the clarified lysates to 15 mL conical tubes and keep on ice.
9. Take 10 μL aliquots for total biotin-labeled protein analysis by Western blot as described in Subheading 3.7, mix with 2 μL 6X protein sample buffer, and store at −20 °C until needed.

3.6 Purifying Biotin-Labeled Proteins

1. Quantify the lysates' protein content using a reducing agent compatible assay (e.g., Pierce™ 660 nm Protein Assay Reagent) (*see* **Note 10**).
2. Add 60 μL of streptavidin agarose beads (6% slurry) to the lysates for every 360 μg of protein in lysates, and incubate at 4 °C with agitation (e.g., on rotator or gyratory rocker) for at least 1 h and up to 16 h (*see* **Note 11**).
3. Pellet the beads by centrifugation at 1000 × g for 1 min and transfer the supernatants to different tubes. These are the "flowthrough" samples containing unbound proteins and are used to evaluate purification efficiency. Take 10 μL aliquots and mix with 2 μL 6X protein sample buffer for analysis by Western blot and store at −20 °C until needed.
4. Wash the beads with 10 mL of the following: twice with RIPA lysis buffer, once with 1 M KCl, once with 0.1 M Na_2CO_3, once with 2 M Urea in Tris-HCl pH 8.0, and then twice with RIPA lysis buffer. All buffers and samples should be kept on ice.
5. Elute the purified proteins by adding 300 μL of 3X protein sample buffer (prepared from 6X) supplemented with 2 mM biotin and 50 mM DTT; mix and incubate at 95 °C for 10 min. Vortex, cool on ice, and centrifuge briefly to bring down condensation.
6. Mix and then pellet the beads by centrifugation at 1000 × g for 1 min, transfer eluates to new tubes, and keep on ice or store at −20 °C until needed.

3.7 Analyzing Biotin Labeling by Western Blot

1. Thaw the clarified lysates and flowthroughs, and eluate samples containing sample buffer at room temperature, and then centrifuge briefly before incubating at 95 °C for 10 min.
2. Centrifuge the heated samples, and then vortex to ensure sample homogeneity before loading and separating on 8% SDS-PAGE gel.
3. Transfer the proteins to a nitrocellulose membrane and block in 3% (w/v) BSA in TBST at 4 °C for 16 hrs.

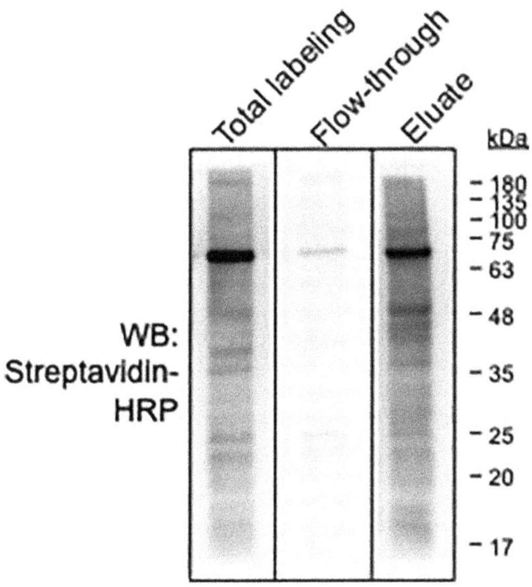

Fig. 2 APEX2 labeling in *Dictyostelium* cells using cAR1 as prototypical POI. The POI-APEX2-mediated protein labeling and efficiency of labeled protein purification should be verified by Western blot prior MS/MS. cAR1-APEX2-mediated labeling is shown as an example

4. Wash the membrane once in TBST and incubate with 0.3 μg/mL streptavidin-HRP in 3% BSA in TBST at room temperature for 1 h.
5. Wash the membrane in TBST for 30 min, followed by four washes of 5 min each.
6. Develop using a standard chemiluminescence reagent and reveal by autoradiography or using a digital imaging system. The clarified lysates and eluates should reveal robust biotinylation, whereas the flowthrough should have minimal biotinylated proteins (Fig. 2).

3.8 Preparing Samples for LC-MS/MS

There are multiple methods to accomplish this. Here, we describe an in-gel trypsin digestion method to prepare the samples for subsequent analysis by mass spectrometry in a core facility.

1. Load the samples on a 12.5% SDS-PAGE gel and perform electrophoresis only until the sample front is at ~1 cm in the gel.
2. Cut gel bands containing the proteins, including migration front with bromophenol blue, using sterile razor blades, and place in 1.7 mL tubes.

3. Incubate the gel bands with destaining solution on a vortex mixer at low speed for 1 h Repeat as necessary to remove all staining, changing the solution by pelleting the gel bands by brief centrifugation in a mini centrifuge and pipetting the supernatant out (*see* **Note 12**).
4. Incubate the gel bands with the reducing solution at 60 °C for 45 min. Pellet the gel bands and discard supernatant.
5. Incubate the gel bands with the alkylating solution in the dark at room temperature for 30 min. Pellet the gel bands and discard supernatant.
6. Wash the gel bands three times with 50 mM ammonium bicarbonate buffer on a vortex mixer at low speed, for 30 min each wash.
7. Transfer the gel bands to a clean glass plate (previously cleaned with methanol or isopropanol), and cut them into small pieces with a sterile razor blade.
8. Place the gel pieces in a 1.7 mL tube and dehydrate by incubating in acetonitrile for 10 min, vortexing a few times. Gel pieces will become white and stick together.
9. Pellet the gel pieces by brief centrifugation in a mini centrifuge, discard acetonitrile, and dry in a Speed Vacuum Concentrator (~30 min).
10. Rehydrate the gel pieces in trypsin digestion solution, enough to cover the gel pieces, for 15 min at room temperature. Vortex to mix.
11. Digest the proteins in the trypsin digestion solution at 37 °C for 16 h.
12. Add 10% formic acid to the in-gel protein digest mixture and vortex to mix. Verify the pH with pH test paper, which should be between 2 and 3. If needed, adjust pH with formic acid.
13. Pellet the gel pieces by brief centrifugation in a mini centrifuge, transfer the supernatant to a new tube, cool on ice, and centrifuge briefly to bring down condensation.
14. Extract the peptides from the gel pieces by adding peptide extraction solution, enough to cover the gel pieces, and sonicate 10 min. Pellet the gel pieces by brief centrifugation in a mini centrifuge. Transfer the supernatant, combining with that from **step 12**. Repeat the extraction procedure a second time.
15. Clarify the peptide solution by centrifugation at 16,000 × g at room temperature for 10 min. Transfer the supernatant to a new tube.
16. Reduce to desired sample volume using a Speed Vacuum Concentrator.

17. The peptides are then separated by standard liquid chromatography followed by tandem mass spectrometry.
18. To identify the proteins, mass spectra are searched against the SwissProt *Dictyostelium discoideum* protein database and an additional common contaminant database (e.g., trypsin, keratins).

4 Notes

1. An alternative buffer that can be used with *Dictyostelium* is the Development Buffer (DB; 5 mM Na_2HPO_4, 5 mM KH_2PO_4; 1 mM $CaCl_2$, 2 mM $MgCl_2$, pH 6.5).
2. The APEX2-mediated labeling of proteins is extremely robust and will lead to the labeling of all proteins that come near the POI, independently of whether they directly interact with it or not [11]. Therefore, ideally, a protein or engineered construct that localizes to the same subcellular compartments as the POI, but that does not interact with it, should be used as an APEX2 control to identify the "bystander" proteins, and distinguish them from true POI-interacting proteins. If stimulus-induced interactions are studied, analyzing the labeling obtained in unstimulated cells can help identify bystander proteins, but depending on the subcellular compartment where the POI localizes, this may not be sufficient.
3. If possible, using null strains of the POI under study is recommended. Using the APEX2-POI corresponding null strain will avoid competition of the POI with its endogenous counterpart for binding signaling proteins. Alternatively, APEX2 can be fused to the endogenous POI using CRISPR [16].
4. APEX2 is 27 kDa and thus the fusion proteins are expected to migrate proportionally slower when analyzed by Western blot. Verify that the APEX2 fusion proteins migrate at their expected molecular weight and that there are no other specific protein species, which would indicate proteolysis. Note that we have not had great success with anti-APEX antibodies and thus recommend using antibodies directed against the POI or tag used.
5. It is important to confirm that the APEX2-POI protein localizes to its expected cellular location, to ensure the significance of the identified interactome. It is also important that the APEX2 control localizes to the same cellular location as the APEX2-POI to identify the location-specific resident proteins that will be labelled by APEX2 even though they do not functionally interact with the POI (bystander proteins).

6. When performing the small-scale protocol for verifying the functionality of APEX2 in the fusion constructs, it is recommended to test conditions with and without BP and H_2O_2 treatments, as well as using cells that do not express APEX2, to distinguish APEX2-mediated labeling from endogenously biotinylated proteins [5].
7. To capture protein interactions involved in early signaling events that are triggered rapidly upon stimulation by a ligand such as cAMP, it is best to add H_2O_2 at the same time as the stimulus. This is achieved using a multichannel pipette.
8. Quenching of the labeling reactions must be done as fast and as thoroughly as possible to ensure best time resolution if comparing different time points after stimulation is desired.
9. Alternative lysis methods may be used, such as mechanical lysis by passing cells through filters, which prevents release of proteases from lysosomes. However, we find that with adequate protease inhibitors and keeping samples on ice and/or 4 °C at all times and performing all the subsequent purification steps in the cold room, proteolysis is minimized.
10. A 1:2 dilution of the samples is usually necessary before performing the protein quantification assay to obtain samples that will be within the linear range of the assay.
11. Use a pipette tip for which you have cut the end with a razor blade to make the opening wider and mix the bead solution well to ensure homogeneity right before pipetting the amount desired. If proteolysis is an issue during a longer incubation, a 1 h incubation can be used and is sufficient to pull down most of the biotinylated proteins.
12. The amount of solution used with the gel bands is minimal, just to cover them, except for the washing steps where 250 μL is suggested, or more depending on the gel band size.

References

1. Trinkle-Mulcahy L, Poterszman A (2019) Recent advances in proximity-based labeling methods for interactome mapping. F1000Research 8135(8):135
2. Roux KJ, Kim DI, Raida M, Burke B (2012) A promiscuous biotin ligase fusion protein identifies proximal and interacting proteins in mammalian cells. J Cell Biol 196:801–810
3. Sears RM, May DG, Roux KJ (2019) BioID as a tool for protein-proximity labeling in living cells. Methods Mol Biol 2012:299–313
4. Branon TC, Bosch JA, Sanchez AD, Udeshi ND, Svinkina T, Carr SA, Feldman JL, Perrimon N, Ting AY (2018) Efficient proximity labeling in living cells and organisms with TurboID. Nat Biotechnol 369(36): 880–887
5. Batsios P, Meyer I, Gräf R (2016) Proximity-dependent biotin identification (BioID) in Dictyostelium amoebae. Methods Enzymol 569:23–42
6. Martell JD, Deerinck TJ, Sancak Y, Poulos TL, Mootha VK, Sosinsky GE, Ellisman MH, Ting AY (2012) Engineered ascorbate peroxidase as a genetically encoded reporter for electron microscopy. Nat Biotechnol 3011(30): 1143–1148

7. Rhee H-W, Zou P, Udeshi ND, Martell JD, Mootha VK, Carr SA, Ting AY (2013) Proteomic mapping of mitochondria in living cells via spatially restricted enzymatic tagging. Science (80-.) 339:1328–1331
8. Udeshi ND, Pedram K, Svinkina T, Fereshetian S, Myers SA, Aygun O, Krug K, Clauser K, Ryan D, Ast T, Mootha VK, Ting AY, Carr SA (2017) Antibodies to biotin enable large-scale detection of biotinylation sites on proteins. Nat Methods 1412(14): 1167–1170
9. Lam SS, Martell JD, Kamer KJ, Deerinck TJ, Ellisman MH, Mootha VK, Ting AY (2015) Directed evolution of APEX2 for electron microscopy and proteomics. Nat Methods 12: 51–54
10. Hung V, Udeshi ND, Lam SS, Loh KH, Cox KJ, Pedram K, Carr SA, Ting AY (2016) Spatially resolved proteomic mapping in living cells with the engineered peroxidase APEX2. Nat Protoc 11:456–475
11. Lobingier BT, Hüttenhain R, Eichel K, Miller KB, Ting AY, von Zastrow M, Krogan NJ (2017) An approach to spatiotemporally resolve protein interaction networks in living cells. Cell 169:350–360.e12
12. Huang M, Lin W, Chang J, Cheng C, Wang HY, Mou KY (2019) The cysteine-free single mutant C32S of APEX2 is a highly expressed and active fusion tag for proximity labeling applications. Protein Sci pro.3685:1703
13. Singer-Krüger B, Fröhlich T, Franz-Wachtel M, Nalpas N, Macek B, Jansen R-P (2020) APEX2-mediated proximity labeling resolves protein networks in Saccharomyces cerevisiae cells. FEBS J 287:325–344
14. Takashima JA, Woroniecka HA, Charest PG (2023) Proximity protein labeling In Dictyostelium with engineered ascorbic acid peroxidase 2. J Biol Methods 10:e99010002. https://doi.org/10.14440/jbm.2023.396
15. Fey P, Dodson RJ, Basu S, Chisholm RL (2013) One stop shop for everything Dictyostelium: DictyBase and the Dicty stock center in 2012. Methods Mol Biol 983:59–92
16. Sekine R, Kawata T, Muramoto T (2018) CRISPR/Cas9 mediated targeting of multiple genes in Dictyostelium. Sci Rep 81(8):1–11

Chapter 10

GRminusRD: A Sensitive Assay to Detect Activation Processes at the Plasma Membrane in Living Cells

Peter J. M. van Haastert, Ineke Keizer-Gunnink, and Arjan Kortholt

Abstract

Activation processes at the plasma membrane have been studied with life-cell imaging using GFP fused to a protein that binds to a component of the activation process. In this way, PIP3 formation has been monitored with CRAC-GFP, Ras-GTP with RBD-Raf-GFP, and Rap-GTP with Ral-GDS-GFP. The fluorescent sensors translocate from the cytoplasm to the plasma membrane upon activation of the process. Although this translocation assay can provide very impressive images and movies, the method is not very sensitive, and amount of GFP-sensor at the plasma membrane is not linear with the amount of activator. The fluorescence in pixels at the cell boundary is partly coming from the GFP-sensor that is bound to the activated membrane and partly from unbound GFP-sensor in the cytosolic volume of that boundary pixel. The variable and unknown amount of cytosol in boundary pixels causes the low sensitivity and nonlinearity of the GFP-translocation assay. Here we describe a method in which the GFP-sensor is co-expressed with cytosolic-RFP. For each boundary pixels, the RFP fluorescence is used to determine the amount of cytosol of that pixel and is subtracted from the GFP fluorescence of that pixel yielding the amount of GFP-sensor that is specifically associated with the plasma membrane in that pixel. This GRminusRD method using GFP-sensor/RFP is at least tenfold more sensitive, more reproducible, and linear with activator compared to GFP-sensor alone.

Key words *Dictyostelium*, Activation membrane proteins, GFP-sensor, Cytosolic-RFP, Ras, Rap, PIP_3

1 Introduction

Cells are surrounded with a plasma membrane, which not only serves to keep the cell content together but is also the interface between the interior and extracellular world. The plasma membrane is instrumental during secretion and endocytosis. Extracellular signals are detected by receptor proteins that activate other proteins in the plasma membrane to transduce the message to the intracellular compartment. In addition, the plasma membrane can function as an integrator of local information. Proteins and lipids in the plasma membrane diffuse ten- to hundred-fold slower than in

Alan R. Kimmel (ed.), *Dictyostelium discoideum: Methods and Protocols*, Methods in Molecular Biology, vol. 2814,
https://doi.org/10.1007/978-1-0716-3894-1_10,

the extracellular or intracellular liquid and therefore can preserve local information on a second-to-minute time scale [1]. Methods to visualize local activation processes at the plasma membrane are critical for our understanding of signal transduction. Excellent methods have been developed such as FRET, FLIM, and BiFC to specifically detect conformational changes of activated proteins or formation of protein complexes during signal transduction [2]. However, these methods are difficult to design and make and require complex instrumentation and analysis. A fusion of GFP to a sensor protein that binds to activated proteins in the plasma membrane is much easier to design and make. Furthermore, the analysis of the microscopy data is relatively simple, because the GFP-sensor translocates from the cytoplasm to the plasma membrane upon activation of the membrane process. Very useful sensors have been developed to detect local Ras-GTP formation using the Ras-binding domain of Raf fused to GFP [3]. Similarly, Ral-GDS-GFP detects activated Rap-GTP and PH-Crac-GFP detects the phospholipid PI(3,4,5)P_3 [4, 5].

Although the GFP-sensors mentioned above are very useful and have provided critical information on local activation processes, they are far from perfect, because translocation from the cytosol to plasma membrane is not very sensitive, is not linear with the amount of activator formed, and is not easily quantifiable in a reproducible manner. The main reason is that the pixels at the boundary of the cell not only contain the plasma membrane with GFP-sensor associated with the activator but also contains some cytoplasm with free GFP-sensor. Each individual boundary pixel may have a different amount of cytoplasm, and this varies in time and place due to cell movement; on average 50% of a boundary pixel is cytoplasm. In unstimulated cells where no GFP-sensor is associated with the plasma membrane, the fluorescence intensity of boundary pixels is between 0% and 100% with an average of 50% relative to the fluorescence intensity of cytosolic pixels. Upon stimulation of the cell, a fraction of the GFP-sensor (say 15%) translocates from the cytoplasm to the plasma membrane, thereby the fluorescence intensity in the cytoplasm decreases from 100% to 85% and the intensity on the boundary pixels increases on average from 50% to 65%. With this example, it is clear that after stimulation, the increase to 65% at the boundary is still smaller than the 85% intensity of the adjacent cytosolic pixels. Experiments reveal that an increase of GFP-sensor at the cell boundary is detectable only when it leads to higher intensities than the adjacent cytosolic pixels. This will require either translocations of some cytosolic GFP-sensor to small but intense areas at the plasma membrane or translocation of at least 30% of cytosolic GFP-sensor to the entire plasma membrane.

The problem of the GFP-sensor translocation assay to detect activation processes at the plasma membrane is the variable and unknown amount of cytoplasm in the boundary pixels. So we reasoned that if we co-express the GFP-sensor with cytosolic-RFP, the RFP signal can be used to determine the amount of cytoplasm in boundary pixels. Subtraction of the RFP signal from the GFP signal of a boundary pixel will yield the GFP signal that is specifically associated with the plasma membrane in that boundary pixel [6]. This GRminusRD method was developed for CRAC-GFP/cyt-RFP to detect local PI(3,4,5)P_3 accumulation in *Dictyostelium* [6]. In the design of the method, both GFP-sensor and cyt-RFP are expressed from the same extrachromosomal plasmid, by which nearly all cells exhibit similar green and red fluorescent levels. The method has been used successfully for other *Dictyostelium* sensors to detect Ras-GTP with RBD-Raf-GFP/cyt-RFP and to detect Rap-GTP with Ral-GDS-GFP/cyt-RFP [3, 7, 8]. Here we describe the GRminusRD method with instructive details on recording of movies and especially on obtaining the GRminusRD information with ImageJ and analysis of the data in Excel (*see* **Note 1**). The method will be illustrated using published experiment on Ras activation.

2 Materials

2.1 General Materials

1. *Dictyostelium* liquid growth medium HL5-C and SM-agar, Formedium.
2. Culture plates, tissue culture treated: Nunc Plate, Thermo Fisher Scientific.
3. Selection markers, Hygromycin B; Gibco. Neomycin and Blasticidin; Thermo Fisher.
4. Experiment with buffer and cAMP gradients: μ-Dish, 35 mM, low; Ibidi.
5. Experiments with buffer and uniform cAMP: Channel slide perfusion chamber, μ-Slide VI 0.5 Glass Bottom; Ibidi.
6. Non-nutrient agar (1 L): 15 g agar in PB, autoclaved.
7. PB: 10 mM Na_2HPO_4/KH_2PO_4 buffer, pH = 6.5.
8. cAMP; Boehringer.
9. LatrunculinA (LatA); Sigma.
10. Zeiss LSM800 confocal laser scanning microscope equipped with a Zeiss plan-apochromatic 63× numerical aperture 1.4 objective.
11. Microinjector with femtotip filled with 1 μM cAMP.
12. Program ImageJ (Fiji) https://imagej.net/software/fiji/.
13. EMBL ImageJ plug-in Bleach corrector https://www.embl.de/eamnet/html/bleach_correction.html.

2.2 Dictyostelium Culture and Preparation of Csells

1. Axenic *Dictyostelium* cells are grown on liquid HL5-C medium [9].
2. Non-axenic cells are grown on solid SM medium in association with *E. coli* or *K. aerogenes.*
3. Starved *Dictyostelium* cells. Collect *Dictyostelium* cells from growth medium by centrifugation for 3 min at 500 g, wash cells three times with PB, and resuspend in PB at a density of 10^7 cells/mL. Starve cells for 5 h by shaking in a rotary shaker at 150 revolutions/minute. Some mutants develop better when starved on non-nutrient agar (2.5×10^6 cells/cm^2). These plates are incubated at 22 °C for 4–6 h or at 8 °C for 16 h followed by an additional 10–30 min at 22 °C.

2.3 Plasmids

The plasmids consist of five parts flanked by restriction sites, thereby the parts are exchangeable as cassettes [10]: part (a) is for selection and amplification in *E. coli*, part (b) is for extrachromosomal replication in *Dictyostelium*, part (c) is a resistance gene for selection in *Dictyostelium*, part (d) is a gene expressing cytosolic-RFP, and part (e) is a gene expressing the protein of interest fused to GFP. The selection marker for expression in *Dictyostelium* in part (d) is driven by an actin-15 promoter and actin-8 terminator; it consists of a gene providing resistance to hygromycin, to blasticidin, or to neomycin.

1. Protein-GFP was cloned in the bacterial NgoMIV shuttle vector pDM329 [10]. This plasmid contains a module of actin-15 promoter, BglII/SpeI multiple cloning site, C-terminal GFP, and actin-8 terminator. Both sites of this module are flanked by an NgoMIV restriction site.
2. The protein-GFP fragments was digested with NgoMIV and ligated in the NgoMIV site (part E) of the mRFP-expressing *Dictyostelium* pDM318 vector [10].
3. For detection of Ras activation: Part (e) consists of the Ras-binding domain of Raf (amino acids 50–134 of RAF1) fused to GFP (RBD-Raf-GFP) [11].
4. For detection of Rap1 activation: Part (e) consists of Ral-GDS (amino acids 786–884) fused to GFP (Ral-GDS-GFP) [5].
5. For detection of PI(3,4,5)P_3 formation: Part (e) consists of the PI(3,4,5)P_3-binding protein Crac (amino acids 1–126) fused to GFP (Crac-GFP) [4].

3 Methods

3.1 *Dictyostelium* Culture Conditions and Preparation of Cells

The strains used are axenic wild-type strains AX2 or AX3 and several mutant strains mainly made in an AX3 background. The mutant strains have their own selection marker, and for the current analysis with GRminusRD, the cells are transformed with the sensor plasmid containing another selection marker. Cells are grown at 22 °C in liquid HL5-C medium supplemented with antibiotics that allows growth of cells that express the selection marker(s). Cells are harvested in the late logarithmic phase, washed three times in PB by repeated centrifugation at 500 g for 3 min, and are resuspended in PB at the required cell density.

Most experiments are performed with starved aggregation-competent cells that have become sensitive to the chemoattractant cAMP (*see* Subheading 2.2). Some mutants have difficulty to develop on their own but are stimulated by wild-type cells (e.g., *gα2*-null cells); then 10% of GFP-labelled mutant cells are starved in the presence of 90% un-labelled wild-type cells at a total density of 2.5×10^6 cells/cm^2. With the confocal microscope, only mutant cells are detected in the fluorescence channels, while wild-type and mutant cells are observed with the brightfield detector.

In some experiments, cells are stimulated with folate; these cells are starved on non-nutrient plates at 22 °C for 1 h.

3.2 Recording of Time Series Movie by Confocal Microscopy

1. Harvest starved cells by centrifugation for 3 min at 500 g, wash with PB, and resuspend in PB at a density of 10^7 cells/mL. The experiment is performed within 1 h after harvest.
2. Place cells in one of the following observation chambers.
   - 2a. For observation in buffer or stimulation with a cAMP gradient using a micropipette: Place 0.5 mL cells (1.5×10^6 cells/mL) in the center of a 35 mM μ-Dish such that about 10% of the bottom is covered by cells.
   - 2b. For observation in buffer or stimulation with uniform cAMP: Place 0.1 mL cells (1.5×10^6 cells/mL) in an Ibidi channel slide.
3. After the cells are settled, aspirate most of the buffer leaving the cells in a thin layer of buffer. Place the chambers in the confocal fluorescence microscope, and start recording a movie with three channels in one track (important for exact overlay of the green and red channel): the GFP channel (488 nM laser, detection wavelength 510–546 nM), the RFP channel (561 nM laser, detection wavelength 600–700 nM), and the brightfield channel.
4. After 1–5 min of recording in buffer, stimulate the cells with cAMP, either with a cAMP gradient from a microcapillary containing 1 μM cAMP (*see* [12] for geometry of the cAMP

concentration profile) or with uniform cAMP by pipetting 0.1 mL 1 μM cAMP into the left reservoir of the channel slide. Continue recording of the movie during cAMP stimulation.

5. Optimal recording of movies depends on the aim of the experiment, the magnification, and the microscope. Microscope settings for the Zeiss LSM800 microscope are optimized with respect to resolution (Nyquist resolution) and pinhole size (1 Airy Unit). Other settings such as field size and recording time are optimized for the experiment, with initial settings of 1024 × 1024 pixels (pixel size of 199 × 199 nM in x, y-direction; the confocal slice in the z-direction is about 700 nM) at a rate of 4 seconds per frame; the field of observation contains about 20 cells. For higher time resolution (1 s per frame), a smaller field of observation is used (512 × 512 pixels). For higher x,y-resolution (67 × 67 nM pixel size), one or two cells are observed in a field of 34 × 34 μm. Recordings in this small field for a few minutes require that cells hardly move in the x,y-directions (after incubation with LatA). Save the movie as <xx_m > (where xx is a specific name related to the project); settings are saved with the images as metadata. *See* **Note 2** for nomenclature of files and operations.

3.3 Recording Three-Dimensional Z-Stack Movie by Confocal Microscopy

1. Place 0.5 mL starved cells (1.5×10^6 cells/mL) in the center of a 35 mM μ-Dish.
2. After the cells are settled, replace the buffer with buffer containing 5 μM latA. After 10 min, place the dish in the confocal fluorescence microscope, and start recording a movie in the GFP channel, the RFP channel, and the phase contrast channel. Record 16 images of 0.5 μm Z-stacks with pixel dimensions of 67 × 67 (x,y). Save file <xx_Z>.

3.4 Conversion of Movie to GRminusRD in ImageJ

1. In ImageJ, open the xx_m or xx_z movie with "Bio-Formats Import Options." Most parameters are standard; select the "Color mode" option to choose "Custom" and put the sliders such that the colors are the same as the channels (green, red, gray).
2. Select a cell to be analyzed (*see* **Note 3**) and crop the movie to the selected area (a) and frames (b). Split green and red channel (c), and convert each to 32 bit movies (d). In ImageJ, these actions use the tab <<Image>>: (a) <<Rectangle Tool>>, <<Crop>>; (b) <<Stacks>>, <<Tools>>, <<Make Substack>>, (c) <<Color>>, <<Split Channels>>, (d) <<Type>>, <<32bit>>.
3. Correct for bleaching using the EMBL ImageJ plug-in Bleach corrector. Usually, the GFP signal does not require correction,

but the RFP signal may show 10–30% bleaching during prolonged imaging.

4. Correct for the difference in expression levels of the GFP and RFP markers: select large areas of the cytoplasm (excluding nucleus and vacuoles) in several images of the movie, yielding the mean fluorescence intensity in the cytoplasm of the red (RD) and green channels (GR), respectively. This provides the correction factor, c = GR/RD. In ImageJ: Tool Bar: <<Freehand Selections>>, Tab <<Analyze>>, <<Set Measurements>>, select <<Mean gray value>>, then <<Analyze>>, <<Measure>>; this opens a new window "Results" with data that can be copied into Excel and used for further calculations.
5. Multiply the red channel by the correction factor c. In ImageJ: <<Process>>, <<Math>>, <<Multiply>>. Save movies as TIFF files: <xx_GR_m_tiff> and < xx_cRD_m_tiff>.
6. Subtract Red movie <xx_cRD_m_tiff> from Green movie <xx_GR_m_tiffi>. In ImageJ: <<Process>>; <<Image calculator>>; Image 1 = <xx_GR_m_tiff>, operation <<Subtract>>, Image 2 = <xx_cRD_m_tiff>, select <<Create new window>>. The movie will appear as Result of xx_GR_tiff.
7. Divide result movie by average green signal of the cytosol obtained in step 4. In ImageJ: <<Process>>; <<Math>>; <<Divide>>, fill in value. Save obtained movie as TIFF with name <xx_GRminusRD_tiff>. This provides the Ψ-values for all pixels of the entire movie. The movie is best observed with fire colors, because this has the same appearance as the kymographs and Z-stacks made below in Subheadings 3.5 and 3.6: In ImageJ: <<Image>>; <<Lookup tables>>; <<Fire>>, and then optimize colors with <<Image>>; <<Adjust>>; <<Brightness/Contrast>>.

3.5 Determine GRminusRD at the Boundary of the Cell

1. Make new Excel file <xx_GRminusRD_exl>.
2. Open file <xx_GRminusRD_tiff> in ImageJ.
3. Select line width: in ImageJ: Tool Bar: <<double click line icon>> select 5 (*see* **Note 4** for line width).
4. Select freehand line In ImageJ: <<right mouse click on line icon>>; <<select freehand line>>.
5. Draw flexible line at the boundary of the cell and make a plot profile of the fluorescent intensity along the line. In ImageJ: <<Analyze>>, <<Plot Profile>>. This produces a graph of the Ψ-values as function of the line length. In the window "Plot of xx," press <<List>>. This produces a window "Plot Values": two columns with X the line length and Y the values of Ψ, respectively. Copy the list to the Excel file.

6. Repeat **step 5** for all images of the movie. Copy all lists in the Excel file next to each other. Keep the first column with the line length of the first image; delete all other columns with line length. For a movie with 40 frames, this produces the data with the line length in column A, and the values of Ψ of 40 frames in the subsequent columns B till AO. Save the file <xx_GRminusRD_exl>.

3.6 Visualization GRminusRD in Excel

During the analysis of the confocal images, it is often convenient to visualize the obtained large data set in Excel with the appearance that is later used in publications.

1. Open file <xx_GRminusRD_exl>.
2. Make the cells of Excel smaller. Select all data (including length of column A). With this selection, do the following steps: (a) reduce the size of columns from the standard 64 pixels to 12 pixels and of the rows from the standard 20 pixels to 12 pixels, and (b) reduce the font size to 2.
3. Color code the Y-values. Select all Ψ data (so excluding length of column A), color code: <<Conditional Formatting>>, <<Color Scales>>, <<Red-Yellow-Green>>. With these operations, the data are visible as squared colors.
4. Zoom out the Excel worksheet such that all data are visible on the screen.
5. For presenting the figure in a report or manuscript, optimize the figure in two steps. The length of the circumference of the cell is not constant with time (the cell of Fig. 1 varies between 37 and 83 μm, between 188 and 418 pixels). Often it is preferred that the kymograph is symmetric in size. Add in Excel empty cells to the shorter columns such that the data go downward and the columns become symmetric as shown in Fig. 1D. Make print screen and save as <xx_GRminusRD_image_exl > (Fig. 1D).
6. Import the figure in drawing program. For a kymograph, the image has not yet an appealing format, because there are many pixels in the circumference and less in the number of images (in Fig. 1, there are 418 pixels in circumference and 93 in time). For a kymograph rescale, the x- and y-dimensions of the image in the drawing program to produce an attractive figure (*see* Fig. 1D).

3.7 Visualization GRminusRD in ImageJ

1. Open the Excel file <xx_GRminusRD_fig_exl>. As described above in 3.6.5, make the kymograph symmetric in circumference by adding empty cells to the shorter columns. Fill the empty cells with NaN (not a number). Add an extra row at the top; fil the row with Y1, Y2, etc.

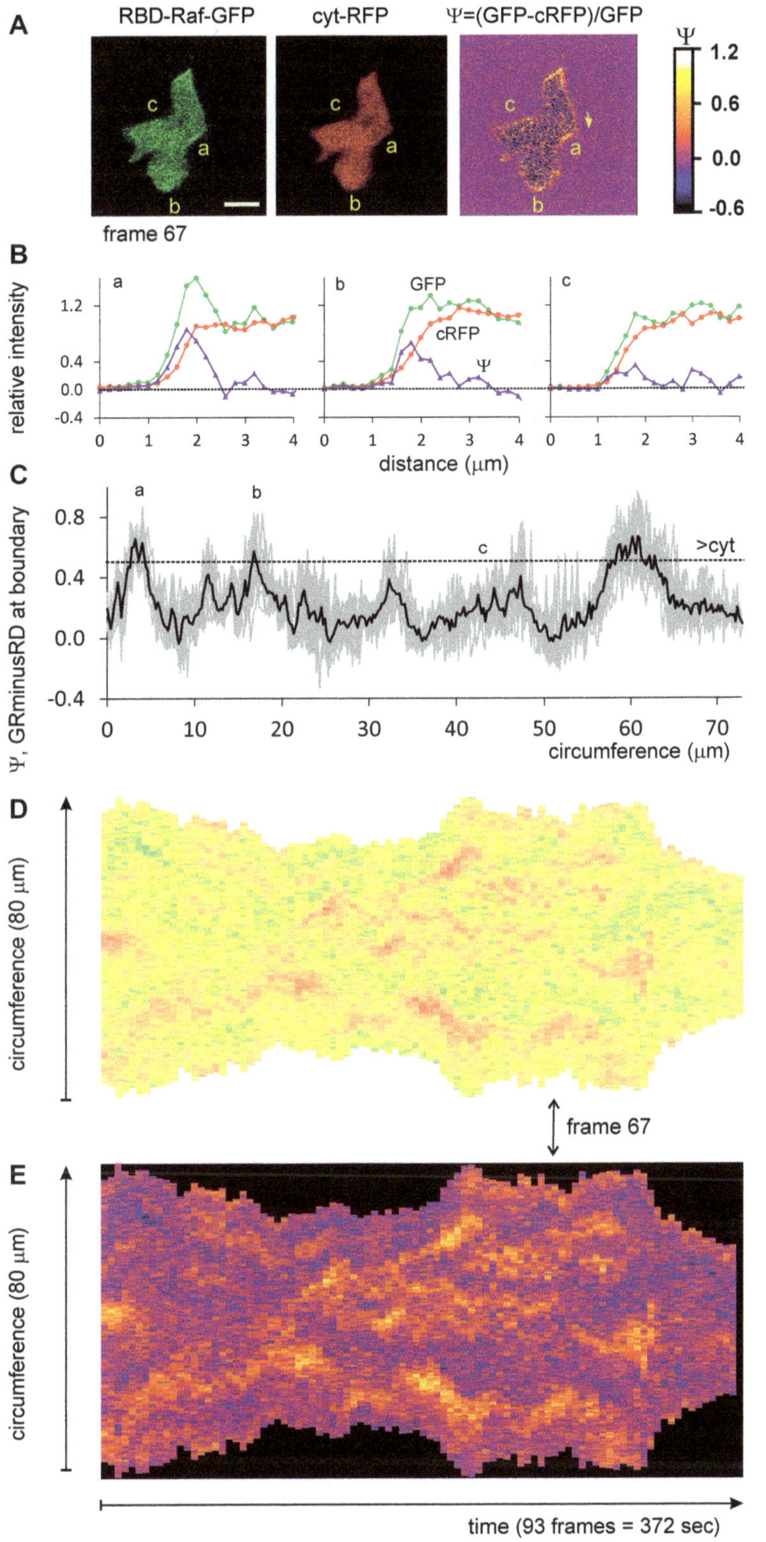

Fig. 1 GRminusRD method used to detect Ras-GTP patches in moving cells. A movie was recorded of unpolarized *Dictyostelium* cells in buffer-expressing

Save the file as tab or comma delimited text file. In excel: <<File>>; <<Save As>>; at "save as" select: text (tab delimited). Save file as <xx_GRminusRD_fig_txt > .

2. Open in ImageJ the text file. In ImageJ: <<File>>; <<Import>>; <<Results>>; at file name select: <xx_GRminusRD_fig_txt>. This opens a new window named Results with a table containing the data.
3. Convert the data in Results table to a figure. In ImageJ: <<Image>>; <<Transform>>; <<Results to Image>>; select 32 bit. This opens a new window named Results Table with a figure of the data.
4. The image is best observed with fire colors, because this has the same appearance as the GRminusRD images of the cell in Subheading 3.6: In imageJ: <<Image>>; <<Lookup Tables>>; <<Fire>>, and then optimize colors with <<Image>>; <<Adjust>>; <<Brightness/Contrast>>.
5. Save file as <xx_GRminusRD_image_ImJ>. The format can be TIFF or BMT (bitmap) depending on the drawing program used to make figures.
6. Import the figure in a drawing program. Optimize the kymograph by rescaling the x- and y-dimensions as described in 3.6.6 (*see* Fig. 1E).

For a Z-stack, the image has not yet the correct format, because the confocal elements have a different size in the x- and z-direction. (In Fig. 2, the size is 67 nM in the x-direction and 500 nM in the z-direction.) For a Z-stack rescale, the x- and y-dimensions of the image in the drawing program have the correct ratio of x-direction and z-direction of the confocal element (*see* Fig. 2d).

Fig. 1 (continued) RBD-Raf-GFP and cyt-RFP. Panel (**A**) shows the images of frame 67 in the green channel, the red channel, and the calculated GFP minus RFP signal Ψ, which is proportional to Ras-GTP; scale bar is 5 μm. (**B**) From the positions indicated by the letters a, b, and c in Figure (**A**), a line is drawn perpendicular to the cell surface to determine the fluorescence intensity in the GFP image (green lines) and the cRFP image (red lines). The data are presented relative to the mean intensity of GFP in the cytoplasm; the pink lines show the difference between green and red (Ψ = GRminusRD). (**C**) The values of Ψ are recorded for the right image of panel (**A**) using a line scan at the boundary of 5 pixels wide (1 μm), starting at the arrow indicated in panel (**A**). The gray lines are ten replica line scans all starting from the same position; the black line is the average. The dashed line at $\Psi = 0.5$ indicates that details below this line are not visible in the GFP channel. (**D**, **E**) Kymograph of the values of Ψ for the entire movie made in Excel (**D**, *see* Subheading 3.6) and in ImageJ (**E**, *See* Subheading 3.7). Figure modified from [8]

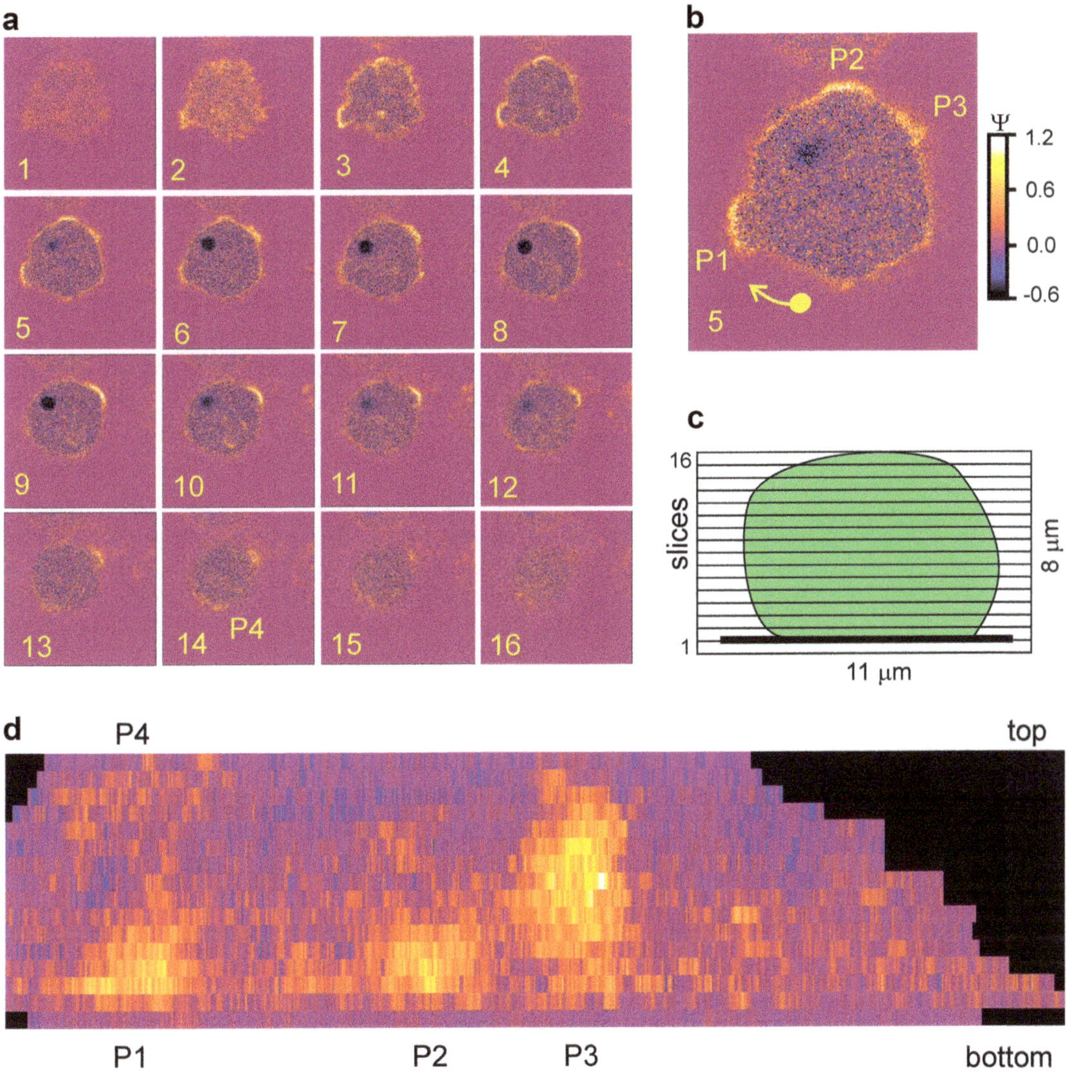

Fig. 2 GRminusRD method used to detect Ras-GTP patches on the entire surface of a single cell. A series of confocal images in the Z-direction were recorded for wild-type cell-expressing RBD-Raf-GFP and cytosolic RFP; cells were incubated with 5 μM LatA for 10 min. The GRminusRD values (Ψ) were calculated for each pixel in each slice (*see* Subheading 3.4). Panel (**a**) presents 16 slices of 0.5 μm of a typical cell. Panel (**b**) shows slice 5 (2.0–2.5 μm above the glass surface). (**c**), Side view of the cell, which is about 11 μm in maximal diameter and 8 μm in height. The glass-cell interface is present in slice 2. Panel (**d**), reconstruction of GRminusRD (Ψ) at the entire cell surface of the cell. The values of Ψ were recorded with a line scan at the boundary of 5 pixels wide (1 μm), starting at the arrow indicated in panel (**b**). The data of the 16 line scans were assembled in Excel and converted to the image made in ImageJ (*see* Subheading 3.7). The figure shows three major RBD-Raf-GFP large patches P1–P3, a weaker large patch P4, and several smaller patches. Figure modified from [7]

3.8 Results

Figure 1A shows a cell in buffer expressing both RBD-Raf-GFP and cyt-RFP. This GFP-sensor binds to activated Ras-GTP at the plasma membrane. In Fig. 1B, the fluorescence intensity was measured at three positions (a, b, and c) with line scans perpendicular to the cell surface. At position a in the image of the green channel, a patch is visible (Fig. 1A left), and the fluorescence intensity of the line scan shows at the boundary of the cell an overshoot above the fluorescence intensity of GFP in the cytosol (Fig. 1B left). At positions b and c, patches are not visible in the image of the green channel, and no overshoot of fluorescence intensity of GFP at the boundary is observed in the line scan. However, at position b, the fluorescence intensity of GFP at the boundary increases much faster than the intensity of RFP (Fig. 1B middle), leading to a clear patch of activated Ras in the GRminusRD image (Fig. 1A middle). At position c, the fluorescence intensity of GFP and RFP at the boundary is nearly same (Fig. 1B right) and no or weak activation of Ras is detectable in the GRminusRD image (Fig. 1A right).

Figure 1C shows the GRminusRD signal of a 5-pixel wide line drawn over the entire boundary of the cell, which reveals at least five areas with increased Ras activation. To stress the sensitivity of the GRminusRD signal, theoretical considerations and experiments reveal that the observed Ras activation in the green minus red signal below Ψ of 0.5 is undetectable in the GFP channel [6, 8]. The GRminusRD signal at position (a) is well above this value, and the patch is clearly visible in the GFP image, whereas the GRminusRD signal at position (b) is around this value and the patch is not visible in the GFP image.

The traces of GRminusRD at the boundary of all images are combined in a kymograph (Fig. 1D, E), demonstrating that about 50 Ras-GTP patches are detectable at the boundary of this cell during the 372 seconds of observation [8]. These patches were analyzed in detail showing many small- and short-living spikes of Ras activation and a few spikes that grow to long-lived large patches of Ras activation. These large patches are often associated with extending pseudopods. These experiments with the sensitive GRminusRD method were instrumental to uncover a mutual positive feedback loop between Ras-activation and F-actin formation responsible for the large patch of Ras activation and strong F-actin formation in the extending pseudopod [8].

Figure 2a shows a Z-stack of images of a cell expressing the same sensors RBD-Raf-GFP and cyt-RFP; the cell was treated with LatA to reduce cell mobility [7]. The GRminusRD signal shows little Ras activation in the lowest Z-stack representing the interaction of the cell with the substratum. Strong Ras-GTP patches are detected in the GRminusRD images of the other Z-slices. The value of Ψ at the cell boundary was measured for each image and combined in the Z-stack of Fig. 2b. This image therefore shows Ras activation at the entire surface of the cell, that is, it is a sensitive complete two-dimensional representation of Ras activation at the inner surface of the plasma membrane of a single cell. This cell has three strong large patches, one less intense large patch and several smaller areas that probably represent the spikes of Fig. 1 [8].

The kymograph and the Z-stacks have been used to quantify and analyze the Ras-GTP patches in detail and were combined with timing of extension of protrusions at that position of the cell [8, 13].

3.9 Reproducibility and Pitfalls

Determining the fluorescence intensity at the boundary of the cell by tracking a 5-pixel yields the average intensity over these five pixels. In the GRminusRD cells, the fluorescent signal in the cytosol is zero. Therefore tracking a 5-pixel wide line over the boundary does not have to be extremely accurate, because the pixels outside the cell and the pixels in the cytosol are zero. As long as the boundary is present in the 5-pixel line, the results will be the same. This is in strong contrast to a cell expressing only the GFP sensor, because the cytosol has a very significant GFP signal, by which the result of line tracking will be higher or lower depending where exactly the line is drawn. This means that boundary tracking of a GFP-sensor is inherently not very reproducible (observed relative SD of 10 traces of the same cell is 24%), in strong contrast to boundary tracking of the GRminusRD sensor (observed relative SD is 8%).

In the absence of Ras activation, the GRminusRD signal at the boundary of the cell is zero, and the GRminusRD signal increases linear with the amount of GFP-sensor that binds to Ras-GTP at the membrane. Since RBD-Raf-GFP has a very high affinity for Ras-GTP, nearly all Ras-GTP molecules are occupied with RBD-Raf-GFP, and the GRminusRD signal will be linear with the amount of Ras-GTP, even at very low amounts of Ras-GTP. This is in strong contrast to a cell expressing only the GFP sensor, because Ras activation leading to increases of GFP-sensor below the GFP value of cytosol pixels are undetectable. Therefore, the intensity at the boundary with a GFP-sensor is only linear with Ras-GTP at very high levels of Ras activation, and low levels of Ras activation cannot been seen.

Because the GRminusRD signal at the boundary of the cell is zero in the absence of Ras activation, the sensitivity of the assay to detect small amounts of Ras-GTP depends mainly on the sensitivity of the cameras to detect green and red light. A value for GRminusRD of $\Psi = 0.05$ is easily detected ($\Psi = 0.05$ means that the GFP signal at the boundary is 5% of the cytosolic GFP signal). In contrast, in cells expressing only GFP-sensor, the GFP signal at the boundary must be at least half the cytosolic GFP signal to become detectable. Therefore the GRminusRD sensor is at least tenfold more sensitive than the GFP-sensor only.

Pitfalls. One of the control experiments in the development of the GRminusRD method was the co-expression of cytl-GFP and cyt-RFP. It is expected that the GRminusRD signal at the cell boundary and in the cytosol will be zero, which is observed when the normal GRminusRD Ψ scale is used (−0.2–1.2). However, at very high sensitivity of the GRminusRD Ψ scale (−0.1 to +0.1), we have observed two deviations that may be related to physical processes. First, the most outward pixels of the cell contain slightly more red signal than green signal, that is, the cell is surrounded by a very faint halo of negative GRminusRD. We attribute this effect to the point-spread function of fluorophores. That is, when a fluorescent molecule emits photons, the photons spread in space; not all photons are captured in the pixel below the fluorophore but also in the adjacent pixels. Since the point-spread function depends on the wavelength, red light is emitted further than green light, explaining a faint red halo, leading to a small negative value of Ψ at the entire cell boundary of the cells expressing cyt-GFP and cyt-RFP (we estimate that this leads to $\Psi = -0.04$). Secondly, we have also observed in some experiments that this faint red halo is stronger at one side of the cell than at the opposite site of the cell where it then is a green halo. We attribute this effect the slight misalignment of the red and green lasers and/or detectors. Since the GRminusRD method subtracts two large numbers from each other (about 100 in 8-bit movies), a misalignment of only 0.1 pixel will lead to an over- or underestimate of GRminusRD equivalent to $\Psi = 0.03$.

In conclusion, the GRminusRD method is very sensitive and reproducible and has an output Ψ that is linear with the fraction of the GFP-sensor that translocate to the plasma membrane. Because the method is both linear and very sensitive, very small signals can have meaningful information. However, at these extreme small signals, the method approaches physical and instrumental limits of differences of the point spread function between GFP and RFP and instrumental limitations on alignment and stability of lasers and cameras. Therefore, we are extra careful to draw conclusions from data with $\Psi \leq 0.05$.

4 Notes

1. The method has been developed initially as part of Quimp3 for the automatic detection of deformations (protrusions) and fluorescence at the boundary of the cell [14]. For good analysis of GRminusRD, this automatic method needs careful selection of cells and correction of membrane "flaws" (such as organelles that touch the membrane, boundary pixels with very low fluorescence, etc.). Here we describe a more simple and hand-operated method with full control at each step.

2. The following nomenclature has been used: <> indicates a file names. <<>> indicates an operation in computer programs. A sequence of operations are separated by semicolon (;).
3. Selection of a cell to be analyzed. The normal field of observation is 204 × 204 μm and contains about 20 cells; usually about 4 cells are analyzed from one movie. Criteria for selection of cells to be analyzed are as follows: (i) the cell stays in the field of observation during the entire movie, (ii) the cell does not touch other cells for prolonged times, (iii) the cell expresses red and green fluorescence to similar levels, and (iv) the cell moves over the substratum. From the remaining cells, the speed was recorded, and the three or four cells closest to the mean speed were used.
4. The output for a 5-pixel wide line is the magnitude of the measured values averaged over the five pixels. Because in GRminusRD the fluorescence intensity in extracellular space and the cytoplasm is close to zero, the selection of the line width is not very critical. Select a width so that the line can be easily drawn over the cell boundary.

References

1. Postma M, Van Haastert PJ (2001) A diffusion-translocation model for gradient sensing by chemotactic cells. Biophys J 81: 1314–1323
2. Kim H, Ju J, Lee HN, Chun H, Seong J (2021) Genetically encoded biosensors based on fluorescent proteins. Sensors (Basel) 21:795
3. Kortholt A, Keizer-Gunnink I, Kataria R, Van Haastert PJM (2013) Ras activation and symmetry breaking during Dictyostelium chemotaxis. J Cell Sci 126:4502–4513
4. Parent CA, Blacklock BJ, Froehlich WM, Murphy DB, Devreotes PN (1998) G protein signaling events are activated at the leading edge of chemotactic cells. Cell 95:81–91
5. Plak K, Keizer-Gunnink I, Van Haastert PJM, Kortholt A (2014) Rap1-dependent pathways coordinate cytokinesis in dictyostelium. Mol Biol Cell 25:4195–4204
6. Bosgraaf L, Keizer-Gunnink I, Van Haastert PJM (2008) PI3-kinase signaling contributes to orientation in shallow gradients and enhances speed in steep chemoattractant gradients. J Cell Sci 121:3589–3597
7. Van Haastert PJM, Keizer-Gunnink I, Kortholt A (2018) The cytoskeleton regulates symmetry transitions in moving amoeboid cells. J Cell Sci 131:jcs208892
8. Van Haastert PJM, Keizer-Gunnink I, Kortholt A (2017) Coupled excitable Ras and F-actin activation mediates spontaneous pseudopod formation and directed cell movement. Mol Biol Cell 28:922–934
9. Sussman R, Sussman M (1987) Cultivation of dictyostelium discoideum in axenic medium. Biochem Biophys Res Commun 29:53–55
10. Veltman DM, Akar G, Bosgraaf L, Van Haastert PJM (2009) A new set of small, extrachromosomal expression vectors for Dictyostelium discoideum. Plasmid 61:110–118
11. Kortholt A, Kataria R, Keizer-Gunnink I, Van Egmond WN, Khanna A, Van Haastert PJM (2011) Dictyostelium chemotaxis: essential Ras activation and accessory signalling pathways for amplification. EMBO Rep 12:1273–1279
12. Postma M, Van Haastert PJM (2009) Mathematics of experimentally generated chemoattractant gradients. Methods Mol Biol 571: 473–488
13. Van Haastert PJM (2020) Symmetry breaking during cell movement in the context of excitability, kinetic fine-tuning and memory of pseudopod formation. Cells 9:1809
14. Bosgraaf L, Van Haastert PJM (2010) Quimp3, an automated pseudopod-tracking algorithm. Cell Adhes Migr 4:46–55

Chapter 11

Analyzing the Micro-topography Guidance of *Dictyostelium* Cells Using Microfabricated Structures

Gen Honda and Satoshi Sawai

Abstract

Over the last decade, the use of microfabricated substrates has proven pivotal for studying the effect of substrate topography on cell deformation and migration. Microfabrication techniques allow one to construct a transparent substrate with topographic features with high designability and reproducibility and thus well suited to experiments that microscopically address how spatial and directional bias are brought about in the cytoskeletal machineries and hence cell motility. While much of the progress in this avenue of study has so far been made in adhesive cells of epithelial and mesenchymal nature, whether related phenomena exist in less adhesive fast migrating cells is relatively unknown. In this chapter, we describe a method that makes use of micrometer-scale ridges to study fast-migrating *Dictyostelium* cells where it was recently shown that membrane evagination associated with macropinocytic cup formation plays a pivotal role in the topography sensing. The method requires only basic photolithography, and thus the step-by-step protocol should be a good entry point for cell biologists looking to incorporate similar microfabrication approaches.

Key words Microfabrication, Cell migration, Contact guidance, Actin waves

1 Introduction

Surface topography provided by the aligned fibrous proteins in the extracellular matrix is thought to serve as an important cue that guides directional cell movement or "contact guidance" in development [1, 2], immunity [3, 4], and cancer metastasis [5]. Contact guidance and the associated cell deformation are often selective to certain dimension of ridges/grooves on the substrate; therefore it is a critical parameter that needs to be taken into consideration for experimental analysis. The surface that cells likely encounter in vivo can be curved or flat with defined edges having spatial scales that range from tens of nanometers as in the fibril diameters of ECM polymers to micrometers as in the size of cells and blood vessels [2, 6, 7]. Oriented cell movement that depends on the surface geometry features has been extensively studied in vitro especially

Alan R. Kimmel (ed.), *Dictyostelium discoideum: Methods and Protocols*, Methods in Molecular Biology, vol. 2814, https://doi.org/10.1007/978-1-0716-3894-1_11,

in adhesive cells such as fibroblasts and epithelial cells [8, 9]. These studies have elucidated mechanisms that depend on the formation of stable focal adhesion and strong contraction by actin stress fibers [10]. Spatial rearrangement of these machineries is at the basis of generating anisotropic contractile forces across the entire cell to align cell orientation along the ridges/grooves [11]. Fast-migrating cells, such as immune cells, and *Dictyostelium* amoeba that form only weak and transient adhesion sites and no stress fibers are also known to migrate along certain substrate topography features [12, 13]; however, the mechanism remains relatively unexplored.

Study of cell contact guidance in vitro is greatly facilitated by microfabrication techniques. They allow one to reproducibly construct with well-defined structures that are transparent and thus compatible with microscopy. Most techniques including ones described in this chapter involve micrometer-scale photolithography [14] and would typically require a few hours work in a clean room (class 1–1000) with orange light. Epoxy-based photoresist SU-8 that is initially coated uniformly on top of a glass substrate is etched or cross-linked by applying UV light to regions of choice through a mask or by a digital micromirror device (DMD). Not only is SU-8 biocompatible, but its high optical transparency above 360 nm also makes it ideal to etch near vertical structures of 1–100 μm height. In this chapter, we provide a practical guide to fabricate ridge structures of micrometer-scale that consisted of two layers of SU-8 and demonstrate how they can be used to image and quantify cytoskeletal dynamics. The method should be easy to follow by biologists after a few basic trainings in photolithography.

2 Materials

2.1 Reagents

1. Mask blanks. 4 × 4 inch soda-lime glass substrate pre-coated with chromium (100-nM-thick) and AZP1350 photoresist (540-nm-thick) (Clean Surface Technology Co., CBL4006Du-AZP).
2. Reagents for photolithography: Photoresist developer NMD-3 (Toyko Ohka Kogyo Co.). Chromium etcher MPM-E350 (DNP Fine chemicals Co.). Acetone. Photoresist SU-8 2 and SU-8 developer (KAYAKU Advanced Materials, Inc., Y020100; 1-Methoxy-2-propanol acetate, > 99.5%). Store SU-8 at 4–21 °C in the dark.
3. Glass coverslips (Matsunami,φ25 mM, 24 × 60 mm or 40 × 50 mm).
4. Reagents for washing the coverslips: 10% alkaline cleaner Contaminon LS-II (Wako, 035-22,251) diluted with milliQ water, 99.5% EtOH and 0.08 M NaOH, 250 mL, respectively.

2.2 Equipment

1. Required to wash glass coverslips: A sonicator bath. A dry oven. A glass vessel (a 100 mL beaker or a 250 mL staining vessel) (Matsunami, BT-210, 50 × 100 × 100 mm). A coverslip rack (Sigma, Wash-N-Dry, Z688568 or Matsunami, B-20).
2. A laser direct imaging device (LDI) (NEOARK, DDB-201-TW, Fig. 1A(a)) and a CAD software (Photron, Zuno Rapid).
3. A mask aligner (MIKASA, MA-20, Fig. 1A(b)).
4. An alternative to 2 and 3. A maskless aligner (Heidelberg Instruments, μMLA).
5. An O_2 plasma etcher (Samco, FA-1). A spin coater (MIKASA, MS-A150). Hotplates (IKA, C-MAG HS 4; 100 × 100 mm, at least three).
6. A draft chamber. An air blow gun connected to a compressor or in-house nitrogen supply.
7. A metallurgical microscopes. Electronic microscopes or Scanning Probe Microscopes (SPMs), AFM (JPK, Nanowizard3) mounted on an inverted microscope (Olympus) or Dektak XT (Bruker, Fig. 1A(c)). AFM cantilever: ACTA (AppNano). SPMs should be operated in a vibration-free environment. The use of an active vibration isolator (Hertz) is recommended.
8. A cell culture dish (Mattek, P35G-0-14-C) and 9 × 9 mm Frame-seals (BIO-RAD, SLF0201).
9. A glass cutter (Mitsuboshi diamond industrial, GCC-P-M17P).

2.3 Reagents for Cell Culture

1. *Dictyostelium* cell culture medium: PS medium (modified HL5 medium). Solution A: 10 g special peptone (OXOID, LP0072), 7 g yeast extract (OXOID, LP0021), 0.12 g $Na_2HPO_4 \cdot 7H_2O$, and 1.4 g KH_2PO_4. Dissolve in milliQ water to 900 mL final volume and autoclave. Solution B: 10 mg cyanocobalamin (vitamin B12) and 20 mg folic acid in 10 mL of milliQ water. Store solution B at −30 °C. Solution C: 15 g Dextrose dissolved in milliQ water to 90 mL final volume, 10 mL of antibiotic-antimycotic (100×) (Gibco, 15240-062), and 40 μL of solution B. Filter sterilize. Mix solutions A and C to obtain 1 L medium.
2. Developmental buffer (DB). Solution A: 10 × Na/K-phosphate buffer (0.2 M, pH 6.5). Dissolve 8.16 g KH_2PO_4 and 10.7 g $Na_2HPO_4 \cdot 7H_2O$ in milliQ water to 500 mL final volume. Check pH using a pH meter. Solution B: 10 mM $CaCl_2$ aq. Solution C: 100 mM $MgSO_4$ aq. Autoclave solution A, B, and C. Mix 25 mL of solution A, 10 mL of solution B, 10 mL of solution C, and 455 mL of milliQ to 500 mL final volume.

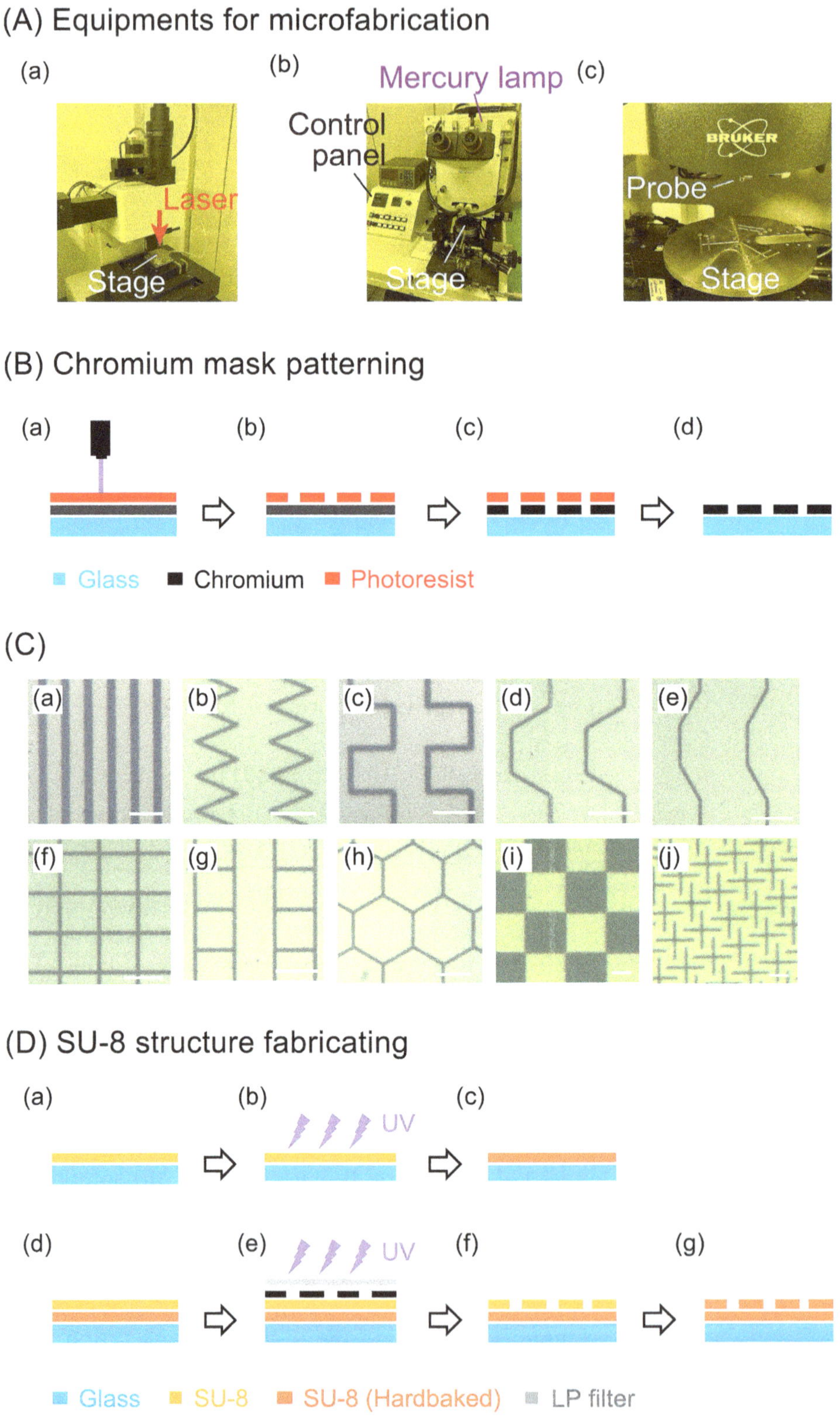

Fig. 1 Microfabrication overviews. (**A**) Devices employed. (*a*) A laser direct imaging device (DDB-201-TW, NEOARK). (*b*) A mask aligner MA-20 (MIKASA). (*c*) A stylus profiler Dektak XT (Bruker). (**B**) Mask patterning schematics. (*a*) Laser drawing on mask blanks. (*b*) Photoresist etching. (*c*) Chromium etching. (*d*) Washing. (**C**) Mask pattern examples on the chromium substrate. (*a*) Straight lines. (*b–e*) Zigzag lines with edges of (*b*)

2.4 Live-Cell Imaging

An inverted microscope (Olympus, IX83), confocal scanning unit (Yokogawa, CSU-W1), piezo Z stage (molecular device).

3 Methods

The following step-by-step protocol describes fabrication of two layers of SU-8 on glass coverslips. The dual layer design is employed so that all sides of the structures that the cells come in contact with consist of the same material. The fabrication protocol is followed by a demonstration experiment where one uses the microfabricated substrates for live-cell imaging analysis of the F-actin dynamics in *Dictyostelium*. The photolithography processes should be performed under yellow illumination to prevent photoresist from reacting with blue and UV light. It is recommended that all procedures from Subheadings 3.2, 3.3 and 3.4 are performed in a class 1–1000 dust-free environment with HEPA filter ventilation. If one has access to a maskless aligner, skip the mask production (Subheading 3.2) and go directly to Subheading 3.4.

3.1 Washing Coverslips

Cleaning the coverslip surface is critical for uniform coating of the photoresist and to reduce defects.

1. Insert 20 coverslips in a rack and place it in a 250 mL glass vessel (Matsunami, BT-210). Sonicate in 10% Contaminon LS-II for 60 min. Rinse coverslips by filling the vessel with 250 mL of tap water and draining it. Repeat the process with tap water ten times and then with milliQ water three times.
2. Sonicate in milliQ water for 3 min. Rinse with milliQ water three times.
3. Sonicate in 99.5% EtOH for 30 min. Rinse with milliQ water three times.
4. Sonicate in milliQ water for 3 min. Rinse with milliQ water three times.
5. Sonicate in 0.08 M NaOH for 30 min. Rinse with milliQ water three times.
6. Sonicate in milliQ water for 3 min. Rinse with milliQ water three times.
7. Dry the coverslips at 140 °C for 1 h by placing them in the vessel in an oven. Seal the vessel with a parafilm and store at room temperature.

Fig. 1 (continued) 45, (*c*) 90, (*d*) 120, and (*e*) 150°. (*f*) Grids. (*g*) Ladders. (*h*) Honeycombs. (*i*) Checkers. (*j*) Crosses. Bars: 30 μm. (**D**) Schematics of fabricating a two-layered SU-8 structure: (*a*) Spin-coating of SU-8, (*b*) uniform UV application, and (*c*) hardbaking. (*d*) Spin-coating of the second SU-8 layer, (*e*) UV application with a chromium mask, (*f*) baking and development of the second layer, and (*g*) hardbaking

3.2 Creating the Patterned Mask

1. Using a CAD software, prepare a design file in DXF format. The file specifies regions of laser application.
2. Load the design file to the laser direct imaging (LDI) system.
3. Place the mask blanks on the LDI system stage.
4. In the LDI software, set the laser power to 1.0 V, the scan speed to 100 μm/s, and the focus offset from the substrate surface to −0.8 V. These parameters should be good starting points; however, optimal parameters that achieve a desired line width etched on the mask will vary from instrument to instrument and thus should be determined beforehand by running a test.
5. Run the program to apply laser illumination (Fig. 1B(a)).
6. Place four stainless square bats in a fumed hood. Pour 200 mL each of the photoresist developer (NMD-3), chromium etcher (MPM-E350), acetone, and milliQ water.
7. Immerse the mask blanks into the developer for 1 min (Fig. 1B(b)).
8. Rinse in water and then immerse it in the chrome etcher for 1 min (Fig. 1B(c)).
9. Rinse again in water and then transfer it to acetone and gently agitate for a few minutes to remove the remaining resist (Fig. 1B(d)).
10. Dry the mask surface with a blower.
11. Wipe carefully with acetone using an industrial-grade cotton wiper (Bemcot).
12. Check the pattern by a metallurgical microscope (Fig. 1C).

3.3 Fabrication of the SU-8 Structure Using a Mask Aligner

The protocol below uses photoresist SU-8 2 to fabricate ridge structures with ~1 μm height (i.e., 1 μm-thick SU-8 2). For other thickness, one may choose other SU-8 formulations with different viscosity and determine the appropriate parameter for spin-coating, UV dose, and post exposure baking time. For the choice of the photoresist, refer to the datasheet from the manufacturer (https://kayakuam.com/products/su-8-photoresists/) and optimize the parameters indicated therein.

1. Turn on the plasma chamber and put it on hold.
2. Prewarm three hot plates each at 65 °C, 95 °C, and 150 °C, respectively.
3. Place a coverslip into the plasma chamber and treat it with O_2 plasma for 10 s at RF power 125 W under 20 mL/min O_2 flow. This step is important to improve attachment between the coverslip surface and SU-8.
4. Set the coverslip on the spin coater stage. Dispense 100 μL of SU-8 2 using a micropipette to the center of the coverslip.

5. Rotate at 500 rpm for 5 s and then 2000 rpm for 30 s.
6. To evaporate the solvent, heat the substrate on the hotplate for 1 min at 65 °C and then for 3 min at 95 °C. Place the substrate on a horizontal stage and allow it to cool down to room temperature (Fig. 1D(a)).
7. Load the substrate to the mask aligner. Irradiate UV uniformly for 1 min without a mask (Fig. 1D(b)). This step is performed to make the SU-8 surface flat.
8. Heat the substrate on the hotplates for 1 min at 65 °C and then for 3 min at 95 °C and for 10 min at 150 °C (Fig. 1D(c)). Remove the substrate and allow it to cool down to room temperature. The last so-called hardbake process at 150 °C is important when shaping sharp edges that stand near vertically in the second layer.
9. Treat the substrate with O_2 plasma for 10 s.
10. To form the second layer, dispense 100 μL of SU-8 2 to the center of the substrate and spin-coat at 500 rpm for 5 s then at 2000 rpm for 30 s.
11. Heat the substrate on the hotplates for 1 min at 65 °C and then for 3 min at 95 °C. Remove the substrate from the hotplate and allow it to cool down to room temperature (Fig. 1D(d)).
12. Place the patterned chrome mask and a long-pass filter (> 350 nM) in the mask aligner. Look into the eyepiece and adjust the focus to the mask surface. Place the substrate on the stage and lift the stage to bring the substrate in contact with the chrome mask.
13. Apply UV light for 2 s (the duration will depend on the resist thickness and UV lamp power) (Fig. 1D(e)). Repeat two more times.
14. Heat the substrate on the hotplates for 1 min at 65 °C and then for 1 min at 95 °C. Remove the substrate from the hotplate and allow it to cool down to room temperature.
15. Place two stainless steel bats (φ100 mM, 20 mM of depth) in a draft chamber and pour 50 mL SU-8 developer each.
16. Immerse the substrate in the first tray for more than 1 min and gently agitate at times. Rinse in the next tray and dry the surface with a blower (Fig. 1D(f)).
17. Check the patterned surface by a microscope. If incomplete, return to the developer step (**step 16**). Streaks remaining on the substrate after blowing air are the sign of incomplete etching. As a check, the surface can also be rinsed with isopropanol which leaves a white film when the surface is incompletely developed. The developer can be reused a few times.
18. Hardbake the substrate for 10 min at 150 °C (Fig. 1D(g)).

19. Measure the surface profile by AFM or Dektak (*see* **Note 1**).
20. To assemble the observation plate, detach the coverslip of glass bottom dish (φ35 mm) or drill a hole (14 mm) at the center of a plastic plate. If required, cut out the fabricated coverslip using a glass cutter to the size of 11 × 11 ~ 20 × 20 mm. Stick a 9 × 9 mm adhesive film with a square frame (Frame-seal; Biorad) around the hole from the bottom of the plate. Peel off the cover film to reveal the adhesive surface of the square frame. Attach the fabricated coverslip to the frame by gently pressing it against the frame with a tweezer. Make sure that the SU-8 structure is facing the inside of the plate through the hole. *See* also **Note 2**.

3.4 Fabrication Using a Maskless Aligner

1. Using the LayoutEditor software (https://layouteditor.com), prepare the binary design file in the DXF format.
2. Wash a coverslip of 40 × 50 mm dimension (Subheading 3.1). Coat with SU-8 following the **steps 1–11** in Subheading 3.3.
3. In the μMLA software, start a new job. Select the exposure mode to "Series." Input the size and shape of the loading sample, focusing method, the type of resist, and its thickness. Select the design file used for exposure. For a new design, click "Convert design" to start the conversion software. In the "Conversion" window, select the source file in the DXF file format. Check the layers to be used and set the "DXF units [nm]" according to the grid unit in the design. Click "Complete Tasks" to save the conversion settings, and click "Finish." The new design file is now listed in the job setting window.
4. Load the substrate on the stage. Allow the stage to hold the sample down pneumatically. In the μMLA software, press "Load Substrate" and follow the instructions for automatic focusing.
5. In the μMLA software, select the "Series" mode to duplicate the pattern. Set the "Number of Fields" to 2 × 2 and "Step size in X/Y" to 20 mm. Set the "Start value" of the dose to 90 mJ/cm^2 ("Dose") and defocusing to 0 ("Defoc"). The "Step size" should be set to 0 for both the dose and defocusing. Begin applying UV by pressing "Start exposure" in the μMLA software. Heat and develop the substrate following the **steps 14–19** in Subheading 3.3.
6. Prepare the chamber for cell observation following the **step 20** in Subheading 3.3.

3.5 Cell Preparation

1. Grow *Dictyostelium discoideum* AX4-expressing GFP-Lifeact in PS medium containing a 10 μg/mL G418 at 22 °C to a cell density of 1–2 × 10^6 cells/mL. Leave sufficient room for aeration. Typically, 10 mL culture is shaken in a 50 mL conical tube at 155 rpm in a rotary shaker.

2. Collect the 2.5×10^7 cells and remove the nutrient medium by pelleting by centrifugation at $700 \times g$, decanting the supernatant and resuspending the cells in DB. Repeat this three times.
3. Adjust the cell density in DB to 5×10^6 cells/mL and place 5 mL of the cell suspension in a 50 mL conical flask. Incubate on a rotary shaker for 1 h at 155 rpm.
4. After 1 h, begin applying 50 μL of the 5 μM cAMP solution (50 nM, final concentration) every 6 min using a peristaltic pump. Continue incubation for 4.5 h.
5. Pellet cells by centrifugation at $700 \times g$ and then resuspend in fresh 2.5 mL DB (1×10^7 cells/mL). Adjust cell density to 3×10^4 cells/mL in DB. Transfer 1 mL of the diluted cell suspension to a 1.5 mL tube. Arrest cells from further differentiation by placing the tube on ice and proceed immediately to **step 6**.
6. Treat the assembled observation plate (**step 20** in Subheading 3.3) with O_2 plasma for 10 s at RF power 125 W under 20 mL/min O_2 flow to improve cell attachment. Note that this should be performed immediately before plating the cells.
7. Pipette up and down the suspension ten times to dissociate agglutinated cells. Introduce the cell suspension to the plasma-treated dish. To reduce cell-cell encounter, cell density of ~3×10^3 cells/cm^2 is recommended. Incubation for >2 h increases cells exhibiting PIP3/F-actin-enriched patches [15].

3.6 Image Acquisition

Acquire timelapse confocal Z-stack images. Use the multidimensional acquisition in the imaging software of choice (Micromanager, Metamorph, etc.). Set the parameter for timelapse three-dimensional image acquisition. Typically, the thickness of the lower SU-8 layer should be 2 μm or less so that the cells will be within the working distance of a high magnification objective: 60×. For three-dimensional acquisition, scan in the z-direction from $z = -1$ to 2 μm ($z = 0$ μm at surface) at an interval of 0.5 μm. Optimize the exposure time based on the trade-off between signal/noise, acquisition time, and damage to the cells. A typical exposure time per image is 50 ms. The use of piezo driven Z-stage is critical to shorten the acquisition time. If available in the laser launcher employed, it is also important to minimize time of laser illumination by synchronizing it to the movement steps of the piezo-z stage using TTL signal. Timelapse interval should be 10 s or shorter in order to sufficiently track the cortical F-actin dynamics.

3.7 Image Analysis

The sample data is from a cell migrating along a zigzag ridge showing F-actin enriched patch at the leading front (Fig. 2A). F-actin accumulation is observed along ridges of 1.5 μm height and 4 μm width [15]. The following procedure demonstrates quantification of differential F-actin accumulation between the

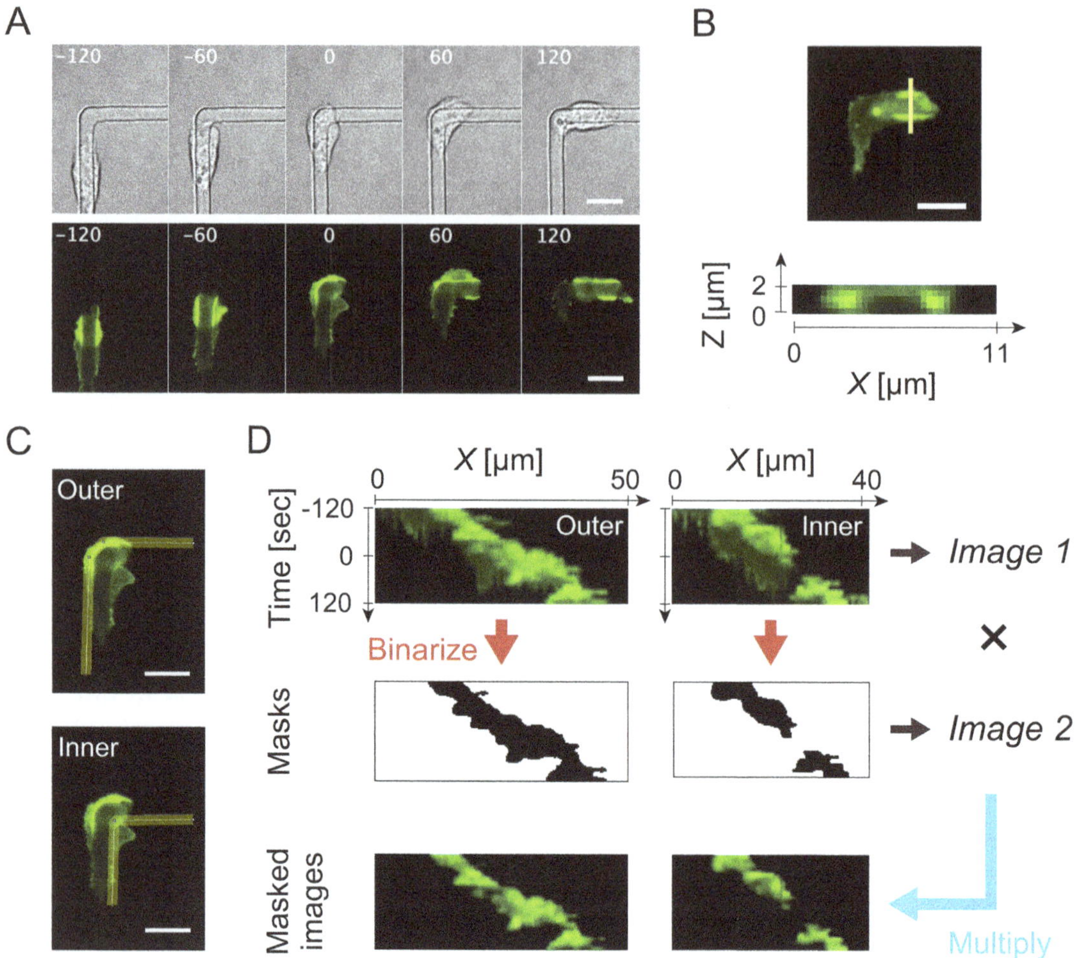

Fig. 2 (**A**) Transmitted-light (*top*) and confocal (*bottom*) images of a *Dictyostelium* cell-expressing GFP-Lifeact on the turning ridge (1.5 μm high and 4 μm wide). The confocal slice taken at z = 1 μm from the surface. Time in sec. (**B**) *Top panel*: Maximum intensity projection of Z-stacked confocal images of GFP-Lifeact. *Lower panel*: An XZ cross-section along the yellow line in the *top* panel. (**C**) ROIs (yellow lines, width = 2 μm) along the outer (*left*) and inner (*right*) corners. Bars, 10 μm. (**D**) The regions of actin patches are extracted from the kymographs (*top*) taken along the linear ROIs in **C**. *left*: Outside, *right*: Inside. Binarized images (*middle*) are obtained by smoothing and thresholding, which are then multiplied with the raw images to obtain the masked images (*bottom*). Images are enlarged four times in *t*-axis for visibility

outside and inside of a ridge corner (Fig. 2A, *lower*, t = −120 to 0 s) which evens out as the cell exits the corner (Fig. 2A, *lower*, t = 0 to 120 s). The ImageJ commands are indicated by square brackets [ ].

1. Open the image file "sample_488nm_t-Zstack.tif" using ImageJ [File > Import > Image Sequence…]. To reconstruct the three-dimensional image, create a hyperstack from the image

stack [Image > Hyperstacks > Stack to Hyperstack. . .]. Specify the number of slices for the substacks. For the demo file, ("channels":1, "*z*-slices": 5, "time frames": 401).

2. (Optional) To have a quick glance at the dynamics, it is convenient to obtain two-dimensional projected image of the three-dimensional data. To obtain the maximal intensity projection on the XY plane, choose [Image > Stacks > Z Project. . .] and select the projection type "Max Intensity."
3. (Optional) Using the Line Selection Tool (Straight Line) in ImageJ, specify a linear region-of-interest (ROI) (Fig. 2B, *top panel*). In the ImageJ menu, select [Image > Stacks > Reslice. . .] to obtain an XZ cross-section along the line specified (Fig. 2B, *lower panel*). The cross-section can be used to obtain the fluorescent intensity distribution in the *z*-direction (Fig. 1D in [15]; *see* **Note 3** for *z*-calibration).
4. Execute [Image > Duplicate. . .] to extract a single-plane time-lapse images at a specified z-plane. In the pop-up window, check "Duplicate hyperstack" and enter the number of the z-slice (in the demo file, "Slices (z): 3" which is at approximately $z = 1$ μm from the bottom plane).
5. On this re-sliced substack, run again "Duplicate. . ." function, this time to extract frames from the time the F-actin-enriched patch entered the ridge corner until it exited. In the pop-up window, enter the time frames to be extracted. In this example, set "Frames (t): 6-46" (Fig. 2A; 240 s total).
6. Using the Line Selection Tool (segmented line) in ImageJ, define ROIs for fluorescence intensity measurements. In this example, linear ROIs are specified along the inner and outer corners of the ridge, respectively (Fig. 2C). Obtain kymographs using [Image > Stacks > Reslice. . .] function (Fig. 2D, *top*). In the kymographs, F-actin patches appear as regions of high fluorescence intensity of GFP-Lifeact.
7. Smooth by applying Gaussian filter [Process > Filters > Gaussian Blur. . .]. For the value of sigma, enter 1 (pix).
8. Convert the data type to 32-bit [Image > Type >32-bit] and binarize the images by [Image > Adjust > Threshold. . .]. Use "Otsu" method with the option "Set to NaN" (Fig. 2D, *middle*) enabled so that the value NaN (not a number) is assigned to the background. Select [Process > Math > Divide. . .] and enter 255 to normalize the pixel intensity of the patch region to 1.
9. In the ImageJ menu, select [Process > Image Calculator. . .]. In the pop-up window, set the raw image as "Image1" and the binarized mask image as "Image2." Run the "Multiply" operation to obtain an image in which only patch regions contain values (Fig. 2D, *bottom*).

10. Export the results by choosing [File > Save as > "Text Image..."].
11. Repeat **steps 5–10** for all other turning events at the corner (in the demo file, t -slice 53–93, 108–148, 166–206, 222–262, and 285–325). To obtain the average each for the inner and outer corners, align the time frame so that the 21st frame is set to $t = 0$.
12. To further quantify the fluorescence intensity of GFP-Lifeact from the text image files of the kymographs obtained above, use a computational tool of choice (e.g., Matlab, Python, or Excel). Spatially integrate the intensities (along the x-axis) at each position in the y-axis (timepoint) and normalize against the initial value at $t = -120$ s. Obtain the average over all events and plot it (y-axis) against time (x-axis) for the outer and inner corner, respectively (*see* Fig. 2C, D in [15]).

4 Notes

1. SPMs can be substituted by three-dimensional confocal imaging of fluorescently stained surface. Apply 30 μg/mL Cy5 solution to the plasma-treated SU-8 substrate for 30 min. Wash with buffer solution. Submerge the stained substrate in buffer solution and acquire *z*-slices using a confocal microscope. The advantage of this method over SPMs is that it allows profiling of surfaces difficult to access by SPM probes such as sharp structures that stand at an angle of 90° or higher from the basal plane. Note that surface profiling with co-axial illumination is still limited by light diffraction which distorts the image in the z-direction. Lightsheet microscopy may be employed to circumvent this issue. *See* also **Note 3**.
2. The plate with the center hole can be reused by removing the SU-8 substrate. This requires peeling off the "frameseal" by soaking the dish in 99.9% ethanol overnight. The substrate is for one-time use.
3. Fluorescent beads of several micrometers in diameter are useful to calibrate the Z-motor. Acquire the *z*-scan of fluorescent beads, and in the XZ cross-section, measure the diameter in *x*- and *z*-direction, respectively. Assuming the bead is a true sphere and given the calibrated length in *x*-direction with micrometer, use its ratio to calibrate the *z*-step length.

Acknowledgments

We thank Yuichi Wakamoto, Okura Reiko, and Akihiko Nakajima for technical assistances. This work was supported by grants from the Japan Science and Technology Agency (CREST JPMJCR1923) and the Japan Society for Promotion of Science (JSPS); the Ministry of Education, Culture, Sports, Science and Technology (MEXT; KAKENHI JP19H05801 to S.S.); and the JSPS fellowship (JP18J14678 to G.H.).

References

1. Reig G, Pulgar E, Concha ML (2014) Cell migration: from tissue culture to embryos. Development 141:1999–2013
2. Hatten ME (1990) Riding the glial monorail: a common mechanism for glial guided neuronal migration in different regions of the developing mammalian brain. Trends Neurosci 13: 179–184
3. Wolf K, Müller R, Borgmann S et al (2003) Amoeboid shape change and contact guidance: T-lymphocyte crawling through fibrillar collagen is independent of matrix remodeling by MMPs and other proteases. Blood 102:3262–3269
4. Salmon H, Franciszkiewicz K, Damotte D et al (2012) Matrix architecture defines the preferential localization and migration of T cells into the stroma of human lung tumors. J Clin Invest 122:899–910
5. Provenzano PP, Inman DR, Eliceiri KW et al (2008) Contact guidance mediated three-dimensional cell migration is regulated by Rho/ROCK-dependent matrix reorganization. Biophys J 95:5374–5384
6. Bajénoff M, Egen JG, Koo LY et al (2006) Stromal cell networks regulate lymphocyte entry, migration, and territoriality in lymph nodes. Immunity 25:989–1001
7. Bovetti S, Hsieh Y-C, Bovolin P et al (2007) Blood vessels form a scaffold for neuroblast migration in the adult olfactory bulb. J Neurosci 27:5976–5980
8. Mathur A, Moore SW, Sheetz MP et al (2012) The role of feature curvature in contact guidance. Acta Biomater 8(2595):2601
9. Teixeira AI, Abrams GA, Bertics PJ et al (2003) Epithelial contact guidance on well-defined micro- and nanostructured substrates. J Cell Sci 116:1881–1892
10. Leclech C, Barakat AI (2021) Is there a universal mechanism of cell alignment in response to substrate topography? Cytoskeleton 78:284–292
11. Ray A, Lee O, Win Z et al (2017) Anisotropic forces from spatially constrained focal adhesions mediate contact guidance directed cell migration. Nat Commun 8:14923
12. Kwon KW, Park H, Song KH et al (2012) Nanotopography-guided migration of T cells. J Immunol 189:2266–2273
13. Driscoll MK, Sun X, Guven C, Fourkas JT, Losert W (2014) Cellular contact guidance through dynamic sensing of nanotopography. ACS Nano 8:3546–3555
14. Ermis M, Antmen E, Hasirci V (2018) Micro and nanofabrication methods to control cell-substrate interactions and cell behavior: a review from the tissue engineering perspective. Bioact Mater 3:355–369
15. Honda G, Saito N, Fujimori T, Hashimura H, Nakamura MJ, Nakajima A, Sawai S (2021) Microtopographical guidance of macropinocytic signaling patches. Proc Natl Acad Sci 118: e2110281118

Chapter 12

Evaluating the Activity of Ras/Rap GTPases in *Dictyostelium*

Stephen F. Smith and Pascale G. Charest

Abstract

Ras and Rap small GTPases of the Ras superfamily act as molecular switches to control diverse cellular processes as part of different signaling pathways. *Dictyostelium* expresses several Ras and Rap proteins, and their study has and continues to greatly contribute to our understanding of their role in eukaryote biology. To study the activity of Ras and Rap proteins in *Dictyostelium*, several assays based on their interaction with the Ras binding domain of known eukaryotic Ras/Rap effectors have been developed and proved extremely useful to study their regulation and cellular roles. Here, we describe methods to assess Ras/Rap activity biochemically using a pull-down assay and through live-cell imaging using fluorescent reporters.

Key words *Dictyostelium discoideum*, Ras, Rap1, Activity assays, Ras binding domain, Pull-down, Imaging

1 Introduction

Ras and Ras-related protein (Rap) GTPases play key roles in numerous biological processes that are studied in *Dictyostelium*, including cell proliferation, chemotaxis, cell adhesion, and phagocytosis [1]. In *Dictyostelium*, there are 11 Ras and 3 Rap proteins, but only 7 of them have been studied thus far: RasB, RasC, RasD, RasG, RasS, RapA (Rap1), and, more recently, RapC [1–4]. RasB, RasC, RasD, and RasG are most closely related to each other and to human Ras proteins whereas RasS is more divergent; and Rap1 was found to be the homologue of human Rap1 whereas RapC is more divergent.

Ras/Rap GTPases act as molecular switches, cycling between an inactive GDP-bound form and an active GTP-bound form. To study the activity of Ras/Rap in cells, methods using the Ras binding domain (RBD) of known Ras/Rap effectors have been developed, which allow detecting GTP-bound Ras/Rap proteins both biochemically by pull-down (Fig. 1a) and in live cells with fluorescent reporters and imaging approaches (Fig. 1b), and these

Alan R. Kimmel (ed.), *Dictyostelium discoideum: Methods and Protocols*, Methods in Molecular Biology, vol. 2814,
https://doi.org/10.1007/978-1-0716-3894-1_12,

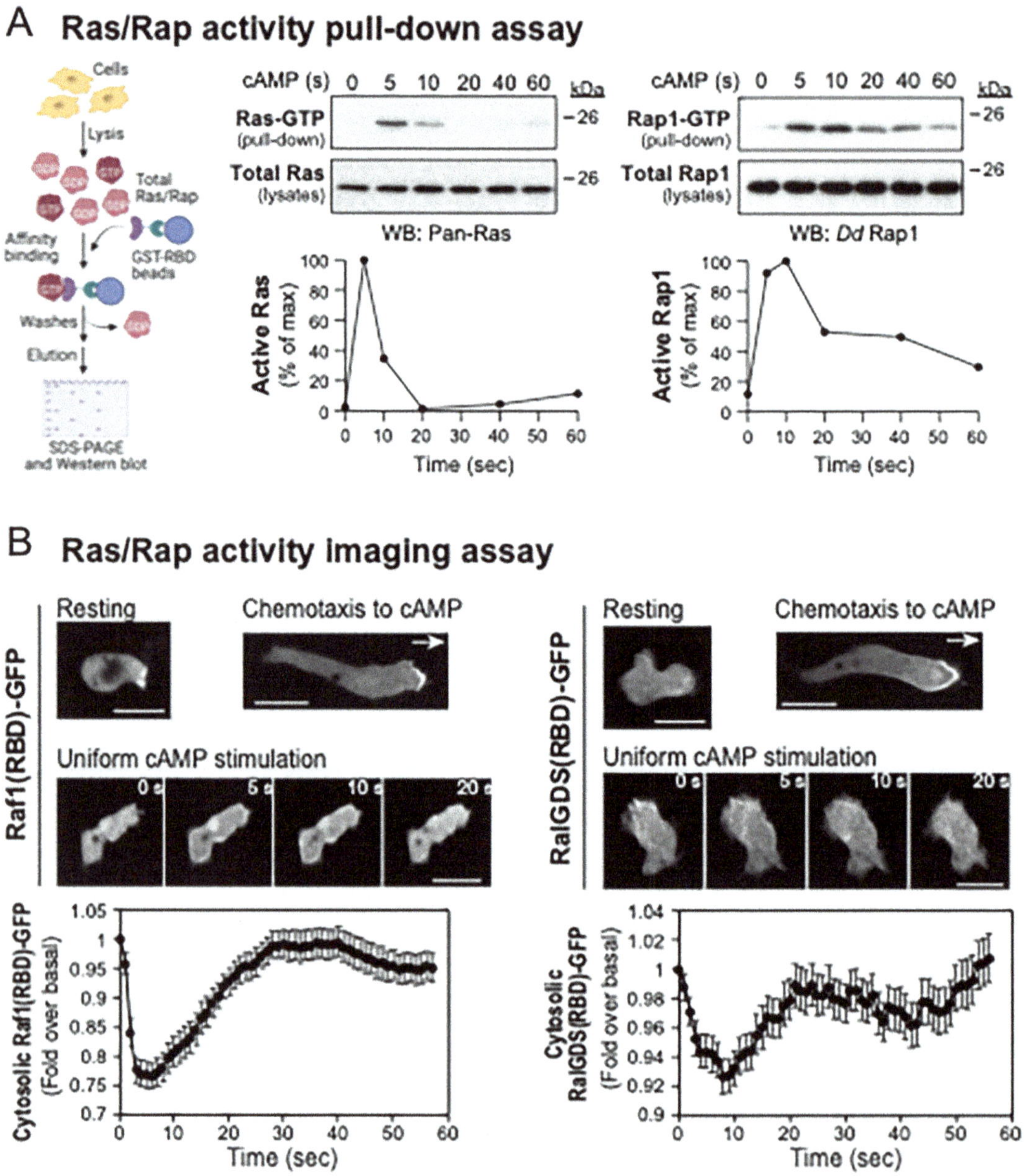

Fig. 1 Ras/Rap activity assays. Examples of data obtained with 5.5 h developed AX3 cells are shown. (**a**) The Ras/Rap activity pull-down assay. An outline of the assay is depicted on the left. Cells were stimulated with cAMP for the indicated time before lysing the cells and purifying the GTP-bound Ras or Rap proteins using GST-Raf1(RBD) for Ras and GST-RalGDS(RBD) for Rap1, followed by their analysis by Western blot using a Pan-Ras antibody or *Dictyostelium* (Dd) Rap1-specific antibody. The graphs illustrate the relative cAMP-induced activation of the Ras and Rap1 proteins as quantified by densitometry, normalized to the total Ras and Rap1, and expressed as percentage of the maximal response for each. (**b**) The Ras/Rap activity imaging assay. The Raf1(RBD)-GFP and RalGDS(RBD)-GFP were used as reporters of Ras and Rap1 activity, respectively. Images show localization of the reporters and, thereby, active Ras and Rap1, in resting/randomly moving cells, cells chemotaxing to cAMP in a pipette chemotaxis assay, and upon the uniform stimulation of cells with cAMP

have been successfully adapted for use in *Dictyostelium* [5, 6]. Commonly, the RBD of the human Rap1 effector RalGDS is used to assay *Dictyostelium* Rap1 activity [6]; the RBD of human Ras effector Raf1 is used to assay RasB, RasD, and RasG activity; and the RBD of the yeast Ras effector Byr2 is used to assay RasC activity [5, 7]. Of note, RasC has a slightly different effector domain sequence, and because of that, it does not interact effectively with Raf1-RBD [5]. On the other hand, Byr2-RBD interacts with RasC as well as the other Ras proteins and Rap1, but because its production in bacteria produces much lower yields than Raf1-RBD and RalGDS-RBD, those are still preferentially used with the other Ras/Rap proteins. Moreover, while Byr2-RBD is a great tool in Ras/Rap activity pull-down assays, especially for RasC, it does not work well as a fluorescent reporter in live cells due to its too high affinity to the GTP-bound Ras/Rap proteins, which leads to Ras/-Rap function inhibition. Here, we describe the biochemical and imaging methods to spatiotemporally detect and quantify Ras/Rap activity in *Dictyostelium* (Fig. 1).

2 Materials

Prepare all solutions with ultrapure deionized water (ddH_2O) and sterilize the ones that are used with cells.

2.1 Ras/Rap Activity Pull-Down Assay

2.1.1 GST-RBD Expression and Purification

1. BL21 DE3-competent bacteria.
2. GST-Byr2(RBD), GST-Raf1(RBD), and/or GST-RalGDS (RBD) bacterial expression plasmids, depending on the Ras/Rap proteins to be assayed (*see* **Note 1**).
3. LB media and LB agar plates with appropriate selection antibiotic.
4. Incubator shaker capable of maintaining temperatures of 16 and 37 °C.
5. GSH agarose beads and Sephadex G-10 beads.
6. 6× Protein sample buffer: 0.375 mM Tris-HCl pH 6.8, 30% glycerol, 6% SDS, 600 mM DTT, and 6% bromophenol blue, in ddH_2O. Aliquot and freeze at −20 °C.
7. 1 M IPTG: dissolve in ddH_2O, aliquot, and store at −20 °C.

Fig. 1 (continued) for the indicated time. The graphs show the relative fluorescence of the RBD-GFP reporter in the cytosol in cAMP-stimulated cells, as mean ± S.E.M of more than 20 cells from three different experiments, expressed as fold over the basal level. The fluorescence intensity in the cytosol is inversely proportional to that at the plasma membrane and, as such, inversely reflects Ras and Rap1 activation at the plasma membrane. The data shown here is provided as an example and was originally published in Scavello et al. [11]

8. Phosphate-buffered saline (PBS): 116 mM NaCl, 13.2 mM Na_2HPO_4, and 1.47 mM KH_2PO_4, in ddH_2O. Adjust pH to 7.4 with HCl. Store at 4 °C. Before use, supplement PBS with a bacteria-specific protease inhibitor cocktail.
9. 50% glycerol/PBS: mix 1 part glycerol with 1 part PBS. Store at 4 °C.

2.1.2 Pull-Down of Active Ras/Rap

1. Tabletop centrifuge and microcentrifuge.
2. Gyratory shaker for flasks.
3. Rotator or gyratory rocker for sample tubes.
4. 100 mg/mL BSA: dissolve in ddH_2O, aliquot, and store at −20 °C.
5. 2× Ras/Rap lysis buffer: 100 mM Tris pH 7.5, 300 mM NaCl, 50 mM $MgCl_2$, 20% glycerol, and 1% NP-40 or IGEPAL CA-630, in ddH_2O. Store at 4 °C. Before the assay, prepare a working solution by adding 2 mM DTT, 2 mM Orthovanadate, 4 μg/mL aprotinin, and 4 μg/mL leupeptin (*see* **Note 2**).
6. 12 mM Na/K phosphate buffer: 2.5 mM Na_2HPO_4, 9.5 mM KH_2PO_4, and pH 6.1. To prepare a 1 L10× solution, dissolve 3.4 g Na_2HPO_4 and 13 g KH_2PO_4 in ddH_2O. Autoclave to sterilize and store at 22 °C (*see* **Note 3**).
7. 30 mM cAMP stock: dissolve cAMP-Na salt monohydrate in 12 mM Na/K phosphate buffer. Aliquot and store at −20 °C.
8. 10 μM cAMP: dilute from 30 mM cAMP stock with Na/K phosphate buffer.

2.1.3 Analysis of Ras/Rap Activity by Western Blot

1. Tris-buffered saline (TBS): 50 mM Tris base, 150 mM NaCl, and 2.7 mM KCl, in ddH_2O. Adjust pH to 7.4 with HCL. Store at 4 °C.
2. TBS-Tween (TBST): Dilute Tween-20 to 0.1% in TBS.
3. Appropriate antibody: Pan-Ras, *Dictyostelium* Ras- or Rap-specific, or anti-epitope antibodies depending on goal of assay (*see* **Note 4**).
4. Software to perform densitometry analyses (e.g., ImageJ).

2.2 In Vivo Ras/Rap Activity Imaging Assay

2.2.1 Preparation of Cells

1. *Dictyostelium* expression plasmids for Rafl(RBD)-GFP (for detecting active RasB, RasD, and RasG) or RalGDS(RBD)-GFP (for detecting active Rap1) (*see* **Note 5**).
2. Coverslip-bottom dishes: Can be purchased or made from regular dishes. To make them, use a hole puncher to make holes of ~1 cm in diameter in the bottom and lid of 35 mM dishes. The bottoms can be used for imaging resting cells or for random cell motility and uniform stimulation assays, and the

lids (making shallower "dishes") are used for imaging cells in the micropipette chemotaxis assay because they allow more room to properly position the micropipette. Use vacuum grease or epoxy to secure 22 × 22 mM glass coverslips #1.5 over the holes, from the outside (*see* **Note 6**).

2.2.2 Live Cell Imaging

1. 150 μM cAMP: dilute from 30 mM cAMP stock in 12 mM Na/K phosphate buffer.
1. Spinning disk confocal imaging system equipped with a 488 laser, DIC capability, CCD camera, and 20×, 40×, and, nice but not necessary, 63× objectives.
2. Eppendorf TransferMan® micromanipulator.
3. Eppendorf Microloader™ tips and Femtotip® micropipettes.

2.2.3 Image Analysis

1. Confocal images analysis software (e.g., ImageJ/FIJI).

3 Methods

Ras/Rap activity can be measured in *Dictyostelium* cells at any stage of their life cycle and in many different conditions. We describe a protocol for assaying Ras/Rap activities in response to cAMP in 5.5 h. developed, aggregation-competent cells that are highly responsive to cAMP, but the method can be easily applied to other cellular contexts and conditions. Cells should be prepared using standard *Dictyostelium* cell culture and development methods [8].

3.1 Ras/Rap Activity Pull-Down Assay

3.1.1 GST-RBD Expression and Purification

1. Transform the desired GST-RBD-encoding plasmid into competent bacteria using standard procedures.
2. Pick a colony and grow in 10 mL LB overnight at 37 °C with shaking at 300 rpm in an incubator shaker.
3. Transfer the overnight culture to 1 L of LB in a 2.5 L flask (*see* **Note 7**).
4. Grow bacteria at 37 °C with shaking at 300 rpm in an incubator shaker until the culture reaches an O.D. of 0.6–0.8.
5. Induce protein expression by adding 1 mM IPTG and incubate at 16 °C with shaking at 300 rpm in an incubator shaker for 16 h.
6. Collect the bacteria and pellet by centrifugation at 5000 × *g* for 20 min at 4 °C (*see* **Note 8**).
7. Resuspend cells in 20 mL cold PBS containing protease inhibitors.
8. Sonicate on ice, at 50% amplitude and duty cycle 5, 10 × 1 min with 2 min cooldown between each sonication.

9. Add NP-40 (or IGEPAL CA-630) to 1% final concentration, mixing for ~10 min on ice until complete dissolution.
10. Transfer the lysates to appropriate centrifuge tubes and clarify by centrifugation at 20,000 × *g* for 30 min at 4 °C.
11. Transfer the clarified lysates to a new tube with 1 mL of GSH Sepharose beads and incubate for at least 1 h at 4 °C with agitation on rotator or gyratory rocker.
12. Pellet the beads by centrifugation at 1000 × *g* for 1 min and wash five times with cold PBS.
13. Pellet the beads, remove PBS, and add 250 μL of 50% glycerol/PBS.
14. Take a 20 μL aliquot, mix with 4 μL 6× protein sample buffer, and quantify approximately by SDS-PAGE and Coomassie staining against 1, 2, 5, and 10 μg BSA standards using standard procedures.
15. Dilute the GST-RBD beads to 0.5 μg/μL with 50:50 hydrated Sephadex G-10 beads in 50% glycerol/PBS. Store at −20 °C.

3.1.2 Pull-Down of Active Ras/Rap

To measure the total Ras activity or the activity of overexpressed Ras/Rap proteins, we suggest using 1×10^8 cells, and to measure the activity of a single Ras/Rap protein expressed at endogenous or low levels, we suggest using 5×10^8 cells.

1. Collect for 5.5 h developed cells by centrifugation at 500 × *g* for 5 min at 4 °C.
2. Wash the cells twice with cold 12 mM Na/K phosphate buffer.
3. Resuspend the cells in 5 mL of 12 mM Na/K phosphate buffer, and incubate for 30 min at 22 °C with shaking at ~140 rpm on a gyratory shaker (*see* **Note 9**).
4. Transfer one 450 μL aliquot of cells to the "0" time point tube containing 500 μL 2× lysis buffer +50 μL 12 mM Na/K phosphate buffer and keep on ice.
5. Stimulate the remaining cells with 500 μL of 10 μM cAMP (1 μM final), and collect 500 μL aliquots at 5, 10, 20, 40, and 60 s, mixing the cells and changing the pipette tip between each sample, stopping the stimulation by transferring the aliquots to the corresponding pre-prepared tubes containing 500 μL 2× lysis buffer on ice (*see* **Note 10**).
6. Vortex the samples thoroughly and incubate on ice for 10 min, vortexing every 2 min.
7. Clarify the lysates by centrifugation at 20,000 × *g* for 10 min at 4 °C.
8. Transfer the supernatants to new tubes on ice and vortex thoroughly.

9. Collect 20 μL samples of the total cell lysates and mix with 20 μL 2× protein sample buffer (diluted from 6× protein sample buffer with ddH_2O) for total Ras Western blot. Store at −20 °C.
10. Add 20 μL of 100 mg/mL BSA (2 mg/mL final) and 20 μL of 0.5 μg/μL GST-RBD beads (10 μg final) to each lysate sample and incubate for 30 min–2 h at 4 °C with agitation on rotator or gyratory rocker (*see* **Note 11**).
11. Pellet the beads at 1000 × *g* for 1 min and wash three times with 1× lysis buffer (diluted from 2× lysis buffer with ddH_2O).
12. Carefully remove all lysis buffer and elute the Ras/Rap proteins from the beads by adding 30 μL of 2× protein sample buffer containing 100 mM DTT. Mix well and incubate at 65 °C for 15 min.
13. Vortex thoroughly and centrifuge at 1000 × *g* 1 min to bring down condensation, and then vortex and pellet beads again.
14. Carefully transfer the supernatants containing eluted Ras/Rap proteins to new tubes, without transferring any beads. Use immediately for Western blot analysis or store at −20 °C until needed.

3.1.3 Analysis of Ras/Rap Activity by Western Blot

1. Denature all proteins by incubating the samples at 95 °C for 10 min.
2. Centrifuge the heated samples to bring down condensation, and vortex to ensure sample homogeneity before loading and separating the proteins on 12% SDS-PAGE gels (*see* **Note 12**).
3. Transfer the proteins to a nitrocellulose membrane, perform the Western blot according to the appropriate antibody's manufacturer protocol, and reveal by chemiluminescence or fluorescence using standard procedures (*see* **Note 13**).
4. Save the Western blot images as grayscale JPEG files and open in an appropriate software for densitometric analyses (e.g., ImageJ).
5. If using ImageJ, set the measurement criteria to "Grey Mean Value" only and use the rectangle tool to draw a frame around the largest band to be measured, and then drag the frame to other bands and adjust the thickness accordingly to make sure that it can be used for all bands.
6. Using the same frame for all the Ras/Rap protein bands on the Western blot, measure the mean gray value (by clicking "Measure" under "Analyze" menu) for each band, one at a time. The first measurement will open a measurement window where all of them will be displayed in order to be taken.

7. Using the same frame, also take background measurements. If the background is very uniform, only one measurement can be taken, but if there are differences in background over the Western blot, take one background measurement next to each band.
8. Transfer the measurements to an Excel spreadsheet for analyses.
9. First, invert the pixel density for all data recorded by ImageJ (x) by subtracting them from 255 (255 – x), and then subtract the inverted background from all inverted band values.
10. Divide each GTP-Ras/Rap value by their corresponding total Ras/Rap loading control.
11. To facilitate comparison between separate experiments, because it is extremely challenging to have equivalent exposures of the GTP-Ras/Rap and total Ras/Rap Western blots between experiments, we suggest expressing the data as percent of the strongest response in the control condition (wild-type cells, no drug treatment, etc.).

3.2 In Vivo Ras/Rap Activity Imaging Assay

3.2.1 Preparation of Cells and Microscope

1. Generate *Dictyostelium* cells transformed with the appropriate RBD-GFP expressing plasmid and develop the cells for 5.5 h using standard procedures.
2. Transfer the cells to coverslip-bottom dishes containing 2 mL 12 mM Na/K phosphate buffer, at ~1×10^6 cells for random motility and uniform stimulation assays in a custom-made dish bottom with coverslip and at ~2.5×10^5 cells for a chemotaxis pipette assay in a custom-made dish lid with coverslip. Spread the cells evenly and let them settle and disaggregate for ~30 min (*see* **Note 14**).
3. Place a cell-containing coverslip-bottom dish on the stage of a spinning disk confocal imaging system and focus on cells that are on the coverslip using a 20× objective.
4. Find an area with multiple cells but that do not touch each other and that express the RBD-GFP reporter at low levels, which will allow for better contrast between the signal at the plasma membrane and the cytosol when the reporter is recruited to active Ras/Rap at the plasma membrane.
5. Set up the fluorescence imaging parameters, keeping the laser power as low as possible to obtain a clear image with a ~ 100 ms exposure, which will limit phototoxicity.

3.2.2 Imaging RBD-GFP in Resting, Randomly Moving, or Uniformly Stimulated Cells

1. For imaging resting and randomly moving cells, set up a time-lapse for capturing DIC (sometimes useful to see whole cells especially when the reporter expression is low) and fluorescence images at 6 s intervals for 30 min.

2. For imaging RBD-GFP in uniformly stimulated cells, set up a time-lapse for capturing fluorescence images at 1 s intervals for 60 s (or longer if desired). Start the acquisition and stimulate the cells on the third capture by adding 100 μL of 150 μM cAMP dropwise directly over the imaged cells (*see* **Note 15**).
3. To use as controls for photobleaching and background fluorescence in later analyses of the uniform stimulation experiment, also image unstimulated cells and cells not expressing any fluorescent reporter using the same time-lapse parameters.

3.2.3 Imaging RBD-GFP in Chemotaxing Cells Using a Micropipette Assay

1. Keeping the protective tip of a micropipette on, use a microloader tip to load a micropipette with 20 μL of 150 μM cAMP (*see* **Note 16**).
2. Set the micropipette holder of the micromanipulator to ~35°, carefully remove the micropipette's protective tip, and secure the micropipette on the holder according to the manufacturer's instructions.
3. Use a thin coverslip-bottom dish with cells (the ones made with the lids), and focus on the cells using a 20× objective in bright-field imaging or by looking through the eye piece directly.
4. Using the micromanipulator's control panel and joystick, position the tip of the micropipette in the middle of the objective, and start to slowly lower in the buffer. Focus up to find the tip of the micropipette in the field of view and gradually, very carefully, lower to ~10 μm above the cells (*see* **Note 17**).
5. Bring the micropipette a bit higher and then hit "home" on the micromanipulator's controller to take it out of the buffer. Change to the 40× objective (or 63× for higher resolution images), and find a good area to image, with several separated cells and where the micropipette can be positioned away from the cells so that they will not reach the tip before the end of the recording (*see* **Note 18**).
6. Hit "home" to bring the micropipette back into position and finely bring down to ~10 μm above the cells and place slightly away from cells, ideally at the edge of the field of view.
7. Set up a time-lapse for capturing DIC and fluorescence images at 6 s intervals for 10 min. To image the response upon introduction of the micropipette and gradient establishment, start the time-lapse before bringing the micropipette back.
8. Export images as RGB TIFF files or the entire recording as series movies for publication and/or presentations, and as 16-bit TIFF files (multi-page TIFF) for subsequent quantitative analyses.

3.2.4 Quantitative Analysis of Uniform cAMP Stimulations

1. Open the 16-bit TIFF files in an appropriate image analysis software such as ImageJ/FIJI.
2. For each condition and cell to be measured, select a cytosolic region omitting organelles with higher or lower fluorescence intensities, making sure the region is good for the entire time-lapse. The average value of all pixels within the selected cytosol region then serves as a measure of cytosolic fluorescence intensity at time t, $I_c(t)$ (*see* **Note 19**).
3. Use the measurements on cells not expressing any fluorescent reporter to determine the background fluorescence intensity at each time point, $I_b(t)$.
4. Average the fluorescence intensity measured in the first frames before cAMP was added to obtain the basal cytosolic fluorescence intensity, I_{c0} and I_{b0}.
5. Calculate the normalized fluorescence signals, $I_{c\ nor}(t)$ as follows: $I_{c,nor}(t) = [I_c(t) - I_b(t)]/(I_{c0} - I_{b0})$ (*see* **Note 20**).
6. To measure the photobleaching effect, first measure the fluorescence intensity of unstimulated RBD-GFP-expressing cells at each time point, $I_a(t)$, and normalize the time series by calculating $I_{a,nor}(t) = [I_a(t) - I_b(t)]/(I_{a0} - I_{b0})$.
7. Correct each fluorescence measurement of the RBD-GFP reporter for the photobleaching effect and determine the final cytosolic signal by calculating $I(t) = I_{c\ nor}(t)/I_{a\ nor}(t)$.
8. Plot the fluorescence intensity in the cytosol as a function of time to generate a graph showing the changes of cytosolic fluorescence intensity upon cAMP stimulation.

4 Notes

1. The RBD of yeast Byr2 can be used to pull down RasB, RasC, RasD, RasG, and Rap1; the RBD of human Raf1 can be used to pull down active RasB, RasD, and RasG (it does not interact with RasC); and the RBD of human RalGDS can be used to pull down Rap1. Of note, Raf1-RBD agarose and RalGDS-RBD agarose are commercially available, so they can be purchased instead of prepared as described in this protocol.
2. Excess 2× lysis buffer can be prepared to be diluted with Na/K phosphate buffer afterward and used as 1× lysis buffer for washing the beads.
3. An alternative buffer that can be used with *Dictyostelium* is the development buffer (DB; 5 mM Na_2HPO_4, 5 mM KH_2PO_4; 1 mM $CaCl_2$, 2 mM $MgCl_2$, pH 6.5).

4. Commercial Pan-Ras antibodies will detect many *Dictyostelium* Ras proteins, including RasB, RasC, RasD, and RasG. Alternatively, custom-made Ras/Rap-specific antibodies can be used to detect the activity of specific Ras/Rap proteins. Finally, if custom-made antibodies are unavailable and cells expressing epitope-tagged Ras/Rap proteins are used, epitope-specific antibodies are then used.
5. The Byr2(RBD) does not work as a reporter of Ras activity in live cells, presumably because its affinity for the Ras/Rap proteins is too strong and, consequently, inhibits their function. Therefore, it is important to note that none of the currently available fluorescent reporters of Ras activity report the activity of RasC. Also note that several versions of the fluorescent reporters exist, with the latest optimized ones by Arjan Kortholt [9, 10].
6. Vacuum grease should only be used when preparing dishes the same day of a planned experiment, to avoid it drying leading to buffer leaking out. Epoxy-secured coverslip-bottom dishes can be prepared in advance and can be used to culture cells before the experiment if needed.
7. For GST-Raf1(RBD) and GST-RalGDS(RBD), 1 L produces sufficient protein for multiple experiments. However, GST-Byr2(RBD) expression and purification from bacteria generate much lower yields, and we recommend using 5 L of culture for this one.
8. Bacteria can be frozen at −20 °C at this stage and then thawed directly into PBS when needed. Especially for GST-Byr2(RBD) that produces low protein yields, it is useful to grow larger volumes of culture, store aliquots of pelleted bacteria at −20 °C, and purify as needed.
9. During this incubation, prepare the tubes for the samples to be collected before (for basal activity level) and after cAMP stimulation: 0, 5, 10, 20, 40, and 60 s. Add 500 μL of 2× lysis buffer in each tube and place on ice. Add 50 μL of 12 mM Na/K phosphate buffer to the "0" time point tube, which is the equivalent of the cAMP solution that will be in the other samples. We find it helpful to use a rack to hold the tubes in place on ice.
10. Collecting the 5 and 10 s time points can be particularly challenging, especially with shaking constantly as much as possible. We hold the tube with the cells in one hand, mixing the cells after addition of cAMP and in between each collection, while using the other hand to collect and transfer samples to the tubes on ice. Between each time point, the tip is discarded, and another tip is rapidly mounted on the pipette from a racked tip box nearby. We strongly suggest changing tips, because for

optimal rapid stopping of the stimulation, it is best to deliver the samples *in* the lysis buffer so that some mixing occurs before all the time points are finished being collected, and if the tip is not changed, then there would be a risk of transferring some lysis buffer to the cells upon collection of the next time point sample. We also strongly suggest practicing doing this procedure to get used to it before doing a first experiment.

11. An incubation time of 30 min with the GST-RBD beads is enough and recommended when assaying total Ras activity or overexpressed Ras/Rap proteins, to limit binding of inactive Ras. However, when assaying the activity of single Ras/Rap protein expressed at endogenous or low levels, we have had better success incubating for longer times.
12. The total Ras/Rap proteins may give a considerably stronger signal, so we advise loading those on different gels than the active Ras/Rap samples. Endogenous Ras/Rap proteins migrate ~21 kDa.
13. For detecting single Ras/Rap1 proteins expressed at endogenous or low levels, we recommend using a super sensitive ECL reagent (e.g., SuperSignal™ West Femto Maximum Sensitivity Substrate) to facilitate detection of the pulled-down proteins, especially.
14. We suggest always preparing several dishes to have backups in case a dish leaks and to allow performing replicates in the stimulation assays.
15. To avoid disturbing the cells, which are not very adherent on the glass substrate, we suggest delivering the cAMP dropwise right above the buffer as quickly as possible. Because the cAMP cannot be mixed after addition, we found that using a higher cAMP concentration leads to more consistent results. Also note that the addition of cAMP can be done at later captures (instead of at the third one), the important thing is to be consistent to facilitate later analyses.
16. To avoid breaking the micropipette, keep the protective tip on during loading and until ready to mount on the micromanipulator. Also, it is important to make sure that there are no air bubbles in the micropipette, especially at the very tip as this will prevent cAMP from diffusing out. To get rid of air bubbles, one can give a good flick of the wrist while holding the micropipette, or centrifuge briefly in an appropriate microcentrifuge (which accommodates the long micropipette with the protective tip). Also of note, we suggest using 150 μM cAMP in the micropipette if simple diffusion of the cAMP is used to create the gradient. If the micromanipulator is equipped with a pump creating positive pressure to push cAMP out of the micropipette, much lower cAMP concentrations can be used.

17. The micropipette is extremely fragile and will break upon touching the coverslip. It is important not to break it because then the cAMP that will come out will cause a stimulation of the cells, and the resulted wider micropipette opening will create a different gradient. Therefore, for reproducibility, one should not use a broken micropipette. To avoid breaking it, keep the focus on the tip while finely lowering it, until the cells start being visible. You can then focus on the cells, and the micropipette tip should remain slightly out of focus.
18. If cells reach the tip and it is low enough, cells can then hit it, burst, and leave pieces of membranes in it and, thereby, block it. Consequently, the micropipette will not be usable anymore. If cells reach the pipette, it is best to stop the recording and bring the micropipette "home." The micropipette can then be reused for another recording in another area of the dish and/or for another dish.
19. We find that measuring the fluorescence in the cytosol, instead of measuring the fluorescence at the plasma membrane using kymographs, provides much more accurate measurements and reproducible results. The recruitment of the RBD-GFP to active Ras/Rap at the plasma membrane is then proportionally reflected by the loss of cytosolic fluorescence.
20. Since the basal intensity mostly reflects the expression level of the RBD-GFP, it is best to normalize the data and express as fold change over basal to compare between the conditions.

References

1. Weeks G, Spiegelman GB (2003) Roles played by Ras subfamily proteins in the cell and developmental biology of microorganisms. Cell Signal 15:901–909
2. Eichinger L, Pachebat JA, Glockner G, Rajandream MA, Sucgang R, Berriman M, Song J, Olsen R, Szafranski K, Xu Q, Tunggal B, Kummerfeld S, Madera M, Konfortov BA, Rivero F, Bankier AT, Lehmann R, Hamlin N, Davies R, Gaudet P, Fey P, Pilcher K, Chen G, Saunders D, Sodergren E, Davis P, Kerhornou A, Nie X, Hall N, Anjard C, Hemphill L, Bason N, Farbrother P, Desany B, Just E, Morio T, Rost R, Churcher C, Cooper J, Haydock S, van Driessche N, Cronin A, Goodhead I, Muzny D, Mourier T, Pain A, Lu M, Harper D, Lindsay R, Hauser H, James K, Quiles M, Madan Babu M, Saito T, Buchrieser C, Wardroper A, Felder M, Thangavelu M, Johnson D, Knights A, Loulseged H, Mungall K, Oliver K, Price C, Quail MA, Urushihara H, Hernandez J, Rabbinowitsch E, Steffen D, Sanders M, Ma J, Kohara Y, Sharp S, Simmonds M, Spiegler S, Tivey A, Sugano S, White B, Walker D, Woodward J, Winckler T, Tanaka Y, Shaulsky G, Schleicher M, Weinstock G, Rosenthal A, Cox EC, Chisholm RL, Gibbs R, Loomis WF, Platzer M, Kay RR, Williams J, Dear PH, Noegel AA, Barrell B, Kuspa A (2005) The genome of the social amoeba Dictyostelium discoideum. Nature 435:43–57
3. Kortholt A, van Haastert PJ (2008) Highlighting the role of Ras and Rap during Dictyostelium chemotaxis. Cell Signal 20:1415–1422
4. Park B, Kim H, Jeon TJ (2018) Loss of RapC causes defects in cytokinesis, cell migration, and multicellular development of Dictyostelium. Biochem Biophys Res Commun 499: 783–789

5. Kae H, Lim CJ, Spiegelman GB, Weeks G (2004) Chemoattractant-induced Ras activation during Dictyostelium aggregation. EMBO Rep 5:602–606
6. Kang R, Kae H, Ip H, Spiegelman GB, Weeks G (2002) Evidence for a role for the Dictyostelium Rap1 in cell viability and the response to osmotic stress. J Cell Sci 115:3675–3682
7. Charest PGG, Shen Z, Lakoduk A, Sasaki ATT, Briggs SPP, Firtel RAA (2010) A Ras signaling complex controls the RasC-TORC2 pathway and directed cell migration. Dev Cell 18:737–749
8. Fey P, Kowal AS, Gaudet P, Pilcher KE, Chisholm RL (2007) Protocols for growth and development of Dictyostelium discoideum. Nat Protoc 2:1307–1316
9. Kortholt A, Keizer-Gunnink I, Kataria R, Van Haastert PJ (2013) Ras activation and symmetry breaking during Dictyostelium chemotaxis. J Cell Sci 126:4502–4513
10. Kortholt A, Bolourani P, Rehmann H, Keizer-Gunnink I, Weeks G, Wittinghofer A, Van Haastert PJ (2010) A Rap/phosphatidylinositol 3-kinase pathway controls pseudopod formation [corrected]. Mol Biol Cell 21:936–945
11. Scavello M, Petlick ARARAR, Ramesh R, Thompson VFVF, Lotfi P, Charest PGPG (2017) Protein kinase A regulates the Ras, Rap1 and TORC2 pathways in response to the chemoattractant cAMP in Dictyostelium. J Cell Sci 130:1545–1558

Chapter 13

Analysis of cGMP Signaling in *Dictyostelium*

Peter J. M. van Haastert, Ineke Keizer-Gunnink, and Arjan Kortholt

Abstract

Biochemical assays are described to analyze signal transduction by the second messenger cGMP in *Dictyostelium*. The methods include enzyme assays to measure the activity and regulation of cGMP synthesizing guanylyl cyclases and cGMP-degrading phosphodiesterases. In addition, several methods are described to quantify cGMP levels. The target of cGMP in *Dictyostelium* is the large protein GbpC that has multiple domains including a Roc domain, a kinase domain, and a cGMP-stimulated Ras-GEF domain. A cGMP-binding assay is described to detect and quantify GbpC.

Key words *Dictyostelium*, cGMP, Guanylyl cyclase, Phosphodiesterase, cGMP-binding protein GbpC

1 Introduction

The intracellular signal molecule cGMP plays an important role in the eukaryotic microorganism *Dictyostelium* to establish cell polarity during basal cell movement and chemotaxis to extracellular cAMP (recently summarized in [1]). Intracellular cGMP is produced by the enzyme guanylyl cyclase that has basal activity and can be activated by surface cAMP receptors through the activation of heterotrimeric G-proteins. The produced cGMP can be degraded by a phosphodiesterase, or it can bind to its target protein. These proteins have been studied in large details.

Dictyostelium contains two guanylyl cyclase enzymes, the soluble protein sGC and the transmembrane protein GCA [2–4]. Degradation of cGMP by phosphodiesterases is complex in *Dictyostelium* [5]: the majority of cGMP is degraded by the cGMP-stimulated cGMP-specific phosphodiesterase PDE5 (also named PdeD or GbpA) [6, 7]. Other routes of degradation are by the high-affinity PDE3 [8] and by secretion followed by degradation by the nonspecific cell surface phosphodiesterase PDE1. *Dictyostelium* has only one target protein that mediates the function of cGMP: the cGMP-binding protein GbpC [9, 10]. Four proteins

Alan R. Kimmel (ed.), *Dictyostelium discoideum: Methods and Protocols*, Methods in Molecular Biology, vol. 2814,
https://doi.org/10.1007/978-1-0716-3894-1_13,

with potential cyclic nucleotide-binding domains have been identified (GbpA-D) [11]. GbpA and GbpB are proteins that combine a cyclic nucleotide-binding domain and a phosphodiesterase domain. GbpA is the cGMP-stimulated cGMP-specific phosphodiesterase PDE5 mentioned above, whereas GbpB is a cAMP/cGMP stimulated cAMP-phosphodiesterase [6]. GbpD combines a cyclic nucleotide-binding domain and a Ras-GEF domain; however, the cyclic nucleotide-binding domain does not bind cAMP or cGMP but appears to bind the phospholipid PI(3,4,5)P_3 [9, 12]. Upon disruption of the *gbpC* gene, all high-affinity cGMP-binding activity is lost [9, 10]. GbpC is a very complex protein with multiple domains, including a Roc domain, a kinase domain, and a cGMP-stimulated Ras-GEF domain. GbpC is activated by an intramolecular cascade: cGMP-binding induces activation of the Ras-GEF domain, leading to the GDP/GTP exchange of the Roc domain and activation of the kinase domain [13]. In addition, GbpC has leucine-rich repeats that are expected to play a role in protein-protein interactions and a GRAM domain that mediates membrane binding [14].

The function of these proteins and their complex regulation has been studied using mutants with deletion or overexpression of the encoding genes or by replacing the encoding gene by a copy with specific mutation in the catalytic or regulatory domains. Biochemical assays are indispensable tools for the analysis of both wild-type and mutant cells. This chapter is an update of previously published assays to analyze cGMP signal transduction [15].

2 Materials

2.1 General Materials

1. [8-^{3}H]cGMP (14.3 Ci/mmoL = 0.53 TBq/mmoL, Perkin Elmer).
2. cAMP, cGMP, ATP, GTP, and GTPγS (Boehringer).
3. Dithiothreitol, DTT (Sigma); 1 M stock solution stored at −20 °C, made fresh each week.
4. Scintillator 299 (Packard).
5. Nuclepore polycarbonate filters, 3 μm pore size, mounted in filter holders (Whatman).
6. Nitrocellulose filters 0.45 μm pore size (Whatman).
7. Herring sperm DNA (Promega), denatured by boiling for 3 min and immediately cooled on ice.
8. Protease inhibitor mixture: mini Complete-tablets, EDTA-free (Roche Applied Science).
9. Calf intestine phosphatase (Sigma).

10. cGMP Enzyme Immunoassay Kit (Sigma-Aldrich).
11. Sheep cGMP-antibody (ab19226, Abcam).

2.2 Dictyostelium Culture and Preparation of Lysates

1. *Dictyostelium* liquid growth medium [16] HL5-C and SM-agar, Formedium.
2. Culture plates, tissue culture treated: Nunc Plate, Thermo Fisher Scientific.
3. Starved *Dictyostelium* cells. Collect *Dictyostelium* cells from growth medium by centrifugation for 3 min at 500 g, wash cells three times with PB, and resuspend in PB at a density of 10^7 cells/mL. Starve cells for 5 h by shaking in a rotary shaker at 150 revolutions/minute.
4. PB: 10 mM Na_2HPO_4/KH_2PO_4 buffer, pH = 6.5.
5. Non-nutrient agar (1 L): 15 g agar in PB, autoclaved.

2.3 cAMP-Induced cGMP Formation

1. Starved *Dictyostelium* cells. Collect starved cells by centrifugation for 3 min at 500 g, wash cells with PB, and resuspend cells in PB at a density of 2×10^8 cells/mL in PB; bubble air through the cell suspension for 10 min at a rate of about 15 mL air per mL suspension per min.
2. Stimulus solution: 5 μM cAMP in PB.
3. 3.5% (v/v) perchloric acid (PCA); store at 22 °C.
4. $KHCO_3$ (50% saturated at 22 °C); store at 22 °C.

2.4 Assay for cGMP

2.4.1 Radioimmunoassay for cGMP

1. cGMP assay buffer: 200 mM K_2HPO_4/KH_2PO_4, 10 mM EDTA, and 10 mM EGTA, pH 7.0; store 4 °C.
2. Tracers: 100,000 cpm [^{3}H]cGMP per mL (5000 Bq/mL) in assay buffer; prepare fresh.
   3a. Commercial cGMP antiserum (part of kit by Amersham)
   3b. Home-made cGMP antiserum AB99 [17]. Dilute AB99 in a preparation of calf serum such that about 30% of [^{3}H] cGMP will bind to the antibody in the radioimmunoassay; prepare fresh (about 12 μL AB99 bleed per mL calf serum). The calf serum is prepared as follows: incubate 1 mL calf serum with 5 mL 60% saturated ammonium sulphate for 5 min on ice, centrifuge for 5 min at 14,000 g, and dissolve the pellet in 1 mL cGMP assay buffer; store at 4 °C.
   3c. Commercial cGMP antiserum (sheep cGMP antibody; Abcam). Dilute antibody in the preparation of calf serum such that about 30% of [^{3}H]cGMP will bind to the antibody in the radioimmunoassay; prepare fresh (30 μL cGMP antibody; Abcam per mL calf serum).

2.4.2 Enzyme Immunoassay Kit for cGMP

The provided components of the immunoassay kit are the following:

- 96-Well Plate Coated with Goat Anti-Rabbit IgG.
- Antibody Rabbit Anti-cGMP (a polyclonal rabbit antibody to cGMP).
- cGMP-Alkaline Phosphatase Conjugate (alkaline phosphatase conjugated to cGMP).
- Wash buffer.
- Stop Solution.
- *p*-Nitrophenyl phosphate substrate solution.

2.4.3 Isotope Dilution Assay for cGMP Using GbpC

1. Cleared lysate of *Dictyostelium* cells. Collect starved cells by centrifugation for 3 min at 500 g; wash and resuspend to a density of 1.5×10^8 cells/mL in lysis buffer on ice (40 mM Hepes-NaOH, pH 7.0, 0.5 mM EDTA, 0.5 mM EGTA, 10 mg/mL protease inhibitor (crushed tablets). Lyse the cells through 3 μm nuclepore filter in the cold room (4 °C); clear by centrifugation for 3 min at 14,000 g and the supernatant for 1 h at 48,000 g.
2. Binding mixture (1 mL). 400 μL of 100 mM HEPES/NaOH pH 7.0, 5 μL of 100 mM EGTA, 6 μL of 1 M $MgCL_2$, 2.5 μL of ten-fold diluted stock [^{3}H]cGMP, 40 μL of 10 mg/mL denatured herring sperm DNA, and 546.5 μL of H_2O.
3. Wash buffer. 40 mM HEPES/NaOH, and 0.5 mM EDTA, pH 7.0.
4. Nitrocellulose filters 0.45 μm pore size (Whatman).

2.5 Guanylyl Cyclase

1. Starved *Dictyostelium* cells. Collect starved cells by centrifugation for 3 min at 500 g; wash and resuspend the cells in 15 mM Tris, pH 8.0, to a density of 2×10^8 cells/mL.
2. Lysis buffer: 15 mM Tris, 500 mM sucrose, and 6 mM EGTA, pH 8.0.
3. Lysis buffer (1 mL): 150 μL of 100 mM Tris, 500 μL of 1 M sucrose, 6 μL of 1 M EGTA pH 8.0, and 344 μL of H_2O.
4. Assay mixture (1 mL): 150 μL of 100 mM Tris, 250 μL of 1 M sucrose, 100 μL of 10 mM GTP, 20 μL of 1 M DTT, 50 μL of 0.1 M $MgSO_4$, and 430 μL of H_2O; prepare fresh from 1 M stock solutions that are stored at −20 °C.
5. 1 M DTT; store −20 °C.
6. 0.1 M GTP; store −20 °C.
7. 0.1 M GTPγS; store −20 °C.
8. PCA 3.5% (v/v); store 22 °C.

2.6 Degradation of cGMP by cGMP-Stimulated cGMP-Phosphodiesterase

1. Starved *Dictyostelium* cells. Collect starved cells by centrifugation for 3 min at 500 g; wash and resuspend the cells in 40 mM HEPES-NaOH and 0.5 mM EDTA pH 7.0 to a density of 2.0×10^8 cells/mL.
2. Lysis buffer (1 mL). 400 µL of 100 mM HEPES/NaOH pH 7.0, 500 µL of 1 M sucrose, 5 µL of 100 mM EGTA, and 95 µL of H_2O.
3. PDE assay mixture (1 mL). 400 µL of 100 mM HEPES/NaOH, pH 7.0, 250 µL of 1 M sucrose, 5 µL of 100 mM EGTA, 10 µL of 1 M $MgCl_2$, 10 µL of 1 M DTT, 10 µL of 10 mM cAMP, 2.5 µL of ten-fold diluted stock [^{3}H]cGMP, and 312.5 µL of H_2O.
4. Calf intestine phosphatase (CIP); Sigma; 10 units/mL in CIP buffer (50 mM Tris/HCl pH 7.9, 100 mM NaCl, 10 mM $MgCl_2$, and 1 mM DTT.
5. Dowex AG1X2; Sigma; 50% (vol/vol) slurry in H_2O: suspend Dowex AG1X2 in H_2O, let stand to settle the Dowex and decant liquid; mix equal amounts of Dowex and H_2O to make the 50% slurry.

2.7 cGMP-Binding to Target Protein GbpC

1. Starved *Dictyostelium* cells. Collect starved cells by centrifugation for 3 min at 500 g; wash and resuspend the cells in 40 mM HEPES-NaOH, pH 7.0, to a density of 1.5×10^8 cells/mL.
2. Lysis buffer (1 mL). 400 µL of 100 mM HEPES/NaOH pH 7.0, 5 µL of 100 mM EDTA, 5 µL of 100 mM EGTA, 100 µL of 100 mg/mL crushed tablets protease inhibitor mixture, and 490 µL of H_2O.
3. Binding mixture (1 mL). 400 µL of 100 mM HEPES/NaOH, pH 7.0, 5 µL of 100 mM EGTA, 6 µL of 1 M $MgCL_2$, 2.5 µL of tenfold diluted stock [^{3}H]cGMP, and 586.5 µL of H_2O.
4. Wash buffer. 40 mM HEPES/NaOH and 0.5 mM EDTA, pH 7.0.
5. Nitrocellulose filters 0.45 µm pore size (Whatman).
6. Protease inhibitor mixture: mini Complete-tablets and EDTA-free (Roche Applied Science).

3 Methods

In Subheadings 3.1 and 3.2, general methods are described for preparation of *Dictyostelium* cells and lysates, respectively. In the Subheadings 3.3, 3.4, 3.5, 3.6 and 3.7, methods are described to investigate cGMP signalling in vivo and in vitro. Proteins that comprise the cGMP signalling cascade and especially their regulation are unstable in vitro. It is important to complete the experiments as soon as possible after preparation of the cells and lysates. Therefore each assay is divided in (a) preparations that should be done before the cells are harvested, (b) the assay itself that should be completed in a short period, and (c) procedures done afterwards to finish the experiment.

3.1 Dictyostelium Culture Conditions

The strains that have been used in cGMP studies are wild-type NC4, axenic strains AX2 or AX3, several chemically mutagenized strains derived from NC4, and transformants derived mainly from AX3. Axenic strains are grown at 22 °C in liquid HG5 medium. Non-axenic strains are grown in co-culture with *Klebsiella aerogenes* on solid SM medium. Axenically grown strains are harvested in the late logarithmic phase and bacterially grown strains just before clearing of the bacterial lawn. Cells are washed three times in PB by repeated centrifugation at 500 g for 3 min and are resuspended in PB at the required cell density.

Most experiments are performed with starved aggregation-competent cells that become sensitive to the chemoattractant cAMP. Cells are generally starved by shaking in suspension for 4–6 h at a density of 10^7 cells/mL in PB at 150 rpm and 22 °C (*see* **Note 1**). Briefly starved cells are sensitive to the chemoattractant folate. For experiments with folate cells are starved in PB for 1 h.

3.2 Lysis of Cells to Assay for Guanylyl Cyclase, Phosphodiesterase, or cGMP-Binding Activity

Guanylyl cyclases, cGMP-phosphodiesterases, and the cGMP-binding activity in lysates are not very stable with half-times of a few minutes (guanylyl cyclase) or about 1 h, in part due to the high protease activity of *Dictyostelium*. The procedure is performed on ice, preferentially in the cold room to have cold pipets, tips, syringes, etc. Cells are lysed by pressing them through the 3 μm holes of a nuclepore filter. The nucleus cannot easily pass the holes thereby shearing the cell; importantly, other organelles such as lysosomes and endosomes easily pass the 3 μm holes and stay largely intact, especially if lysis is performed in isotonic buffer containing 250 mM sucrose (*see* **Note 2**). For intermediate volumes of cells (between 1 and 6 mL), a 25 mM-diameter nuclepore filter is mounted in a filter holder (Whatman) that is connected to a 10 mL syringe with luer-lock. For smaller volumes (between 100 μL and 1 mL), the 25 mM diameter nuclepore filter is cut in four parts, and one part is mounted carefully between the slip-tip of a 1 mL syringe and its needle. The syringes are filled with the appropriate cell suspension in lysis buffer, the plunger is mounted, the syringe is hold over a tube, and cells are lysed by controlled downward pressing the plunger.

3.3 cAMP-Induced cGMP Formation

Stimulation of aggregation-competent *Dictyostelium* cells with cAMP leads to the activation of guanylyl cyclase and the formation of cGMP in vivo. The guanylyl cyclase assay in vitro uses GTP as substrate that is converted to cGMP. The reactions are terminated at the desired time by addition of perchloric acid (*see* **Note 3**). After neutralization, the levels of cGMP are determined by one of the three assays described in Subheading 3.4.

Procedure

1. Prepare. Label 1.5 mL tubes and add 20 μL of stimulus solution. Add 100 μL of PCA to the t_0 samples.
2. Assay. Add 80 μL of the cell suspension to the tubes. Stop the reaction after 3, 6, 9, 12, 15, 20, and 30 s by the addition of 100 μL of PCA. Shake and place samples on ice for about 10 min or store at −20 °C.
3. Finish. Neutralize by adding 50 μL of $KHCO_3$ solution. Let stand to allow CO_2 to escape (shake carefully). Centrifuge 2 min at 14,000 g. Use 20 μL of the supernatant in the isotope dilution assays (see Subheading 3.4).

3.4 Determination of cGMP Levels

Isotope dilution assay has been used to determine cGMP levels in extracts from *Dictyostelium* cells. Classically the experiment was performed with a commercial radioimmunoassay (RIA) in which cGMP from the extract competes with radioactive [^{3}H]cGMP for binding to a limited number of binding sites on a cGMP-specific antibody (e.g., Amersham, [15]); the measured [^{3}H]cGMP that is bound to the antibody is inversely related to the amount of cGMP in the extract. For economic reasons, we have prepared our own antibody (AB99) that exhibits high affinity and specificity for cGMP [17]. Unfortunately, the commercial radioimmunoassay with [^{3}H]cGMP is no longer available. We have carefully analyzed other methods to measure cGMP levels. An alternative to the RIA of Amersham is a RIA with our own antibody AB99 or the commercial cGMP sheep antibody from Abcam. Another alternative is the non-radioactive Enzyme Immunoassay in which cGMP from the extract competes with cGMP-conjugated-alkaline-phosphatase for binding to a limited number of binding sites on a cGMP-specific antibody; the measured alkaline phosphatase activity that is bound to the antibody is inversely related to the amount of cGMP in the extract. A very different alternative is to use GbpC from *Dictyostelium* as the high-affinity cGMP-binding protein; cGMP from the extract competes with radioactive [^{3}H]cGMP for binding to a limited number of binding sites on GbpC. All these assays are described in detail below.

3.4.1 Procedure Radioimmunoassay (Amerham, Abcam, or Home-Made AB99)

1. Incubate at 0 °C in 1.5 mL tubes 20 μL of tracer, 20 μL of H_2O, standard or sample, and 20 μL of cGMP antiserum (Amerham, Abcam, or home-made AB99).
2. Terminate the incubation after 2 h by addition of 0.5 mL of 60% saturated ammonium sulphate. Incubate for 5 min.
3. Centrifuge tubes for 2 min at 14,000 g; aspirate the supernatant.
4. Finish by adding 100 μL of H_2O. Dissolve the pellet and add 1.3 mL of scintillator.

3.4.2 Procedure Enzyme Immunoassay Kit (Sigma-Aldrich)

1. Add to the wells of the provided microtiter plate 100 μL of assay buffer, 100 μL of H_2O/standard/or sample, 50 μL of cGMP-Alkaline Phosphatase Conjugate, and 50 μL of cGMP EIA Antibody. Incubate for 2 h at room temperature on a plate shaker at 500 rpm.
2. Terminate the incubation: Empty the contents of the wells and wash three times with 200 μL wash buffer. Carefully aspirate the last wash buffer completely.
3. Add 5 μL of cGMP-alkaline phosphatase conjugate and 200 μL of the *p*-nitrophenyl phosphate substrate solution to every well. Incubate at room temperature for 1 h without shaking. Add 50 μL of stop solution to every well.
4. Determine the optical density at 405 nM, preferentially in a plate reader.

*3.4.3 Procedure Using GbpC from Dictyostelium (See **Note 4**)*

1. Prepare fresh GbpC lysate.
2. Incubate for 20 min at 0 °C in 1.5 mL tubes: 50 μL of $[^3H]$ cGMP mixture, 50 μL of H_2O/cGMP standard/or sample, and 100 μL of GbpC lysate.
3. Terminate the incubation by filtering the incubation mixture through 0.45 μm nitrocellulose filter. Wash the filters twice with 3 mL ice-cold wash buffer and dry the filters.
4. Finish by putting the filters in scintillation vials and add 5 mL of scintillator.

3.4.4 Calculation

The output of the three assays is the radioactivity bound to the antibody or the GbpC or the absorbance of the enzyme assay. Each assay has tubes with no added cGMP (C_0), tubes with excess cGMP (Bl), and tubes with either known amounts of cGMP or unknown samples (C_x). The relationship between pmole ligand and measured C_x is given by the following equation:

$$\text{pmole} = X^*[(C_0 - Bl)/(C_x - Bl) - 1]$$

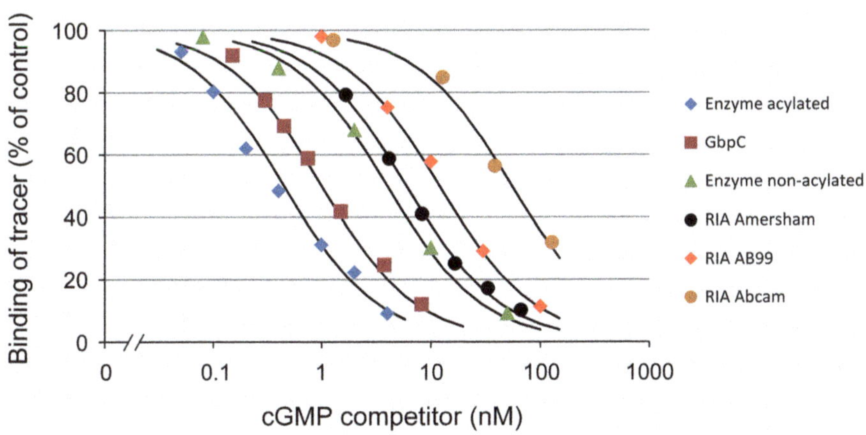

Fig. 1 cGMP assays. A cGMP-binding protein is incubated with tracer ([^{3}H]cGMP for RIA and GbpC or cGMP-alkaline phosphatase for enzyme) in the absence or presence of different concentrations of competitors (acetylated-cGMP for enzyme acetylated). The binding of [^{3}H]cGMP for RIA and GbpC assays was measured or the activity of alkaline phosphatase for enzyme assays. The data are means of three determinations

The value of X is determined from a standard curve with at the abscissa known amounts of ligand in pmole and at the ordinate the observed $(C_0 - Bl)/(C_x - Bl) - 1$. Then X is used to calculate the amounts of unknown ligand in the samples.

3.4.5 Results

A summary of the primary data of the assays is presented in Fig. 1 and Table 1. To develop an alternative to the radioimmunoassay of Amersham, we first determined the concentration of unlabeled cGMP that competes for binding [^{3}H]cGMP or cGMP-alkaline phosphatase to the specific binding-protein (Fig. 1). Each assay has a characteristic affinity for its ligand, which determines the absolute concentrations that can be determined (Table 1). Furthermore, each assay has a characteristic ratio of maximal/minimal binding (C_0/bl); these values determine the range of ligand concentrations that can be determined accurately. The original RIA from Amersham, the RIA with AB99, and the Enzyme-linked assay all have an affinity in the lower nM range and are well suited to detect cGMP in extracts of unstimulated and stimulated *Dictyostelium* cells. The RIA with Abcam antibody has an affinity that is too low for efficient detection of cGMP in extracts. Antibodies against cGMP are produced against succinylated-cGMP coupled with BSA. The commercial enzyme-linked assay kit contains the ingredients to acetylate cGMP in the extract, which increases the affinity for binding to the antibody. Indeed, the affinity for acetylated cGMP is about eightfold higher compared to unacetylated cGMP. Subsequently, we estimated the cGMP concentration in each assay using the known basal and stimulated levels of 1 and 10 pmoL/10^7 cells, respectively, to determine for each assay the optimal cell density during cell stimulation (Table 1). Cells were stimulated with cAMP

Table 1
Optimizing cGMP determinations using different assays

Property	Units	RIA Amersham	RIA AB99	RIA Abcam	Enzyme non-acylated	Enzyme acylated	GbpC
Volume assay	μL	60	60	60	200	200	200
Tracer		[3H]cGMP	[3H]cGMP	[3H]cGMP	cGMP-Alkaline phosphatase	cGMP-Alkaline phosphatase	[3H]cGMP
Competitor		cGMP	cGMP	cGMP	cGMP	Acetylated-cGMP	cGMP
(a) Properties from Fig. 1							
[cGMP] for 80% binding	nM	1.5	3	14	1	0.1	0.25
[cGMP] for 50% binding (Kd)	nM	6	12	55	4	0.45	1
[cGMP] for 20% binding	nM	24	48	220	16	1.8	4
(b) Design of optimal assay[a, b]							
Optimal cell density	cells/mL	2×10^8	2×10^8	2×10^8	1×10^8	1×10^7	5×10^5
Unstimulated cells, expected [cGMP] in assay	nM	2	2	2	1.6	0.16	0.4
Stimulated cells, expected [cGMP] in assay	nM	21	21	21	16	1.6	4
(c) Experimental result assay[c]							
Input [^{3}H]cGMP (cpm)	cpm	2000	2000	2000			2000
No cGMP (C0)	cpm or AU	620 ± 31	649 ± 39	678 ± 47	0.450 ± 0.018	0.350 ± 0.034	764 ± 48
Excess cGMP (bl)	cpm or AU	95 ± 7	33 ± 4	24 ± 5	0.068 ± 0.005	0.067 ± 0.007	54 ± 8
Dynamic rang (C0/bl)		6.5	20	28	6.6	5.2	14.3

(d) Experimental result cells[d]							
Unstimulated cells	cpm or AU	459 ± 18	565 ± 24	672 ± 66	0.337 ± 0.010	0.272 ± 0.036	590 ± 32
Stimulated cells	cpm or AU	208 ± 12	277 ± 21	471 ± 41	0.148 ± 0.008	0.133 ± 0.017	207 ± 17
cGMP of unstimulated cells	pmoL/10^7 cells	1.18 ± 0.18 (16%)	1.03 ± 0.34 (33%)	0.50 ± 3.0 (602%)	1.14 ± 0.14 (13%)	1.39 ± 0.78 (56%)	0.93 ± 0.22 (24%)
cGMP of stimulated cells	pmoL/10^7 cells	9.8 ± 1.4 (15%)	10.0 ± 1.3 (13%)	12.9 ± 3.8 (29%)	10.2 ± 1.2 (12%)	11.9 ± 3.9 (33%)	10.4 ± 1.4 (13%)

[a]The cGMP concentration inducing 80%, 50%, or 20% binding of the tracer is deduced from Fig. 1

[b]Experiment conditions are optimal when the cGMP levels in the samples will yield between 20% and 80% of binding of the tracer

[c]Data are means and SD of four incubations

[d]Cells were stimulated at the optimal cell density; data are the means and SD (and relative SD) of four experiments

at these optimal conditions, and the cGMP concentration was determined in cell extracts from unstimulated and stimulated cells (Table 1). All assays yield similar cGMP levels, except the RIA from Abcam; the observed cpm bound is too close to the C_0 and the assay becomes inaccurate and insensitive. For the other assays, the relative error for stimulated cells is around 14%, except for the enzyme assay with acetylated samples (33% error). In conclusion, a good alternative for the RIA from Amersham are the non-acetylated enzyme assay, GpbC, or our AB99. The cost for the assays is predominantly the enzyme kit (for 96 samples) and the radioactive [^{3}H]cGMP (for many samples with GbpC or AB99). Therefore, the enzyme assay is optimal for determination of a few samples, whereas GbpC of AB99 are optimal if thousands of samples have to be determined. The enzyme assay has the additional advantage that very low cGMP levels can be determined upon acetylation of the samples. This allows cGMP determination when the number of cells is limited (e.g., upon cell sorting or experiments with monolayers) or when the sample has to be diluted because it contains inhibitors of cGMP-binding to the binding protein.

3.5 Guanylyl Cyclase Assay

3.5.1 Principle

The activity of guanylyl cyclase can be measured with Mn^{2+}-GTP or Mg^{2+}-GTP as substrate. The intracellular concentrations of Mn^{2+} and Mg^{2+} in combination with the concentrations required to support enzyme activity indicate that Mg^{2+}-GTP is the physiological substrate. As with many mammalian adenylyl and guanylyl cyclase, the activity measured with Mn^{2+} represents the intrinsic cyclase activity, whereas with Mg^{2+}, the regulated activity can be measured that is modified by surface receptors, G-proteins, or Ca^{2+} ions (see [1]). Unfortunately, in *Dictyostelium* lysates, the Mg^{2+} GTP-dependent activity is very labile with a half-time of a few minutes, while the Mn^{2+}-GTP-dependent activity is more stable, indicating the rapid loss of regulation of guanylyl cyclase activity in lysates [17]. Therefore, the enzyme assay has to start immediately after cell lysis and has to be completed within 2 min after cell lysis. Experiments are performed in the cold room with a 22 °C water bath.

3.5.2 Procedure

1. Prepare. Label 1.5 mL tubes (three tubes per lysis for t_0 t_{30} t_{60} incubations). Add 20 μL of PCA to all tubes; add 20 μL of assay mixture to t_0 incubations; place tubes on ice. Prepare 1 mL syringes with nuclepore membrane adjusted between the tip and the needle.
2. Assay. Place a tube with 50 μL assay mixture in the water bath at 22 °C. Mix 100 μL of cells with 100 μL of lysis buffer at 0 °C. Immediately lyse the cells by pressing them through a nuclepore filter with 3 μm pores. Collect the lysates in tubes at 0 °C. Start the guanylyl cyclase assay at 30 s after lysis by addition of

50 μL of lysate to the tube in the water bath containing 50 μL of assay mixture. Transfer 40 μL of the incubation mixture at 30 and 60 s after the start of the incubation to tubes labeled t_{30} and t_{60}, respectively. Add 20 μL of lysate to tube labeled t_0 (*see* **Note 5**). Repeat the experiment with the next lysis/incubation. Store the samples in PCA at −20 °C.

3. Finish. Neutralize the samples with 10 μL KHCO3, and assay cGMP levels using the isotope or enzyme assay (see Subheading 3.4).

3.5.3 Results

Regulators of the enzyme (such as GTPγS) can be included in the lysis buffer so that the enzyme is immediately activated after cell lysis. This requires a new lysate for each experiment, and the control (no regulator) and experimental condition are made with two independent lysates. Alternatively, the regulator can be included in the assay mixture, which allows to incubate the same lysate without and with the regulator; obviously this requires more pipetting in the same amount of time (see legend of Table 2).

Table 2
Guanylyl cyclase assay

Condition		cpm	pmoL cGMP/assay	pmoL/min/10^7 cells
	t_0	529 ± 34	0.18 ± 0.06	
Mn^{2+}-GTP	t_{30}	205 ± 25	1.98 ± 0.37	
	t_{60}	142 ± 15	3.67 ± 0.59	89 ± 14
Mg^{2+}-GTP	t_{30}	287 ± 42	1.23 ± 0.37	
	t_{60}	176 ± 22	2.76 ± 0.52	55 ± 17
Mg^{2+}-GTP + GTPγS	t_{30}	189 ± 19	2.26 ± 0.35	
	t_{60}	124 ± 18	4.69 ± 1.14	108 ± 22

cGMP levels were measured with AB99 antibody (C0 = 650 cpm, bl = 40 cpm). Data shown are the means and SD, $n = 4$
In this experiment, the lysate was incubated simultaneously with three assay mixtures (A) Mn^{2+}-GTP, (B) Mg^{2+}-GTP, and (C) Mg^{2+}-GTP + GTPγS. The actions are the following:
At t = −60 s place three tubes with 50 μL A, B, and C in the water bath
At t = −40 s mix 150 μL cells with 150 μL lysis buffer and lyse through nuclepore filter
At t = −10 s add 20 μL lysates to a tube containing 20 μL PCA for the t_0
At t = 0 s add 50 μL lysates to tube A
At t = 10 s add 50 μL lysates to tube B
At t = 20 s add 50 μL lysates to tube C
At t = 30 s pipet 40 μL from incubations A to a tube containing PCA
At t = 40 s pipet 40 μL from incubations B to a tube containing PCA
At t = 50 s pipet 40 μL from incubations C to a tube containing PCA
At t = 60 s pipet 40 μL from incubations A to a tube containing PCA
At t = 70 s pipet 40 μL from incubations B to a tube containing PCA
At t = 80 s pipet 40 μL from incubations C to a tube containing PCA
The timing before 0 s is approximate, but from 0 s the timing is exact

Primary data of the guanylyl cyclase assay are presented in Table 2. In this experiment, a lysate is incubated simultaneously with Mn^{2+}-GTP, Mg^{2+}-GTP, and Mg^{2+}-GTP + GTPγS. The levels of cGMP were measured by radioimmunoassay using our AB99. Basal enzyme activity is easily detected and is much higher with Mn^{2+}-GTP than with Mg^{2+}-GTP as substrate. Furthermore, GTPγS produces significant activation of guanylyl cyclase in lysates of *Dictyostelium*.

Dictyostelium possesses two guanylyl cyclases: the twelve transmembrane GCA that is similar to mammalian membrane-bound adenylyl cyclases and the soluble sGC that is not the homolog of mammalian soluble guanylyl cyclase but is more similar to soluble adenylyl cyclase involved in sperm maturation [3]. The assay described above was used to characterize these two guanylyl cyclases. Both enzymes appear to be activated by cAMP and GTPγS, but activation of sGC is generally stronger than activation of GCA.

3.6 cGMP-Phosphodiesterase Assay

3.6.1 Principle

In vivo cGMP is degraded for about 60% by the cytosolic cGMP-stimulated cGMP-specific phosphodiesterase PDE5, for about 20% by the cytosolic cGMP-specific PDE3, while about 20% of cGMP is secreted followed by degradation by the cell surface nonspecific PDE1 [5, 18]. Below we show the assay to measure cGMP degradation in vitro and present some results for the predominant PDE5.

3.6.2 Procedure

1. Prepare. Label 1.5 mL tubes. Add 50 μL of [^{3}H]cGMP assay mixtures to all tubes and 20 μL of H2O, cGMP, 8Br-cGMP, or other compounds to be investigated.
2. Assay. Add 30 μL of lysate; incubate for 15 min at 22 °C. Stop the reaction by boiling for 1 min. Store at −20 °C or continue.
3. Finish. Add 100 μL CIP (calf intestine phosphatase) and incubate for 1 h at 37 °C. Then add 300 μL of a 50% slurry of Dowex AG1X2. Incubate for 15 min at room temperature, centrifuge for 2 min at 14,000 g, and determine the radioactivity in 200 μL of the supernatant.

3.6.3 Results

The phosphodiesterase assay is linear with incubation time until about 40% of the substrate is hydrolyzed. The input is about 18,000 cpm [^{3}H]cGMP; since about half of the incubation mixture is used at the end of the assay to determine hydrolyzed [^{3}H]cGMP, the maximal conversion is about 9000 cpm, and 40% conversion is 3600 cpm. Figure 2 shows the typical results of an experiment where the partially purified cGMP-stimulated cGMP-phosphodiesterase is incubated with 10 nM [^{3}H]cGMP in the presence of increasing amounts of non-radioactive cGMP or cGMP analogs. The samples with only [^{3}H]cGMP yield 1306 ± 75 cpm for active enzyme and 382 ± 62 cpm for boiled enzyme (means and SD with $n = 4$). Low concentrations of cGMP stimulate the hydrolysis of

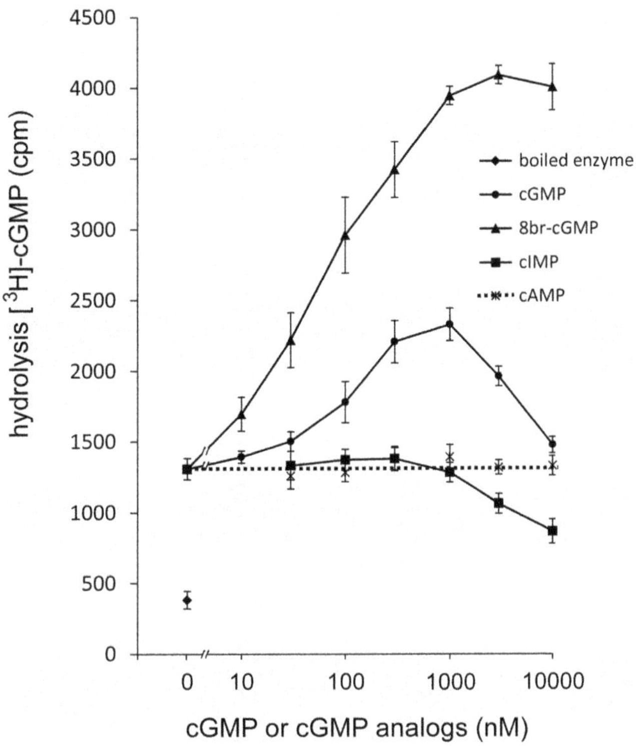

Fig. 2 cGMP phosphodiesterase assay. Partially purified cGMP-stimulated cGMP-PDE was incubated for 30 min with 10 nM [^{3}H]-cGMP (about 18,000 cpm) in the absence or presence of different concentrations of cGMP or cGMP analogs. The data shown are the means and SD of three determinations

[^{3}H]cGMP up to 2328 ± 115 cpm at 1 μM cGMP, whereas higher cGMP concentrations inhibit [^{3}H]cGMP hydrolysis due to competition of cGMP with [^{3}H]cGMP for binding to the catalytic site. The analog 8Br-cGMP strongly activates the hydrolysis of [^{3}H] cGMP while high concentrations do not inhibit [^{3}H]cGMP hydrolysis. On the other hand, cIMP never activates the enzyme, while high concentrations effectively inhibit [^{3}H]cGMP hydrolysis. Finally, cAMP neither activates nor inhibits [^{3}H]cGMP hydrolysis at any concentration. The experiment reveals that the phosphodiesterase assay is relatively simple and robust.

3.7 cGMP-Binding Protein GbpC

3.7.1 Principle

GbpC is the only high-affinity binding protein in *Dictyostelium*. The protein has a complex domain structure and consists of (from N- to C-terminus) a LRR domain, a Roc domain, a dimerization COR domain, a kinase domain, a Ras-GEF domain, and two cGMP-binding domains. Experiments with several mutant proteins suggest an intracellular activation mechanism in which cGMP-binding activates the GEF domain that leads to the exchange of GDP for GTP in the Roc domain, thereby activating the kinase domain. The protein is most easily detected by its cGMP-binding activity (*see* **Note 6**).

3.7.2 Procedure

1. Prepare. Label 1.5 mL tubes. Add 50 μL of [^{3}H]cGMP assay mixture, 50 μL of H_2O, cGMP, or other components of interest.
5. Assay. Add 200 μL of lysate; incubate for 20 min at 0 °C. Stop the incubation by filtering the incubation mixture through 0.45 μm nitrocellulose filter. Wash the filters twice with 3 mL ice-cold wash buffer and dry the filters.
6. Finish by putting the filters in scintillation vials and add 5 mL of scintillator.

3.7.3 Results

The binding of [^{3}H]cGMP to GbpC is shown in Table 1, where GbpC was used as binding protein in an isotope dilution assay. Typical binding data for [^{3}H]cGMP are 764 ± 48 cpm in the absence of unlabeled cGMP and 54 ± 8 cpm in the presence of excess unlabeled cGMP (means and SD, $n = 4$). By measuring [^{3}H]cGMP binding in the presence of different amounts of cGMP, the affinity and number of binding sites can be determined using a Scatchard plot. The specificity of the [^{3}H]cGMP-binding pocket of GbpC has been determined by including different concentrations of cGMP analogs in the [^{3}H]cGMP-binding mixture [19].

4 Notes

1. Some mutants develop better when starved on non-nutrient agar (2.5×10^6 cells/cm^2). These plates are incubated at 22 °C for 4–6 h or at 8 °C for 16 h followed by shaking the cells for an additional hour at 22 °C at a density of 10^7 cells/mL in PB.
2. Sucrose helps to keep the organelles intact (which is especially relevant for lysosomes). However, cell lysis becomes increasingly more difficult when cells are resuspended in sucrose-containing buffer for prolonged periods of time (more than 2 min).
3. The assay described above uses independent stimulations. This procedure is optimal for the determination of the magnitude of the response. In experiments where the time course is more important than the magnitude of the response, we stimulate 900 μL of cells with 100 μL of stimulus (twice as concentrated as described), and at the desired times, 100 μL samples are transferred to tubes containing 100 μL of PCA. In these experiments, the t_0 sample is taken just before stimulation. In order to catch the early time points, cells are stimulated while vortexed, and the same pipette is used to add the stimulus and to draw the samples.

4. The assay follows the protocol for cGMP-binding to GbpC (*see* Subheading 3.7), with the exception that denatured sperm DNA is included in the assay mixture. Denatured sperm DNA strongly reduces the dissociation of the [^{3}H]cGMP-GbpC complex and thereby increases the affinity of GbpC for cGMP and the sensitivity of the assay about threefold. This slow dissociation also facilitates the assay because the timing of washing steps becomes less critical.
5. This sample is the zero-time incubation for basal cGMP levels in the lysate. Preferentially, this sample is collected around the time when the incubation with substrate is started.
6. Studies on the mechanism of GbpC require several assays, including cGMP-binding, GDP/GTP-exchange/activation of the Ras domain, GTPase activity/inactivation of the Ras domain, and kinase activity of the kinase domain. These assays are not always possible with the full-length protein and have been performed with expressed domains; see [13] for details.

References

1. van Haastert PJM, Keizer-Gunnink I, Pots H, Ortiz-Mateos C, Veltman D, van Egmond W et al (2021) Forty-five years of cGMP research in Dictyostelium: understanding the regulation and function of the cGMP pathway for cell movement and chemotaxis. Mol Biol Cell 32: ar8
2. Roelofs J, Snippe H, Kleineidam RG, Van Haastert PJM (2001) Guanylate cyclase in Dictyostelium discoideum with the topology of mammalian adenylate cyclase. Biochem J 354: 697–706
3. Roelofs J, Van Haastert PJM (2002) Characterization of two unusual guanylyl cyclases from dictyostelium. J Biol Chem 277:9167–9174
4. Roelofs J, Meima M, Schaap P, Van Haastert PJM (2001) The Dictyostelium homologue of mammalian soluble adenylyl cyclase encodes a guanylyl cyclase. EMBO J 20:4341–4348
5. Bader S, Kortholt A, Van Haastert PJM (2007) Seven Dictyostelium discoideum phosphodiesterases degrade three pools of cAMP and cGMP. Biochem J 2402:153–161
6. Bosgraaf L, Russcher H, Snippe H, Bader S, Wind J, Van Haastert PJM (2002) Identification and characterization of two unusual cGMP-stimulated phoshodiesterases in dictyostelium. Mol Biol Cell 13:3878–3889
7. Meima ME, Biondi RM, Schaap P (2002) Identification of a novel type of cGMP phosphodiesterase that is defective in the chemotactic stmF mutants. Mol Biol Cell 13: 3870–3877
8. Kuwayama H, Snippe H, Derks M, Roelofs J, Van Haastert PJM (2001) Identification and characterization of DdPDE3, a cGMP-selective phosphodiesterase from Dictyostelium. Biochem J 353:635–644
9. Bosgraaf L, Russcher H, Smith JL, Wessels D, Soll DR, Van Haastert PJM (2002) A novel cGMP signalling pathway mediating myosin phosphorylation and chemotaxis in Dictyostelium. EMBO J 21:4560–4570
10. Van Egmond WN (2010) Biochemical and functional aspects of GbpC and other Roco proteins in Dictyostelium discoideum. PhD thesis, Groningen University
11. Goldberg JM, Bosgraaf L, Van Haastert PJM, Smith JL (2002) Identification of four candidate cGMP targets in Dictyostelium. Proc Natl Acad Sci USA 99:6749–6754
12. Kortholt A, Bolourani P, Rehmann H, Keizer-Gunnink I, Weeks G, Wittinghofer A et al (2010) A rap/phosphatidylinositol 3-kinase pathway controls pseudopod formation. Mol Biol Cell 21:936–945
13. van Egmond WN, Kortholt A, Plak K, Bosgraaf L, Bosgraaf S, Keizer-Gunnink I, van Haastert PJ (2008) Intramolecular activation mechanism of the Dictyostelium LRRK2 homolog Roco protein GbpC. J Biol Chem 83:30412–30420

14. Kortholt A, Van Egmond WN, Plak K, Bosgraaf L, Keizer-Gunnink I, Van Haastert PJM (2012) Multiple regulatory mechanisms for the Dictyostelium Roco protein GbpC. J Biol Chem 287:2749–2758
15. Van Haastert PJM (2006) Analysis of signal transduction: formation of cAMP, cGMP, and ins(1,4,5)P3 in vivo and in vitro. Methods Mol Biol 346:369–392
16. Sussman R, Sussman M (1967) Cultivation of Dictyostelium discoideum in axenic medium. Biochem Biophys Res Commun 29:53–55
17. Janssens PM, De Jong CC, Vink AA, Van Haastert PJM (1989) Regulatory properties of magnesium-dependent guanylate cyclase in Dictyostelium discoideum membranes. J Biol Chem 264:4329–4335
18. van Haastert PJM, van Lookeren Campagne MM, Kesbeke F (1983) Multiple degradation pathways of chemoattractant mediated cyclic GMP accumulation in Dictyostelium. Biochim Biophys Acta 756:67–71
19. van Haastert PJ, van Walsum H, van der Meer RC, Bulgakov R, Konijn TM (1982) Specificity of the cyclic GMP-binding activity and of a cyclic GMP-dependent cyclic GMP phosphodiesterase in Dictyostelium discoideum. Mol Cell Endocrinol 25:171–182

Chapter 14

Quantitative Monitoring of GPCR-Mediated Spatiotemporal IP_3 Dynamics Using Confocal Fluorescence Microscopy

Xuehua Xu, HyunGee Ha, Joseph Brzostowski, and Tian Jin

Abstract

Activation of G protein-coupled receptors upon chemoattractant stimulation induces activation of multiple signaling pathways. To fully understand how these signaling pathway coordinates to achieve directional migration of neutrophils, it is essential to determine the dynamics of the spatiotemporal activation profile of signaling components at the level of single living cells. Here, we describe a detailed methodology for monitoring and quantitatively analyzing the spatiotemporal dynamics of 1,4,5-inositol trisphosphate (IP_3) in neutrophil-like HL60 cells in response to various chemoattractant fields by applying Förster resonance energy transfer (FRET) fluorescence microscopy.

Key words Chemotaxis, Neutrophils, G protein-coupled receptors (GPCRs), Phospholipase C (PLC), 1,4,5-inositol trisphosphate (IP_3), Confocal fluorescence microscopy, Förster resonance energy transfer (FRET)

1 Introduction

Chemotaxis, chemoattractant gradient-directed cell migration, plays critical roles in many physiological processes, such as recruitment of neutrophils to the sites of inflammation, neuron patterning, immune response, angiogenesis, metastasis of cancer cells, and development of the model organism *Dictyostelium discoideum*. Eukaryotic cells detect chemoattractants by various receptors, including G protein-coupled receptors (GPCRs). The engagement of various chemoattractants with their GPCRs activates heterotrimeric G protein, which directly or indirectly activates phospholipase C (PLC) in *D. discoideum* [1, 2] or β2/β3 isoforms of phospholipase C (PLCβ2/β3) in neutrophils [3–6]. Recently, we found that chemoattractant stimulation also triggers membrane recruitment and subsequent activation of PLCγ2, the most abundantly expressed PLC isoform in neutrophils, that plays an essential role in PLC signaling and chemotaxis [7, 8]. All the above indicates

Alan R. Kimmel (ed.), *Dictyostelium discoideum: Methods and Protocols*, Methods in Molecular Biology, vol. 2814, https://doi.org/10.1007/978-1-0716-3894-1_14,

a complex in chemoattractant-induced PLC activation. Activation of PLC isoforms generates two potent second messengers, diacylglycerol (DAG) and 1,4,5-inositol trisphosphate (IP_3). Both mediate multiple signaling pathways pivotal for the functions of neutrophils [6, 7, 9, 10].

IP_3 initiates calcium release (calcium response) from endoplasmic reticulum (ER) stores followed by a later calcium influx into cells [11]. Calcium response in diverse cell types plays critical roles in cell migration and chemotaxis [12–15]. However, the roles and relationship among PLC, IP_3, and subsequent calcium response in the chemotaxis of *D. discoideum* and mammalian neutrophils largely remain unclear. In *D. discoideum*, the deletion of *plc* gene in *D. discoideum* has little effect on chemoattractant cAMP-mediated chemotaxis nor on folic acid-mediated chemotactic response [1, 2]. Deletion of IP_3 receptor (*iplA*$^-$) diminishes both cAMP and FA-induced calcium response in the cells, while causes no effect on cAMP-mediated chemotaxis in *D. discoideum* [16, 17], leading to an assumption that calcium signaling plays no role in chemoattractant-mediated chemotaxis. However, recent findings indicate that some key players require calcium response for their proper localization and subsequent functions in chemotaxis [18, 19]. Similarly, double deletion of PLCβ2/β3 in mouse causes dramatic reduction of chemoattractant-induced IP_3 production, while it has no effect on chemotaxis capability of murine neutrophils and macrophages [6], leading to a same conclusion that calcium signaling plays no role in neutrophil chemotaxis. However, a complete inhibition of PLC signaling impairs chemotaxis capability of human neutrophils [20]. More importantly, key molecules, such as PKB and GSK3, require PLCβ2/β3 for their proper functions in neutrophil chemotaxis [21]. Recently, we also found that calcium response is required for membrane targeting of CAPRI, a key regulator of Ras activation in neutrophil chemotaxis [15]. Importantly, we found that chemoattractant-mediated membrane translocation and subsequent activation of PLCγ2 (the most abundantly expressed PLC isoform in human neutrophils) are required for proper IP_3 production, calcium signaling, and subsequent chemotaxis in human neutrophils [8]. To understand the role of PLC signaling in cell migration, effort has been made to measure the tempospatial dynamics of DAG production and calcium responses at single cellular level in both human neutrophils and *D. discoideum* [8, 20]. It has become increasingly important to measure chemoattractant-mediated spatiotemporal IP_3 dynamics in single live cells. Biochemical assays were developed to measure IP_3 production [6]. In recent years, probes using IP_3 receptors that bind IP_3 in physical conditions to measure IP_3 production have been developed [22–24]. Here, we describe how to apply confocal microscopy to monitor and measure IP_3 dynamics in a single live neutrophil-like HL60 cell using an intramolecular Förster resonance energy transfer (FRET) probe in the hope of a similar probe that can be developed and applied in the model organism *D. discoideum*.

2 Materials

2.1 HL60 Cell Culture

1. RPMI 1640 medium GlutaMAX™ Supplement (Life Technologies Corp., Grand Island, NY).
2. Fetal bovine serum (FBS) (Gemini Bio-Products, West Sacramento, CA).
3. HEPES 1 M sterile solution, pH 7.3 (Quality Biological Inc. Gaithersburg, MD).
4. RPMI 1640 culture media (RPMI 1640 medium GlutaMAX™ Supplement with 10% FBS and 25 mM HEPES).
5. RPMI 1640 transfection culture medium (RPMI 1640 medium GlutaMAX™ Supplement with 20% FBS and 25 mM HEPES).
6. RPMI 1640 starvation medium (RPMI 1640 medium GlutaMAX™ Supplement with 25 mM HEPES).
7. T75 flask (BD Biosciences, Falcon, Bedford, MA).
8. HL-60 cell line (ATCC. Manassas, VA).
9. Cell culture incubator, 37 °C, 5% CO_2.
10. Cell culture hood (NuAir Class II biological safety cabinet type A/type B3, Plymouth, MN).
11. N-Formyl-Met-Leu-Phe (fMLP) (Sigma, St. Louis, MO). A 1 mM stock is made in DMSO and frozen at −20 °C in 10 μL aliquots. The working concentration of fMLP is 100 nM to 1 μM.

2.2 Plasmids for FRET Measurements

1. Plasmids encoding Mem-cerulean (single donor control) and Venus (single acceptor control).
2. Plasmid encoding IP_3 FRET probe (Péter Várnai, Semmelweis University).

2.3 Transfection of HL60 Cells

1. Cell Line Nucleofector® Kit V (Lonza, Basel, Switzerland).
2. Plastic bulb sterile pipette included in Nucleofector® Kit (Lonza, Basel, Switzerland).
3. 1.6 mL tube (Neptune Scientific, San Diego, CA).

2.4 Coating Cover Glass Surface of Four-Well and Single-Well Chamber

1. Gelatin-based coating solution (Cell biologics, Chicago, IL).
2. DEPC-treated water (Quality Biological Inc., Gaithersburg, MD).
3. Single- and four-well Lab-Tek II cover glass chambers (Nalge Nunc International, Naperville, IL).

2.5 Imaging with Zeiss LSM 710 Microscope

1. Four-well Lab-Tek II cover glass chambers (Nalge Nunc International, Naperville, IL).
2. FemtoJet microcapillary pressure supply (Eppendorf, AG, Hamburg, Germany).
3. TransferMan NK2 micromanipulator (Eppendorf, AG, Hamburg, Germany).
4. LSM 710 (Zeiss, Jena, Germany) or equivalent with a 40X 1.3 NA objective lens.
5. Other microscope accessories:
   (a) Heating Insert P Lab-Tek S1 (PeCon, GmbH, Erbach, Germany).
   (b) TempModule S1 (PeCon, GmbH, Erbach, Germany).
   (c) Zen Black Software (Zeiss, Jena, Germany).
   (d) Carl Zeiss Immersol 518 F (Zeiss, Jena, Germany).
   (e) Lasos Ar-Ion Laser Series (Zeiss, Jena, Germany, Model LGN 4000, Remote Control RMC 7812 Z2).
   (f) Controller pad (Wienecke & Sinske GmbH, Gleichen, Germany).

3 Methods

3.1 Cell Culture

1. Culture HL60 cells in RPMI 1640 medium with 10% FBS and 25 mM HEPES.
2. Passage the cells every 2 or 3 days (*see* **Note 1**).

3.2 Coating the Cover Glass Surface of the Four-Well or One-Well Chamber

1. Dilute gelatin-based coating solution with twice volume of DEPC-treated water to make the coating solution.
2. Add 1 mL of the coating solution to the two middle wells of a four-well or a one-well chamber bottomed with cover glass. Incubate the chamber at 37 °C for 1 h.
3. Just before seeding cells, remove the coating solution from the chambers (*see* **Note 2**).

3.3 Transfection of HL60 Cells

1. Differentiate HL60 cells in 1.3% DMSO for 5 days. Measure cell density before transfection (*see* **Note 3**).
2. Centrifuge the cells at 200 × *g* for 5 min.
3. After removing the supernatant, suspend the cells in the transfection reagent according to the manufacturer's recommendation (usually two million cells in 100 μL of reagent per transfection), and add the mixture to an electroporation cuvette.

4. Use the manufacturer's program T-019 to perform electroporation.
5. Without delay, gently add 500 μL of RPMI 1640 transfection culture medium to the electroporated cells in the cuvette.
6. Gently transfer the cells from the cuvette to a 1.6 mL tube with the provided plastic pipettes.
7. Place the tube into a cell culture incubator for 30 min.
8. Seed a 100 μL or 500 μL volume of cells at 3–5 × 10^5 cells/mL into pre-coated wells of four-well or one-well chambers, respectively.
9. Add culture medium up to a final volume of 1 mL or 4 mL to one well of a four-well chamber or a one-well chamber, respectively.
10. After allowing the cells to recover for 4 h in the incubator, gently remove the culture medium from the well of a chamber, and add RPMI 1640 starvation medium (*see* **Note 4**).
11. After an additional 30 min, the cells are ready for imaging experiments.
12. Meanwhile, prepare 100 nM or 1 μM fMLP stimuli.

3.4 Application of Various Fields of Chemoattractant Stimulation

Chemotactic cells respond to various fields of chemoattractant gradient differently. Two different types of chemoattractant fields can be used. Uniform stimulation, that is, applying chemoattractant homogeneously to the cell periphery, is often used to determine the temporal features of a cellular response. In order to determine the spatial dynamics of a cellular response, chemoattractant gradients can be applied. Several different methods have been developed to introduce a chemoattractant gradient to cells, such as the Zegman chamber or the needle (microinjector) assay [25, 26]. Recently, advanced microfluidic chambers and ez-TaxiScan have been developed to facilitate simultaneous monitoring of multiple assays [27, 28]. We have also developed novel methods for visualizing and quantitatively measuring the temporal and/or spatial features of chemoattractant fields [29, 30]. In the current protocol, we add a differential interference contrast (DIC) channel to the FRET imaging acquisition settings to monitor the application of chemoattractant along with the FRET imaging. Below, we describe how to apply two different kinds of chemoattractant fields to neutrophil-like cells:

1. Sudden exposure of a cell to a uniform fMLP stimulation. To generate a uniform stimulation, 200 μL of a mixture of fMLP is applied to cells covered with 800 μL of buffer and placed in a four-well chamber. Under this condition, cells are normally exposed to a uniform fMLP stimulation that is about 2.5 times the final concentration (*see* **Note 5**).

2. Exposure of a cell to a steady fMLP gradient. To establish an fMLP gradient, a micropipette (Eppendorf Femtotip) is filled with 30 μL of fMLP solution and then attached to a FemtoJet and micromanipulator 5171 (Eppendorf). The FemtoJet is set at Pc = 30 and Pi = 30 to inject a constant tiny volume of a mixture of fMLP into a one-well chamber that is covered with 5 mL of buffer. Under this condition, a gradient can be established within 100 mm around the tip of the micropipette, and the gradient is usually stable for more than 1 h (*see* **Note 6**).

3.5 Sensitized Emission FRET Measurement with Confocal Microscopy

Sensitized emission FRET is used to measure actual FRET efficiency. Like spectrally resolved FRET measurement, sensitized emission FRET is widely used to measure protein-protein interactions and/or conformational changes of proteins, especially in fixed cells or tissue samples in which dynamic FRET measurement might not be possible [31] (Fig. 1A). Sensitized emission FRET measurement requires three types of cells: donor cells containing only the donor fluorophore, receptor cells containing only the acceptor fluorophore, and FRET cells containing both donor and acceptor fluorophores. FRET measurement starts with the acquisition of images in three channels under Channel Mode (Fig. 1B):

1. *Donor channel*: Excitation with the donor's excitation wavelength and detection with donor's parameter settings.
2. *FRET channel*: Excitation with donor's excitation wavelength and detection with acceptor's parameter settings.
3. *Acceptor channel*: Excitation with acceptor's excitation wavelength and detection with acceptor's parameter settings.

The two single-labeled samples provide the reference values used to correct the optical bleed-through of the donor (donor emission bleeds into acceptor emission) and the possible excitation of the acceptor through the wavelength of the donor. The third sample is the actual FRET sample on which the measurement is performed.

3.5.1 Acquisition Configuration of Sensitized Emission FRET Measurement by Applying Cerulean/Venus FRET Pairs

Sensitized emission FRET measurement calculates the breed through of donor and cross talk of acceptor which is obtained in cells expressing single fluorophore, as an example here are cells expressing cerulean or Venus, respectively. The key point of sensitized FRET acquisition is that all the imaging parameters (laser output, zoom, pinhole, detector gain, amplifier offset, and so on) should be kept constant.

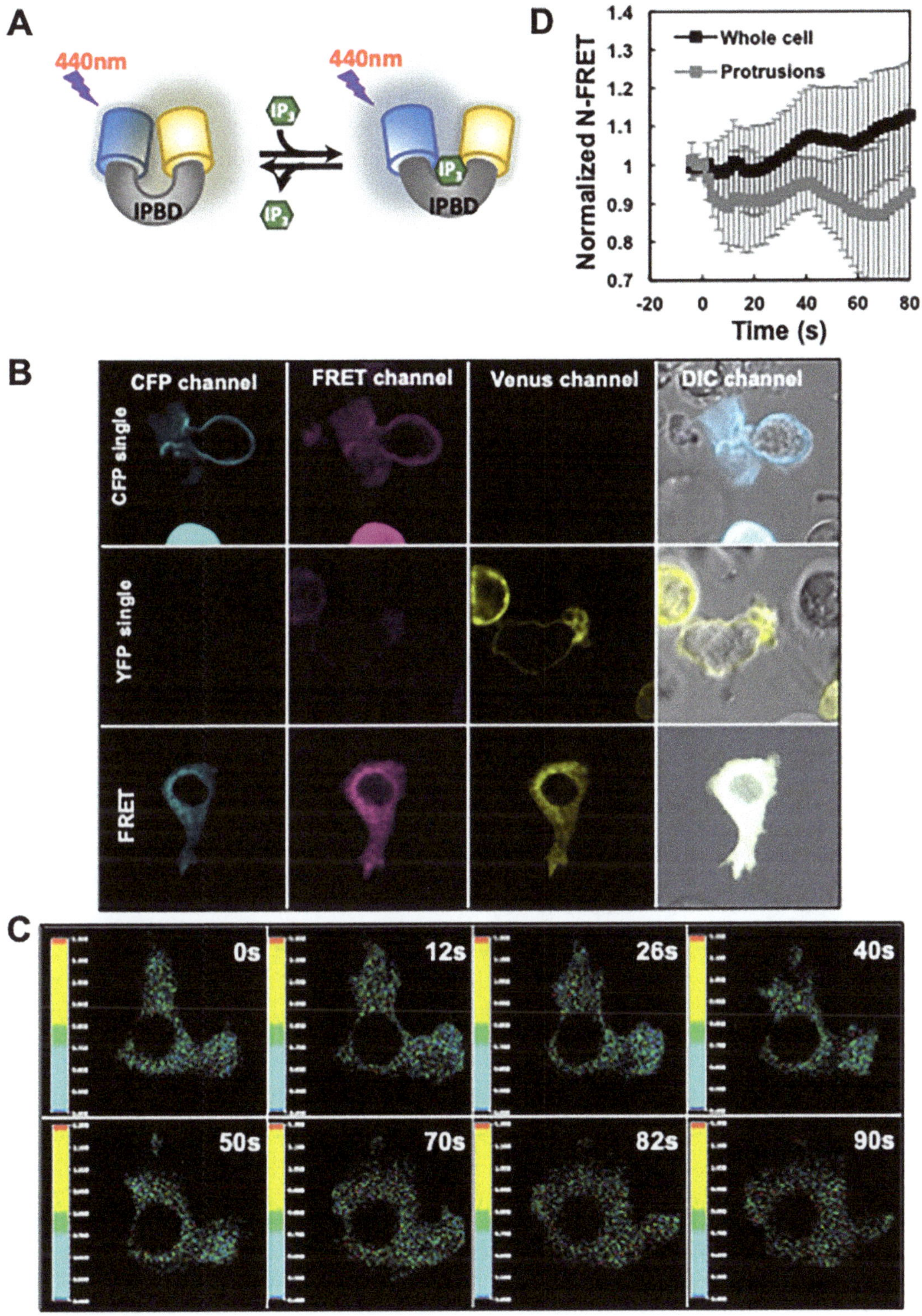

Fig. 1 GPCR-mediated IP_3 production in HL60 cells in response to uniformly applied fMLP stimulation measured by sensitized emission FRET. (**a**) Scheme shows the principle of the applied FRET probe for IP_3. An IP_3 receptor is inserted between the FRET pair cerulean and Venus. (**b**) Montage shows the cells expressing

1. Set the system to the channel acquisition mode and configure the donor cerulean channel, the FRET channel, and the acceptor Venus channel as follows:
   (a) *Cerulean channel:* Excitation with 440 nm laser and application of NFT 515 nm to split the signal and detection with parameter settings of the donor (BP 454–507 nm) (*see* **Note 7**)
   (b) *FRET channel:* Excitation with 440 nm laser and detection with parameter settings of the acceptor (BP 526-592 nm)
   (c) *Venus channel:* Excitation with 514 nm laser and detection with parameter settings of the acceptor (BP 526–592 nm)
2. Adjust the pinhole size for each channel to be an optical slice of the specific type of cells. Here, we set it to about 1–1.5 μm.
3. To image single cerulean expressing cells, adjust the detector gain and amplifier offset of the donor and acceptor channels to obtain a decent signal (Fig. 1b, upper panel). The imaging parameters should be kept constant. The FRET channel will show the bleed-through from the donor cerulean emission channel using donor cerulean excitation wavelength.
4. For single Venus-expressing cells, the FRET channel will show the cross talk from the acceptor Venus emission channel using the donor cerulean excitation wavelength (Fig. 1b, middle panel).
5. For cerulean/Venus double-expressing cells, select those with both cerulean and Venus expression (Fig. 1b, lower panel) (*see* **Note 8**).
6. Add a differential interference contrast (DIC) channel to the above settings to indicate when the stimulation is applied to the cells (Fig. 1b).
7. Set up time-lapse acquisition as detailed in **step 1**.
8. After two or three scans, apply stimuli homogenously to the cells and continue the acquisition. The contrast change in the DIC channel indicates when the stimuli was added to the cells.

Fig. 1 (continued) cerulean single control, Venus single control, or IP_3 FRET probe. Images acquired by cerulean channel, FRET channel, or Venus channel for sensitized emission FRET measurement. A differential interference contrast (DIC) channel is also shown to monitor the application of uniformly applied fMLP stimulation. (**c**) Montage shows the N-FRET images of the IP_3 probe in the cells at the indicated time points after being stimulated by 100 nM uniform fMLP stimulation. Images were obtained by Zeiss LSM 710, analyzed using Zen Black software, and further modified for a better presentation using Adobe photoshop. (**d**) Graph shows the dynamic change in the normalized N-FRET of the IP_3 probe upon uniformly applied fMLP stimulation. Mean ± SD is shown, n = 10

3.5.2 Quantitative Calculation of Sensitized Emission FRET Using Zeiss Zen Black FRET

We followed the data analysis procedures of Zeiss Zen Black FRET. During the data analysis, we found that the following key points should be carefully followed for accurate data analysis:

1. Be aware that whole sets of data for sensitized emission FRET data analysis should be acquired with the same acquisition parameters.
2. To set up the background threshold of donor single, acceptor single, or FRET samples, choose areas without any signal.
3. To set up the bleed-through ratio in donor/FRET/acceptor channels, choose areas without any saturated pixels. We recommend a medium-range intensity for this setup.
4. We recommend choosing the regions of interest based on the biology of the protein of interest. In the current protocol, we show the FRET measurements in whole cells and protrusions (Fig. 1c, d).

3.5.3 FRET Loss Indicates IP_3 Production in HL60 Cells Upon Uniform Stimulation

We used an intramolecular FRET probe designed by inserting an IP_3 receptor between cerulean/Venus FRET pairs [22] (Fig. 1a). Consistent with the above report, we measured a FRET loss in HL60 cells in response to fMLP stimulation (Fig. 1d).

3.6 Monitoring Spatiotemporal IP3 Dynamics in Chemotaxing HL60 Cells

Same FRET imaging acquisition in Subheading 3.5.1 with a DIC channel for the direction of the gradient (the position of micro-injector) was applied to monitor the FRET change in the chemotaxing HL60 cells. We then quantitatively measured the FRET in the front-leading edge and the back-trailing edge of the chemotaxing cells (Fig. 2).

4 Notes

1. It is not recommended that HL60 cells be passaged more than 30 times.
2. About 5–10% of differentiated cells adhere to gelatin-coated cover glass of chambers. We suggest thoroughly removing excess gelatin since we often saw fewer HL60 cells adhere to the cover glass containing floating gelatin.
3. After 5 days of differentiation, cell densities vary from 1 to 2.5×10^6 cells/mL. The best chemotactic cells might be found on day 5 after differentiation. Make your best guess based on morphological changes from round undifferentiated cells to oval polarized chemotactic cells.
4. For the stimulation experiments, cells are much more responsive after a short period of starvation. To starve cells, RPMI1640 culture medium is replaced by the RPMI1640 starvation medium without FBS.

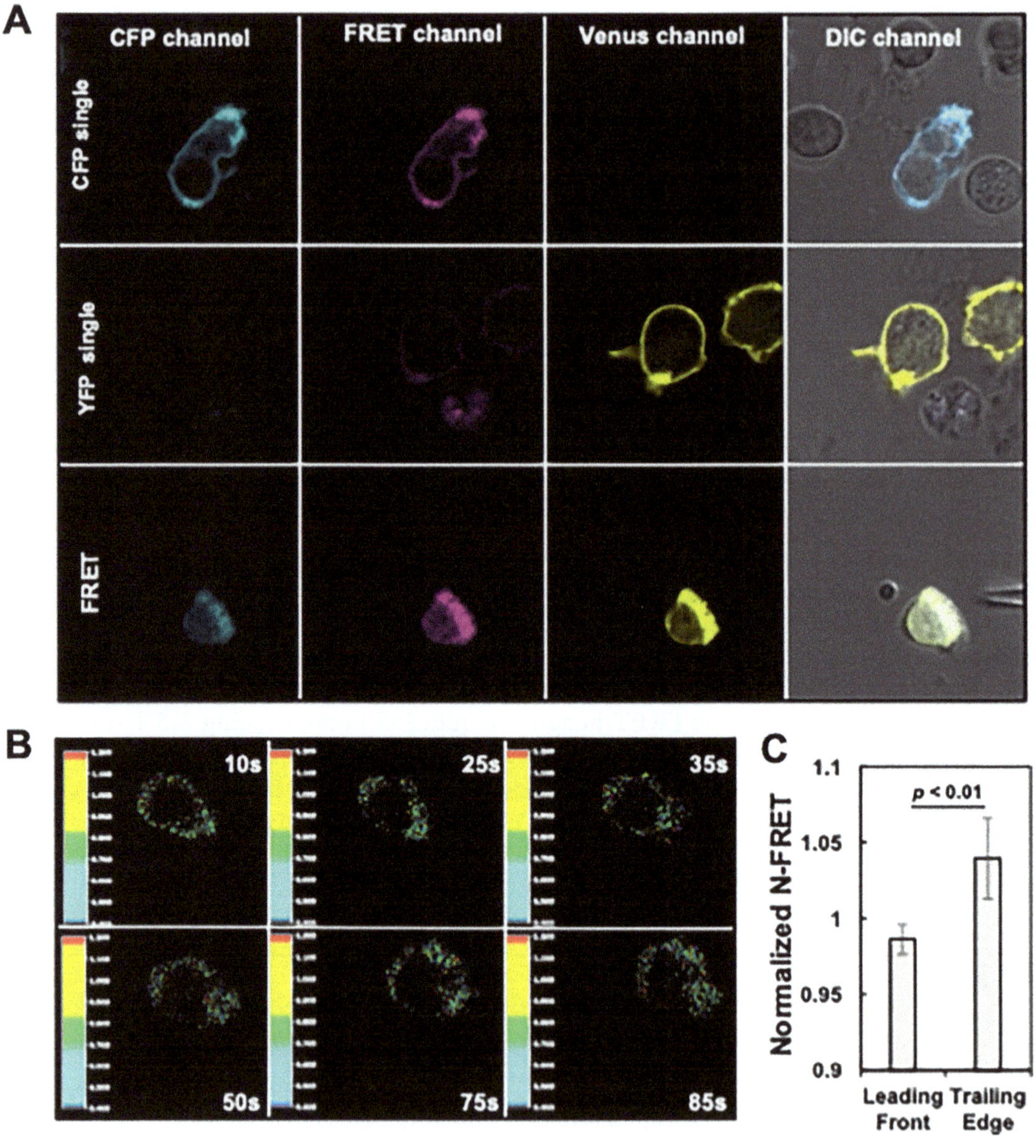

Fig. 2 IP_3 dynamics in chemotaxing HL60 cells. (**a**) Configuration and channel settings of cerulean, FRET, Venus, and DIC (differential interference contrast) for sensitized emission FRET measurement of a cerulean single control, a Venus single control, FRET channel, and a DIC channel using confocal fluorescent microscopy. The DIC channel is to show the microinjector (needle) by which a fMLP gradient is generated. (**b**) Montage shows N-FRET images of the cellular distribution of the IP_3 FRET probe in chemotaxing HL60 cells. Images were obtained by Zeiss LSM 710, analyzed using Zen Black software, and further modified for a better presentation using Adobe photoshop. (**c**) Mean $\pm$ SD of N-FRET in the leading edge and trailing edge of chemotaxing HL60 cells is shown. The time points for the quantitative measurement are shown in (**b**) $N = 6$, *p* value was calculated by student *t* test

5. The addition of fMLP results in a contrast change in DIC images that marks the time point when the uniform stimulation was added. In the FRET imaging with gradient experiment, DIC imaging indicates the source of the fMLP gradient. It is critical that the application of the chemoattractant stimuli be monitored in order to quantitatively measure the FRET change before and after the stimuli.
6. We found that the microinjector (needle) often gets blocked during chemotaxis assay of HL60 cells. To avoid this problem, we suggest removing the non-adhering cells by changing the medium prior to performing the experiments.
7. While a shorter wavelength laser would provide an excitation line closer to the optimal excitation wavelength of CFP/cerulean, effective FRET measurements can be taken using either 404 nm or 458 nm Argon laser lines.
8. Choose cells for FRET analyses that have a balanced cerulean and/or Venus signal. Make adjustments for the relative signal strength of each fluorophore depending on the limitations of the available filter sets on your system. For example, we visually inspect for a strong cerulean signal with a mercury lamp using a CFP/cerulean (or DAPI) filter set. We use a GFP filter set to inspect for cells co-expressing Venus and choose for analysis those that have a slightly higher Venus signal relative to the cerulean signal. Since whole sets of comparable experiments must be performed with the same acquisition conditions, it is recommended that cerulean/Venus double-expressing cells be used to set up the acquisition conditions instead of using single-expressing control cells. We often choose those cells with a similar level of cerulean and Venus expression instead of those highly expressing only one fluorescent protein.

Acknowledgments

The authors would like to thank all members of the Chemotaxis Signal Section of the Laboratory of Immunogenetics (CSS/LIG), NIAID, NIH. This research was supported by the Intramural Research Program of the NIH/NIAID. We also thank Gergő Gulyás and Dr. Péter Várnai of Semmelweis University, Budapest, Hungary, for providing IP_3 probes for this work.

References

1. Srinivasan K, Wright GA, Hames N, Housman M, Roberts A, Aufderheide KJ, Janetopoulos C (2013) Delineating the core regulatory elements crucial for directed cell migration by examining folic-acid-mediated responses. J Cell Sci 126:221–233
2. Kortholt A, King JS, Keizer-Gunnink I, Harwood AJ, Van Haastert PJ (2007) Phospholipase C regulation of phosphatidylinositol 3,4,5-trisphosphate-mediated chemotaxis. Mol Biol Cell 18:4772–4779

3. Camps M, Carozzi A, Schnabel P, Scheer A, Parker PJ, Gierschik P (1992) Isozyme-selective stimulation of phospholipase C-beta 2 by G protein beta gamma-subunits. Nature 360:684–686
4. Jiang H, Wu D, Simon MI (1994) Activation of phospholipase C beta 4 by heterotrimeric GTP-binding proteins. J Biol Chem 269: 7593–7596
5. Park D, Jhon DY, Lee CW, Lee KH, Rhee SG (1993) Activation of phospholipase C isozymes by G protein beta gamma subunits. J Biol Chem 268:4573–4576
6. Li Z, Jiang H, Xie W, Zhang Z, Smrcka AV, Wu D (2000) Roles of PLC-beta2 and -beta3 and PI3Kgamma in chemoattractant-mediated signal transduction. Science 287:1046–1049
7. Xu X, Gera N, Li H, Yun M, Zhang L, Wang Y, Wang QJ, Jin T (2015) GPCR-mediated PLCbetagamma/PKCbeta/PKD signaling pathway regulates the cofilin phosphatase slingshot 2 in neutrophil chemotaxis. Mol Biol Cell 26:874–886
8. Xu X, Wen X, Bhimani S, Moosa A, Parsons D, Ha H, Jin T (2023) G protein-coupled receptor-mediated membrane targeting of PLCγ2 is essential for neutrophil chemotaxis. J Leukoc Biol 114:126–141
9. Tang W, Zhang Y, Xu W, Harden TK, Sondek J, Sun L, Li L, Wu D (2011) A PLCbeta/PI3Kgamma-GSK3 signaling pathway regulates cofilin phosphatase slingshot2 and neutrophil polarization and chemotaxis. Dev Cell 21:1038–1050
10. Liu L, Gritz D, Parent CA (2014) *PKCbetaII* acts downstream of Chemoattractant receptors and mtorc2 to regulate cAMP Production and myosin ii activity in neutrophils. Mol Biol Cell 25:1446–1457
11. Clemens RA, Lowell CA (2015) Store-operated calcium signaling in neutrophils. J Leukoc Biol 98:497–502
12. Wei C, Wang X, Chen M, Ouyang K, Song LS, Cheng H. (2009). Calcium flickers steer cell migration. Nature, 2009. 457:901-905
13. Evans JH, Falke JJ (2007) Ca2+ influx is an essential component of the positive-feedback loop that maintains leading-edge structure and activity in macrophages. Proc Natl Acad Sci USA 104:16176–16181
14. Tsai FC, Meyer T (2012) Ca2+ pulses control local *cycles of lamellipodia retraction and adhesion along the front of migrating cells.* Curr Biol 22:837–842
15. Xu X, Wen X, Moosa A, Bhimani S, Jin T (2021) Ras inhibitor CAPRI enables neutrophil-like cells to chemotax through a higher-concentration range of gradients. Proc Natl Acad Sci USA 118:e2002162118
16. Traynor D, Milne JL, Insall RH, Kay RR (2000) Ca(2+) signalling is not required for chemotaxis in Dictyostelium. EMBO J 19: 4846–48454
17. Traynor D, Kay RR (2017) polycystin-type transient receptor potential (Trp) channel that is activated by ATP. Biol Open 6:200–209
18. Xu X, Pots H, Gilsbach BK, Parsons D, Veltman DM, Ramachandra SG, Li H, Kortholt A, Jin T (2022) C2GAP2 is a common regulator of Ras signaling for chemotaxis, phagocytosis, and macropinocytosis. Front Immunol 13: 1075386
19. Xu X, Bhimani S, Pots H, Wen X, Jeon TJ, Kortholt A, Jin T (2021) Membrane targeting of C2GAP1 enables *Dictyostelium discoideum* to sense chemoattractant gradient at a higher concentration range. Front Cell Dev Biol 9: 725073
20. Xu X, Gera N, Li H, Yun M, Zhang L, Wang Y, Wang QJ, Jin T (2015) GPCR-mediated PLCβγ/PKCβ/PKD signaling pathway regulates the cofilin phosphatase slingshot 2 in neutrophil chemotaxis. Mol Biol Cell 26:874–886
21. Tang W, Zhang Y, Xu W, Harden TK, Sondek J, Sun L, Li L, Wu D (2011) A PLCβ/PI3Kγ-GSK3 signaling pathway regulates cofilin phosphatase slingshot2 and neutrophil polarization and chemotaxis. Dev Cell 21:1038–1050
22. Gulyás G, Tóth JT, Tóth DJ, Kurucz I, Hunyady L, Balla T, Várnai P (2015) Measurement of inositol 1,4,5-trisphosphate in living cells using an improved set of resonance energy transfer-based biosensors. PLoS One 10: e0125601
23. Oura T, Murata K, Morita T, Nezu A, Arisawa M, Shuto S, Tanimura A (2016) Highly sensitive measurement of inositol 1,4,5-trisphosphate by using a new fluorescent ligand and ligand binding domain combination. Chembiochem 17:1509–1512
24. Tanimura A, Nezu A, Morita T, Turner RJ, Tojyo Y (2004) Fluorescent biosensor for quantitative real-time measurements of inositol 1,4,5-trisphosphate in single living cells. J Biol Chem 279:38095–38098
25. Zigmond SH (1978) Chemotaxis by polymorphonuclear leukocytes. J Cell Biol 77:269–287
26. Gerisch G, Albrecht R, Heizer C, Hodgkinson S, Maniak M (1995) Chemoattractant-controlled accumulation of coronin at the leading edge of Dictyostelium cells monitored using a green fluorescent

protein–coronin fusion protein. Curr Biol 5: 1280–1285

27. Wen X, Jin T, Xu X (2016) Imaging G protein-coupled receptor-mediated chemotaxis and its signaling events in neutrophil-like HL60 cells. J Vis Exp 14:54511
28. Li J, Zhu L, Zhang M, Lin F (2012) Microfluidic device for studying cell migration in single or co-existing chemical gradients and electric fields. Biomicrofluidics 6:24121–2412113
29. Xu X, Jin T (2011) Imaging G-protein coupled receptor (GPCR)-mediated signaling events that control chemotaxis of Dictyostelium discoideum. J Vis Exp 55
30. Xu X, Brzostowski JA, Jin T (2006) Using quantitative fluorescence microscopy and FRET imaging to measure spatiotemporal signaling events in single living cells. Methods Mol Biol 346:281–296
31. Xu X, Yun M, Wen X, Brzostowski J, Quan W, Wang QJ, Jin T (2016) Quantitative monitoring spatiotemporal activation of Ras and PKD1 using confocal fluorescent microscopy. Methods Mol Biol 2016(1407):307–323

Chapter 15

Mutation Screening of *Dictyostelium* Restriction Enzyme-Mediated Integration (REMI) Libraries

George Heslop-Harrison, Anthony Goddard, and Robin S. B. Williams

Abstract

Identifying the mechanisms of action of existing and novel drugs is essential for the development of new compounds for therapeutic and commercial use. Here we provide a technique to identify these mechanisms through isolating mutant cell lines that show resistance to drug-induced phenotypes using *Dictyostelium discoideum* REMI libraries. This approach provides a robust and rapid chemical-genetic screening technique that enables an unbiased approach to identify proteins and molecular pathways that control drug sensitivity. Mutations that result in drug resistance often occur in target proteins thus identifying the specific protein targets for drugs and bioactive natural products. Following the identification of a list of putative molecular targets user selected compound targets can be analyzed to confirm and validate direct inhibitory effects.

Key words REMI, BLAST (Basic Local Alignment Search Tool), Drug resistance, CRISPR, Orthologous proteins, Translational science, Mutation recapitulation

1 Introduction

Understanding the molecular mechanisms and targets of drugs and bioactive natural products provides a huge advantage in a range of areas from medicines and pharmaceuticals to toxicology and the development of agrochemicals. Due to the complex nature of animals (and plants), using these models in mechanism of action studies is exceedingly difficult and time consuming without an initial indication of the underlying effects of compounds. The model system *Dictyostelium* provides a validated approach in this research whereby numerous studies have identified mechanism of action for a wide range of structures that have been further validated in translational studies [1, 2]. These studies are based upon two key assumptions: firstly that a library of *Dictyostelium*

Authors George Heslop-Harrison and Anthony Goddard have equally contributed to this chapter.

Alan R. Kimmel (ed.), *Dictyostelium discoideum: Methods and Protocols*, Methods in Molecular Biology, vol. 2814,
https://doi.org/10.1007/978-1-0716-3894-1_15,

insertional mutants contains individual mutants lacking single genes and loss of the protein in each mutant is compensated by cell-specific signaling changes and secondly that a compound of interest acts to modify the activity of a specific protein target and that inhibition of the target protein will cause a phenotype (e.g., reduce cell proliferation). Based upon these assumptions, exposure of a library of mutants to a compound at concentrations necessary to cause a phenotype (e.g., reduce proliferation) will enable the proliferation and survival of mutants having lost the target protein or protein that regulates the target or associated signalling pathway. Thus identifying these proteins through screening a mutant library provides novel insight to molecular mechanisms of action of the compound under investigation.

Site-specific integration of linear DNA using restriction enzymes to stimulate DNA integration into the genome of *Dictyostelium* has been used to develop large libraries of randomly acquired mutants [3, 4]. The REMI protocol involves introducing a restriction enzyme along with linear transforming DNA containing an antibiotic resistance cassette into *Dictyostelium* cells by electroporation. Following selection for stable transformants, these mutants can be pooled into a library with most mutants containing a single mutation due to homologous integration. This library of mutants can be stored and thus shared within the research community.

The REMI library screening protocol has accelerated progress in study of cellular and molecular processes within *Dictyostelium* leading to numerous discoveries applicable to many systems. This approach has identified the GTP-binding protein *racE* involved in cytokinesis [5] and the identification of cheater mutants to understand the genetic complexity of multicellular cooperation [6] in *Dictyostelium*. This screening system has also allowed researchers to identify and test cellular mechanisms of orthologous proteins such as those involved in neuronal ceroid lipofuscinoses (NCL) deficiencies as *Dictyostelium* shares 11 homologues of the 13 human NCL genes [7]. Early drug development has also benefited from this technique including understanding the effect of naringenin treatment on kidney cysts through polycystin-2 inhibition [8], the role of cannabinoids in the treatment of epilepsy and multiple sclerosis [9, 10], and previously the uncharacterized mechanisms of seizure control and tuberous sclerosis treatment through the ketogenic diet-associated medium chain fatty acids [11, 12]. This approach has also been used to investigate allelochemicals mechanisms related to plant function [13]. Recent approaches to screen pools of mutants combined with genome sequencing provide another mechanism for library screening with increase cost but providing a more detailed genome-wide analysis of genes controlling compound function [3].

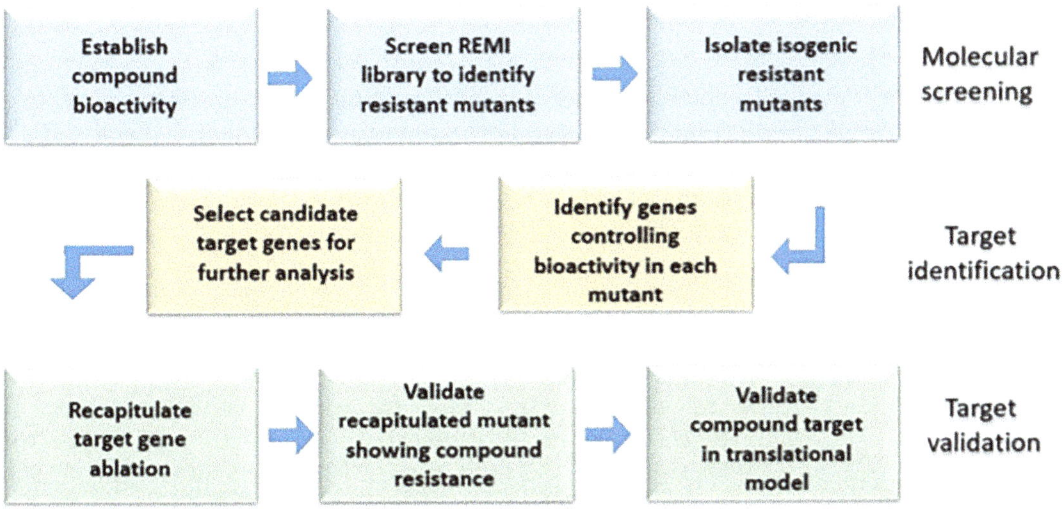

Fig. 1 Workflow schematic for the identification of targets for bioactive compounds using *Dictyostelium* as a model system. Workflow consists of three stages: (1) molecular screening, (2) target Identification, and (3) target validation

The REMI library screening method can be broken down into three discrete stages comprising molecular screening, target identification, and target validation (Fig. 1). These steps can be further broken down to (i) establishing the concentration of the compound necessary for reducing cell proliferation (i.e., bioactivity) to (ii) analyze cell proliferation of wild type cells and the REMI mutant library cells under solvent only conditions and at a concentration of the compound to reduce cell proliferation by 80–100% to (iii) isolate and propagate isogenic cell lines showing resistance to the cell proliferation inhibition caused by the compound. This will enable the (iv) identification of the interrupted genes controlling the bioactivity of the compound in each compound-resistant mutant and (v) the selection of mutants for further investigation. (vi) Independently ablating the target gene in mutants of interest and (vii) confirming compound-resistance in the recapitulated mutant will validate the target gene in controlling the activity of the compound. At this stage, the resistance phenotype can be restored by expressing the gene attached to a fluorescent marker [14] in the mutant to restore sensitivity to the proliferative inhibitory effect of the compound. Further analysis may involve expressing orthologous genes from other species, for example, human genes [14, 15] using *Dictyostelium* codon bias to evidence the related species protein may also be targeted by the compound (viii). Ultimately the target protein should be confirmed in translational models. As such the REMI mutant library screen has become a powerful tool in a range of biological fields and particularly relevant in biomedical studies.

2 Materials

2.1 Cell Culture

1. HL-5 medium (used for routine axenic growth of *Dictyostelium* cells): 10 g/l proteose peptone number 2 (Becton Dickinson) 5 g/l yeast extract 10 g/l glucose 0.35 g/l Na_2HPO_4 0.35 g/l KH_2PO_4 pH 6.4. Adjust pH with H_3PO_4. Sterilize by autoclaving for 40 min and remove from the autoclave immediately to minimize caramelization. Add antibiotics prior to use from the following stock solution. 100x Pen/Strep: 10000 U/mL penicillin G; 10 mg/mL streptomycin sulphate (Gibco 11,860–038).
2. SM agar medium (used for growth of *Dictyostelium* on *R. planticola* bacterial lawns): 10 g/L Bacto-peptone (Becton Dickinson) 1 g/l yeast extract 10 g/l glucose 1 g/l $MgSO_4$ 1.9 g/l K_2HPO_4 0.6 g/l KH_2PO_4 20 g/l Bacto-Agar (Becton Dickinson) pH 6.4. Adjust pH with H_3PO_4 sterilize by autoclaving and pour about 32–35 mL into each 9 cm Petri dish.

2.2 DNA Isolation and Processing

1. Phosphate buffer (16.5 mM KH_2PO_4 3.8 mM K_2HPO_4 pH 6.2).
2. Wizard DNA extraction Kit (Promega).
3. 1% agarose and 1x TAE buffer (Tris-acetate-EDTA) for gel electrophoresis.
4. 10x buffer TANGO (Thermo Fisher Scientific) RsaI (10 U/μL Thermo Fisher Scientific) 70% ethanol 10 × T4 DNA ligase Buffer (Thermo Scientific) and T4 DNA ligase (5 U/μL Thermo Fisher Scientific) used for DNA processing.

2.3 Inverse PCR

1. 5x GoTaq Flexibuffer (Promega)
2. $MgCL_2$ (25 mM Promega).
3. Deoxynucleoside triphosphates (dNTPs) (2 mM of each dNTP).
4. For primers, *see* Subheading 3.6.

2.4 Sequencing Preparation

1. MiniPrep PCR purification kit (Qiagen) used to prepare samples for Sanger sequencing.

3 Methods

3.1 Establishing Bioactivity of Compounds

The first step of chemical-genetic screening is the analysis of the compound to determine the concentrations necessary for library screening [8–11, 16]. This is essential since resistance screens will only work for compounds that show inhibitory effects on

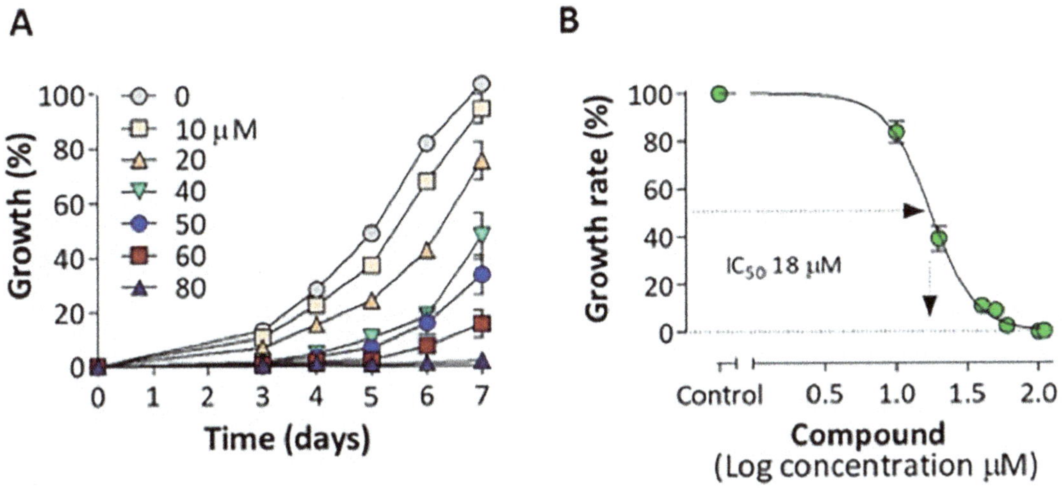

Fig. 2 Defining bioactive compound potency in reducing *Dictyostelium* cell proliferation. (**a**) Initial analysis of a compound on *Dictyostelium* cell proliferation over a seven-day period with cell number normalized to control (solvent only day 7) provides a concentration range to inhibit proliferation by 80–100% for wild-type cells. (**b**) Calculating the rate of cell proliferation (days 5–6) in each condition enables a secondary plot to estimate the compound potency (IC_{50}) of the compound on cell proliferation. (Data derived from Ref. [11])

Dictyostelium cell proliferation. Initial compound concentrations can be based upon therapeutic levels or on that reported in earlier in vitro studies.

1. Using wild-type cells of the background to that used in the mutant library (AX2 AX3 AX4), place 2.5×10^4 cells in individual wells of a 6-well plate in 2 mL HL-5 medium. Choose the number of wells to provide at least duplicate repeats for each condition. Use control (solvent only) and at least six concentrations of compound evenly distributed across the effective concentration range. It is important to have at least one concentration providing minimal inhibition and one concentration providing a block in proliferation. Ensure all wells contain the same concentration of solvent (e.g., DMSO [*see* **Note 1**]). Maintain cells at 22 °C in an aerated box lined with damp paper towels to minimized evaporation.
2. Count cells at days 3–7 using a hemocytometer in duplicate for each well. Ensure cells are thoroughly resuspended in the well before transferring a sample to the hemocytometer for counting.
3. Independent repeats of this analysis can be presented as normalized data (Fig. 2a).
4. Analysis of the proliferation rate in each condition, for example, from days 5 to 6 enables the production of a secondary plot (Fig. 2b) to calculate the IC_{50} value for a compound.

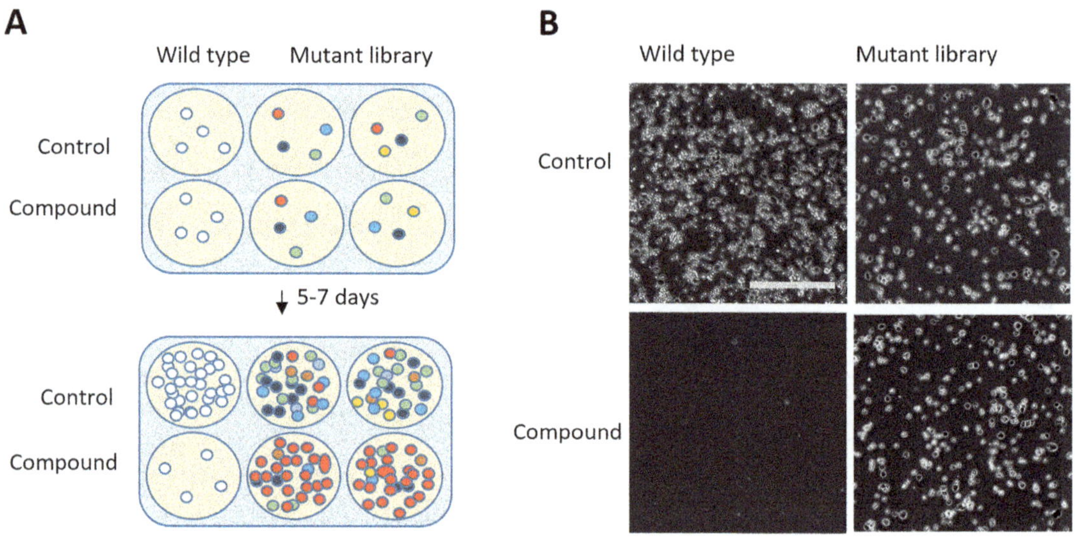

Fig. 3 Identifying compound-resistant *Dictyostelium* REMI mutants. (**a**) Preparing a *Dictyostelium* compound-resistance screen in 6-well plates with wild-type cells (white circle) and distinct REMI insertional mutants (colored circles) in the presence of solvent only (control) or a compound at a concentration to reduce cell proliferation by 80–100% over 5–7-day period. Wild-type cells proliferate under control conditions but are blocked by the compound. REMI mutant library cells show unrestricted cell proliferation for all mutants under control conditions, but only compound-resistant mutants (red circles) proliferate in the presence of the compound. (**b**) Example cell proliferation in a compound-resistant screen with REMI mutants showing proliferation in the presence of a compound that blocks wild-type cell proliferation. (From Ref. [11])

3.2 Screening a REMI Mutant Library

The second step of chemical-genetic screening is the identification of REMI mutants (Fig. 3) that show resistance to the phenotypic effects of the compound. Screening should be performed using a concentration of compound that inhibits *Dictyostelium* cell proliferation by 80–100% to ensure resistant cells outcompete cell retaining compound proliferation-sensitivity.

1. Grow wild-type cells or the REMI mutant library cells for two days in shaking media (using suitable antibiotic selection for the library (10 μg/mL blasticidin).
2. Place 2.5×10^4 wild-type or REMI mutant library cells in 2 mL media into three individual wells per treatment in a 6-well plate in solvent-only conditions or supplemented with the compound at a concentration to inhibit 80–100% of *Dictyostelium* cell proliferation (Fig. 2a).
3. After 5–7 days (depending on rate of cell proliferation to reach near confluent plates), spin down confluent cells and reseed 6-well plates with 2.5×10^4 cells. Add media supplemented with the same concentration of compound. Repeat proliferation and reseeding for a further two rounds of screening.

4. From compound-resistant pools, plate out each cell pool onto individual SM plates seeded with *R. planticola* using serial dilutions to isolate isogenic colonies.
5. Transfer up to 24 clonal isolates from each individual pool into axenic medium (2 mL in 6-well plates), and when confluent, transfer 1.0×10^4 of each isogenic cell line into 500 μL media containing the compound in a 24-well plate. Select those colonies that show resistance after five days for further analysis.
6. Transfer resistant cell lines to axenic medium in 10 cm plates and allow to grow to confluency and freeze 2×10^7 cells in 0.5 mL of freezing medium (sterilized horse serum (Sigma-Aldrich) containing 7.5% (v/v) DMSO) at -80 °C and transfer to liquid nitrogen.
7. Regrow cells for genomic DNA extraction.

3.3 Identification of Disrupted Genes

Following the selection of compound-resistant cell lines, the next step is to identify the location of the REMI insertion site in each cell line. Inverse PCR (Fig. 4) allows for the amplification of the sequences flanking the REMI insert. Amplified sequences can be purified and sequenced allowing for the location of the REMI insert to be identified and thus the ablated gene identified via BLAST (Basic Local Alignment Search Tool) analysis.

1. **Isolation of Genomic DNA.**
   1.1. Remove isogenic resistant cells from plates and wash with phosphate buffer (16.5 mM KH_2PO_4 3.8 mM K_2HPO_4 pH 6.2) twice ($1–2 \times 10^4$ cells/mL).
   1.2. Extract genomic DNA using a genomic DNA extraction kit following the manufacture's protocol (e.g., Wizard Genomic DNA extraction kit (Promega)).
   1.3. Quantify DNA (e.g., using a NanoDrop) and run DNA on a gel to check for quality (ideally ~400–500 ng/μL and 260/280 ration of 1.8–2).

3.4 RsaI Digestion

1. Mix the following in a 1.5 mL Eppendorf tube:
2. 25 μL of genomic DNA (~1 μg)
3. 19 μL water (Sigma-Aldrich).
   - 5 μL 10x buffer for Rsal (e.g., TANGO buffer from Thermo Fisher Scientific)
   - 1 μL RSAI (10 U/mL Thermo Fisher Scientific)
4. Incubate for 1 h at 37 °C.
5. Add 5 μL 3 M sodium acetate and 100 μL ice-cold 100% ethanol.
6. Precipitate for 30 min.

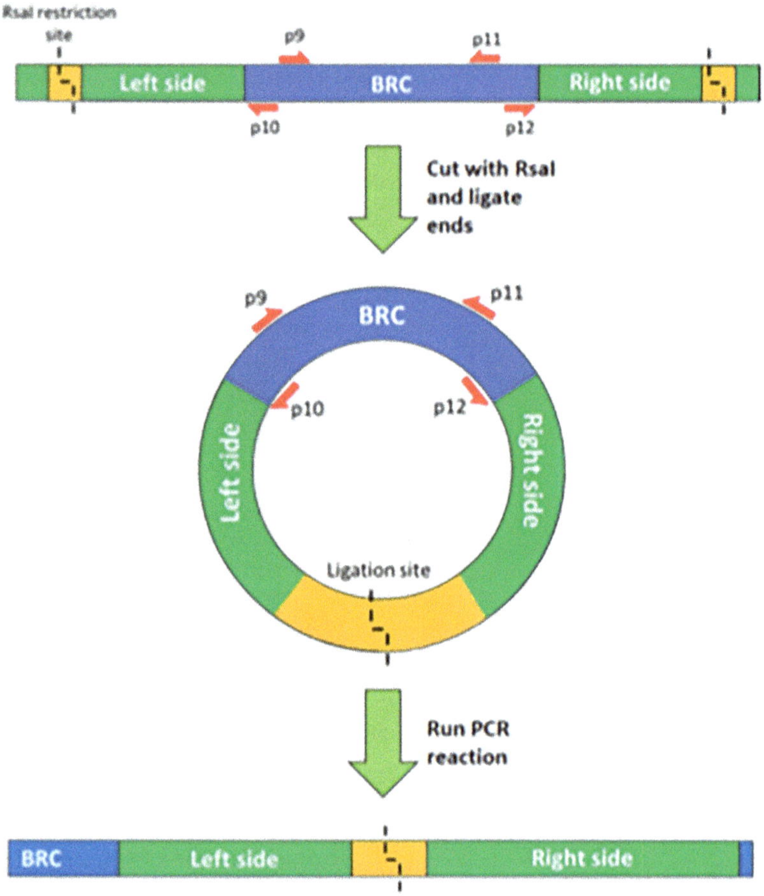

Fig. 4 Using inverse PCR to identify insertionally inactivated genes encoding proteins regulating compound sensitivity. Each compound-resistant mutant will contain the blasticidin resistance cassette (BRC, blue) flanked by the gene of interest (left and right sides, green). Identification of the gene involves the restriction enzyme-dependent cleavage of genomic DNA from each mutant (e.g., using RsaI, yellow) and ligation of the sticky ends to form a plasmid. Primer sites within the insertion cassette (P10 and P12) can be used to amplify the flanking gene sequence in all REMI mutants with control primers (P9 and P11) providing a positive control. Sequencing of the subsequent PCR product will enable the identification of the target gene

7. Centrifuge (16,000 g for 15 min at 4 °C)—remove supernatant.
8. Add 150 μL 70% ethanol (washing).
9. Centrifuge (16,000 g for 2 min at 4 °C)—Discard supernatant.
10. Air-dry the DNA pellet to remove remainder of 70% ethanol.
11. Resuspend pellet in 30 μL sterile H_2O.

3.5 Ligation [See Note 2]

1. Ligate the digested DNA and a negative control using the following reaction mixture:
   - 25 μL RsaI digested DNA
   - 19 μL sterile H_2O (Sigma-Aldrich)
   - 5 μL 10 × T4 DNA ligase Buffer (Thermo Scientific)
   - 1 μL T4 DNA ligase (5 U/μL Thermo Fisher Scientific).
2. Incubate for 1 h at room temperature.
3. Inactivate ligase (74 °C for 10 min).

3.6 Inverse PCR

1. Amplify the flanking restriction sites using the following reaction mixture and PCR protocol:
   - 2 μL ligated DNA
   - 4 μL 5x GoTaq Flexibuffer (Promega)
   - 1.6 μL; $MgCl_2$ (25 mM Promega)
   - 1 μL dNTPs (2 mM of each dNTP)
   - 1 μL forward primer (10 μM)
   - 1 μL reverse primer (10 μM)
   - 0.2 μL GoTaq® G2 Flexi DNA Polymerase (Promega)
   - 9.2 μL water (Sigma-Aldrich).
2. Use PCR condition as detailed:
   - 95 °C 1 min
   - 94 °C 30 seconds, 52 °C 30 seconds, and 68 °C 2 min (34 cycles)
   - 68 °C for 10 min.

     Primers used:

   Pair 1
   - Forward p-09 (5’-TGTATGCTATACGAAGTTATCC-3′).
   - Reverse p-10 (5’-GTAGAAGTAGCGACAGAGAAG-3′).

   Pair 2
   - Forward p-11 (5’-CCATTCGAAACTGCACTACC-3′).
   - Reverse p-13 (5’- TGAATGGTGAAGATAAATATA TGC-3′).
3. Analyze amplified PCR products using gel electrophoresis using 5 μL of PCR product with 1 L loading buffer on a 1% agarose gel.
4. Purify the remainder of the PCR product using a PCR purification Kit (Thermo Fischer Scientific).

3.7 Sequencing of Inverse PCR Product

1. Once the PCR product is purified, quantify DNA concentration (e.g., using a NanoDrop).
2. Prepare PCR sequencing samples, for example, Eurofins Tube-seq services—Mix each sample with one of the forward primers into a 1.5 mL Eppendorf tube (recommended mix of 2 μL primer to 15 μL PCR product to a final concentration of 5 ng/μL).
3. Sequence the PCR product.

3.8 Candidate Target Selection

The analysis of isogenic cell lines showing resistance to a compound often provides a range of mutants indicating multiple proteins involved as potential targets [7, 17]. Although the selection of which candidates to take forward is strongly dependent upon each case, a range of considerations are necessary.

1. If the identified target protein is found in independent mutants (e.g., with distinct insertion sites), this strongly supports the relevance of the target protein in the function of the bioactive compound.
2. Similarly if the mutant is found in independent pools of resistant mutants, this increases the likelihood of the mutant showing resistance to the compound.
3. The identified *Dictyostelium* protein should have orthologous proteins in the relevant translational species (e.g., humans in the context of drug mechanism). BLAST analysis to identify orthologues can quickly identify this. Proteins with comparable size (amino acid number) domain structure and conserved catalytic motifs or amino acids all contribute to this decision.
4. Target proteins identified in the screen may have been proposed in earlier studies from other models. This both increases the likelihood of evolutionarily conserved function of the bioactive compound but reduces the novelty of the discovery.
5. Target proteins identified in the screen should be expressed in the cell during growth. *Dictyostelium* contains many proteins necessary for development and only expressed during development, and these proteins may provide false positives if identified in a growth screen.

3.9 Recapitulation of the Resistance Phenotype

It is important to validate the findings of the mutant library screen by recapitulating identified resistance phenotypes. REMI mutants may have additional off-target effects or multiple unidentified BSR inserts, so it is essential to establish causation between the target gene ablation and resistant phenotype. To do this, independent strains of *Dictyostelium* with candidate genes ablated must be constructed. These genes can be ablated by CRISPR/Cas9 gene editing using an all-in-one plasmid vector pTM1285 described in detail here [18]. This is a relatively uncomplicated process using the following steps:

1. Use CRISPOR or other similar tools to identify suitable guides for a candidate gene.
2. Once oligonucleotide primers for the guide RNA and its complimentary strand have been purchased, anneal them together.
   - 1 μL oligo 1 (100 μM)
   - 1 μL oligo 2 (100 μM)
   - 1 μL 10 × T4 ligation buffer (NEB #B0202)
   - 6.5 μL ddH_2O
   - 0.5 μL T4 PNK (NEB #M0201)
   - Anneal in a thermocycler using the following parameters:
     - 37 °C 30 min
     - 95 °C 5 min and then ramp down to 25 °C at 5°C/min (*total run time 55 min*).
3. Digest annealed oligos using BpiI and ligate into the CRISPR all-in-one-vector:
   - X μL pTM1285 (100 ng).
   - 2 μL phosphorylated and annealed oligo duplex from **step 5.2** (1:250 dilution)
   - 2 μL 10X FastDigest Buffer (or Tango buffer)
   - 1 μL DTT (final concentration of 1 mM) (Sigma #DTT-RO)
   - 1 μL ATP (final concentration of 1 mM) (NEB #P0756S)
   - 1 μL FastDigest *BbsI* or *BpiI* (ThermoFisher Fermentas)
   - 1 μL T4 DNA Ligase (NEB)
   - Y μL ddH_20.
   - 20 μL TOTAL
   - Incubate the ligation reaction in a thermocycler:
     - 37 °C 5 min
     - 23 °C 5 min
     - Cycle the previous two steps for six cycles (*total run time 1 h*).
4. After cloning and purification, this CRISPR vector can be transfected into wild-type cells to ablate the target gene.
   - Resuspend *Dictyostelium* cells and transfer to a 50 mL falcon.
   - Wash cells twice in H50 buffer.

- Spin for 3 min at 500 × g.
- Resuspend in 20 mL of fresh H50 buffer.
- Spin cells for 3 min at 500 × g.
- Resuspend pellet with fresh H50 to a concentration of 5×10^7 cells mL^{-1}.
- Add 100 μL cells to 5 μL plasmid (1–10 μg/3 μg per plasmid) to a chilled 1 mm gap electroporation cuvette.
- Mix and incubate for 5 min.
- Pulse twice at 850 V with 5 s between each pulse.
- Incubate on ice for 5 min.
- Transfer cells to 90 mm tissue culture dish with 10 mL HL5 medium.
- Add 10 μg/mL G418 after 8–16 h.
- Allow to grow for another three days (so that Cas9 can be expressed).

5. Transfected cells are then clonally isolated and PCR is conducted on the edited section of DNA.
6. This PCR product is then purified and then sequenced to confirm gene editing.

Once a candidate gene has been edited and ablation of the gene is confirmed, these independent null cell lines can be tested with the same chemical used in the mutant library screen. Cell proliferation over a range of doses is compared between wild-type and null cell lines to confirm the presence of a resistance phenotype.

4 Notes

1. We extensively use DMSO concentrations up to 2 μL/mL.
2. Clean all workstations with 70% ethanol before and after use to avoid contaminating DNA sequences.
3. BLAST analysis can be used to identify the location of your REMI mutation using the Sanger sequencing data.
4. Any mutants identified to be resistant using this method will need to be recapitulated using independently developed mutants with the same gene ablated.

Acknowledgments

Thanks to Dictybase.org for ongoing support for all researchers in this field.

References

1. Schaf J, Damstra-Oddy J, Williams RSB (2019) Dictyostelium discoideum as a pharmacological model system to study the mechanisms of medicinal drugs and natural products. Int J Dev Biol 63:541–550. https://doi.org/10.1387/ijdb.190228rw
2. Pain E, Shinhmar S, Williams RSB (2021) Using Dictyostelium to advance our understanding of the role of medium chain fatty acids in health and disease. Front Cell Dev Biol 9:722066. https://doi.org/10.3389/fcell.2021.722066
3. Gruenheit N, Baldwin A, Stewart B, Jaques S, Keller T, Parkinson K, Salvidge W, Baines R, Brimson C, Wolf JB, Chisholm R, Harwood AJ, Thompson CRL (2021) Mutant resources for functional genomics in Dictyostelium discoideum using REMI-seq technology. BMC Biol 19:172
4. Kuspa A, Loomis WF (1992) Tagging developmental genes in Dictyostelium by restriction enzyme-mediated integration of plasmid DNA. Proc Natl Acad Sci USA 89:8803–8807. https://doi.org/10.1073/pnas.89.18.8803
5. Larochelle DA, Vithalani KK, De Lozanne A (1996) A novel member of the rho family of small GTP-binding proteins is specifically required for cytokinesis. J Cell Biol 133:1321–1329. https://doi.org/10.1083/jcb.133.6.1321
6. Santorelli LA, Thompson CR, Villegas E, Svetz J, Dinh C, Parikh A, Sucgang R, Kuspa A, Strassmann JE, Queller DC, Shaulsky G (2008) Facultative cheater mutants reveal the genetic complexity of cooperation in social amoebae. Nature 451:1107–1110. https://doi.org/10.1038/nature06558
7. McLaren MD, Mathavarajah S, Huber RJ (2019) Recent insights into NCL protein function using the model organism Dictyostelium discoideum. Cells 8:115. https://doi.org/10.3390/cells8020115
8. Waheed A, Ludtmann MH, Pakes N, Robery S, Kuspa A, Dinh C, Baines D, Williams RS, Carew MA (2014) Naringenin inhibits the growth of Dictyostelium and MDCK-derived cysts in a TRPP2 (polycystin-2)-dependent manner. Br J Pharmacol 171:2659–2670. https://doi.org/10.1111/bph.12443
9. Perry C.J, Finch P, Muller-Taubenberger A, Leung K.Y, Warren E.C, Damstra-Oddy J, Sharma D, Patra P.H, Glyn S, Boberska J, Stewart B, Baldwin A, Piscitelli F, Harvey RJ, Harwood A, Thompson C, Claus SP, Greene NDE, McNeish AJ, Williams CM, Whalley BJ, Williams RSB. (2019). A new mechanism for Cannabidiol in regulating the one-carbon cycle and methionine levels in Dictyostelium and in mammalian epilepsy models. Br J Pharmacol 177::912-928. doi:https://doi.org/10.1111/bph.14892
10. Damstra-Oddy JL, Warren EC, Perry CJ, Desfougeres Y, Fitzpatrick JK, Schaf J, Costelloe L, Hind W, Downer EJ, Saiardi A, Williams RSB (2021) Phytocannabinoid-dependent mTORC1 regulation is dependent upon inositol polyphosphate multikinase activity. Br J Pharmacol 178:1149–1163. https://doi.org/10.1111/bph.15351
11. Warren EC, Dooves S, Lugara E, Damstra-Oddy J, Schaf J, Heine VM, Walker MC, Williams RSB (2020) Decanoic acid inhibits mTORC1 activity independent of glucose and insulin signaling. Proc Natl Acad Sci USA 117: 23617–23625. https://doi.org/10.1073/pnas.2008980117
12. Warren EC, Walker MC, Williams RSB (2018) All you need is fats – for seizure control: using amoeba to advance epilepsy research. Front Cell Neurosci 12:199
13. Heslop-Harrison G, Nakabayash IK, Espinosa-Ruiz A, Robertson F, Baines R, Thompson CRL, Hermann K, Alabadí D, Leubner-Metzger G, Williams RSB (2024) Functional mechanism study of the allelochemical myrigalone A identifies a group of ultrapotent inhibitors of ethylene biosynthesis in plants. Plant Commun 100846. https://doi.org/10.1016/j.xplc.2024.100846.
14. Fischer M, Haase I, Simmeth E, Gerisch G, Muller-Taubenberger A (2004) A brilliant monomeric red fluorescent protein to visualize cytoskeleton dynamics in Dictyostelium. FEBS Lett 577:227–232
15. Ludtmann MH, Otto GP, Schilde C, Chen ZH, Allan CY, Brace S, Beesley PW, Kimmel AR, Fisher P, Killick R, Williams RS (2014) An ancestral non-proteolytic role for presenilin proteins in multicellular development of the social amoeba Dictyostelium discoideum. J Cell Sci 127:1576–1584. https://doi.org/10.1242/jcs.140939
16. Cocorocchio M, Baldwin AJ, Stewart B, Kim L, Harwood AJ, Thompson CRL, Andrews PLR, Williams RSB (2018) Curcumin and derivatives function through protein phosphatase 2A and presenilin orthologues in Dictyostelium discoideum. Dis Model Mech 11:10. https://doi.org/10.1242/dmm.032375

17. Sharma D, Otto G, Warren EC, Beesley P, King JS, Williams RSB (2019) Gamma secretase orthologs are required for lysosomal activity and autophagic degradation in Dictyostelium discoideum independent of PSEN (presenilin) proteolytic function. Autophagy 15:1407–1418. https://doi.org/10.1080/15548627.2019.1586245

18. Sekine R, Kawata T, Muramoto T (2018) CRISPR/Cas9 mediated targeting of multiple genes in Dictyostelium. Sci Rep 8:8471. https://doi.org/10.1038/s41598-018-26756-z

Chapter 16

Single Cell Transcriptome Analysis During Development in *Dictyostelium*

Vlatka Antolović and Jonathan R. Chubb

Abstract

Dictyostelium represents a stripped-down model for understanding how cells make decisions during development. The complete life cycle takes around a day and the fully differentiated structure is composed of only two major cell types. With this apparent reduction in "complexity," single cell transcriptomics has proven to be a valuable tool in defining the features of developmental transitions and cell fate separation events, even providing causal information on how mechanisms of gene expression can feed into cell decision-making. These scientific outputs have been strongly facilitated by the ease of non-disruptive single cell isolation—allowing access to more physiological measures of transcript levels. In addition, the limited number of cell states during development allows the use of more straightforward analysis tools for handling the ensuing large datasets, which provides enhanced confidence in inferences made from the data. In this chapter, we will outline the approaches we have used for handling *Dictyostelium* single cell transcriptomic data, illustrating how these approaches have contributed to our understanding of cell decision-making during development.

Key words Single cell transcriptomics, RNA sequencing, Transcription bursting, Stochastic gene expression, Development, Bioinformatics, Big data

1 Introduction

Dictyostelium is an exceptional system for application of single cell transcriptomic analyses to developmental biology. The developmental program is rapid, with only two major final cell fates, allowing high temporal resolution analysis of cell state transitions and the emergence of cell-type specialization with relatively low cell numbers. In addition, the ease and speed of extraction of high-quality RNA permits very reproducible data, removing the need for complex batch correction procedures and allowing deep insights into how the mechanisms of gene regulation govern cell-type heterogeneity. The data from this system are a clear benchmark for high-quality single cell transcriptome data, used for the development of

Alan R. Kimmel (ed.), *Dictyostelium discoideum: Methods and Protocols*, Methods in Molecular Biology, vol. 2814, https://doi.org/10.1007/978-1-0716-3894-1_16,

reliable quantitative tools for more complex and intractable systems [1].

In this chapter, we will outline the essential approaches for analysis of single cell transcriptome data in *Dictyostelium*, with specific technical and biological examples showing the methods we arrived at to best exploit the data in addition to the gains in knowledge about developmental processes that emerged.

2 Materials

2.1 Cell Preparation

1. Although cells can be cultured on bacteria, HL5 growth media is used for most experiments (Formedium HLG0102), with cells grown in adherent culture on 10 cm tissue culture dishes.
2. Development buffer. Any standard lab development buffer could be used, but we use KK2: 20 mM KPO_4, pH 6.0–6.2.
3. EDTA. We use a 0.5 M stock (pH 8.0) of EDTA, added to the KK2 to a final concentration of 10–20 mM. In practice, we observed little difference in the efficacy of cell disaggregation between 10 and 20 mM.
4. Syringes and needles. For disaggregation of cells, we pass the sample, in KK2, through a 1 ml syringe (BD 303172) attached to a 20 g needle (BD407).
5. Development substrate. We use 1.5–2% agar (BD214010) dissolved in KK2 for most applications. This agar is cast in 3.5 cm petri dishes. For more uniform late development, we plate cells on Whatman #50 filter paper rings (Whatman 1450–090). These rings are kept humid by layering them on rings of Whatman #3 filters (1003–090) saturated with KK2 in 10 cm petri dishes.

2.2 Cell Isolation, Library Preparation, and Sequencing

Two standard strategies have been used for cell isolation: either using the 10× Genomics Chromium Single Cell Gene Expression platform or Fluidigm (now Standard BioTools Inc.) C1 single-cell mRNA sequencing system. Please refer to company websites for more information. Sequencing has been carried out on a NextSeq500 (Illumina). A detailed description of this standard protocol is beyond the scope of this chapter. Please visit the Illumina website.

2.3 Data Analysis

1. Checking the quality of the sequenced reads: FASTQC (Babraham Bioinformatics).
2. Mapping the reads to reference genome and quantifying reads per gene: we used Tophat2 and HtSeq for reads retrieved using C1 sequencing system. Cell Ranger (10× Genomics) was used for reads retrieved using 10× Chromium.

3. Analysis of read count tables (cell transcriptome profiles): MATLAB (MathWorks), Mathematica (Wolfram), and R (R Foundation).
4. Gene ontology analysis: Panther Classification System (National Science Foundation).

3 Methods

3.1 Data Generation Considerations

For completeness, we briefly review the approaches that take us from the biological sample to the sequencing reads. For more details, please visit our recent experimental papers [2–5]. Readers may also wish to consider other studies in which we reused the data based on new scientific questions [6, 7] for insight into the general approaches used.

Single cell isolation in undifferentiated and early developing cells is based on gentle pipetting of cells off their substrates using development buffer. Cells are plated either on 35 mM dishes containing 2–3 mL of 1.5–2% agar made up in KK2 buffer or developed on Whatman #50 filter paper. Generally, 5×10^6–10^7 cells are plated, from a 1 mL cell suspension in KK2 buffer. Cells are allowed to settle for 10–30 min, before buffer is removed, and then the samples are kept in a humidified chamber. At specific time points after this, differentiating cells are removed from the agar surface by gentle pipetting up and down with ice-cold 1 mL KK2 with a standard micropipette (such as a Gilson P1000). Later developmental stages (after 6 hours, when cells have formed aggregates) where cells are more adherent require a slightly more aggressive disaggregation, in which cells are repeat pipetted (20 times) through a syringe and 20G needle. For these later stages, the KK2 also contains 10–20 mM EDTA to inhibit cell-cell adhesions. This is a standard approach for studying more developmentally advanced cells, and the dissociated cells can be used for standard physiological assays after this treatment [4, 6]. The resultant cell suspension can be directly used for transcript extraction but in practice is kept on ice during the transit to the sequencing facility. Gentle pipetting (with Gilson or syringe/needle as appropriate) just prior to loading the suspension ensures single cells are injected into the cell isolation equipment. The absence of a clear stress response in our data, the general concordance of our data with standard population transcriptomics datasets [5] and the overall high predictability of the gene expression behaviors of specific candidate genes arising from our analysis (e.g., [3]) suggest this cold disaggregation approach is minimally invasive, at least as far as the transcriptome is concerned.

The commercial pipelines used for going from single cell suspensions to single cell reads are the 10× Genomics Chromium

Single Cell Gene Expression platform or Fluidigm (now Standard BioTools Inc.) C1 single-cell mRNA sequencing system. The C1 platform benefitted from high sequencing depth but lower cell counts (around 30–60 cells per run was usual although the theoretical maximum was 96). The 10× platform has been used more recently, which has lower read depth but many more cells (in our experiments, usually up to around 5000 per run, however, biological replicates can be loaded at the same time, which greatly facilitates quality control). In addition, the 10× platform can take the cells straight from ice-cold buffer to the chemistry, whereas the C1 required visual inspection of cell counts/well at room temperature, which might permit stress responses to develop—although in our experience, these were not pronounced—with obvious exception of pre-existing spontaneous stress on cells, such as DNA damage [6]. As the RNA conversion to read counts requires amplification steps, final read counts are not linearly related to the initial single cell content of each transcript. However, the option for exact transcript quantification using unique short sequences (unique molecular identifiers: UMIs) can be applied to both C1 and 10× approaches.

3.2 Generating Read Count Tables

The focus of this section is the analysis of the read count data (number of transcripts belonging to each gene in each cell). For this reason, we provide a short overview of what precedes generating a read count table after receiving sequencing read files from the sequencing facility.

Raw sequenced reads are, by standard, stored as FASTQ files. The quality of the FASTQ files should be checked (with FASTQC tool) and then the reads that pass the quality threshold are aligned to the reference genome (in a FASTA file format, together with its correct annotation in GFF format). Based on these alignments (SAM or BAM file format), each transcript for each cell is counted accordingly. Depending on the sequencing technology used, the examples presented in this chapter used data processed in one of the following ways: reads retrieved with the 10× system were processed with Cell Ranger analysis pipelines, while reads from the C1 system were processed with tuxedo suite protocols (previously the Bowtie and Cufflinks packages, now HISAT2 and StringTie tools).

3.3 Data Preprocessing

3.3.1 Basic Data Formatting

After reads have been mapped and counted, the data is in the form of a count table usually in a format of a genes as rows and cells as columns. Depending on the number of samples and the analysis pipeline we used, some basic formatting steps were carried out first, such as merging replicates or naming cells in an appropriate way (e.g., "rep2_cell1" for a first cell from a second replicate, instead of a barcode "AAACGAACATTGACCA-1").

3.3.2 Removing Outlier Samples

We show a simple example where we excluded samples unlikely to be single, whole, properly lysed cells. These were removed to increase the signal-to-noise ratio in later analysis. We excluded these outliers based on the following properties:

1. Low total number of mapped genes per cell. A low number of mapped genes per cell may indicate cell debris or an ineffectively lysed cell. The best way to decide on the low threshold value is to visualize the data. The example in Fig. 1 is from [5]. We would expect the unfragmented, well-lysed cells are less different from each other than from debris. A trough in the plot showing the occurrence of mapped genes per cell (Fig. 1a) indicates that the first peak (with few mapped genes) most likely comes from poor samples.
2. Low total number of transcripts. Extremely low numbers of transcripts also indicate cell debris, an ineffectively lysed cell, or

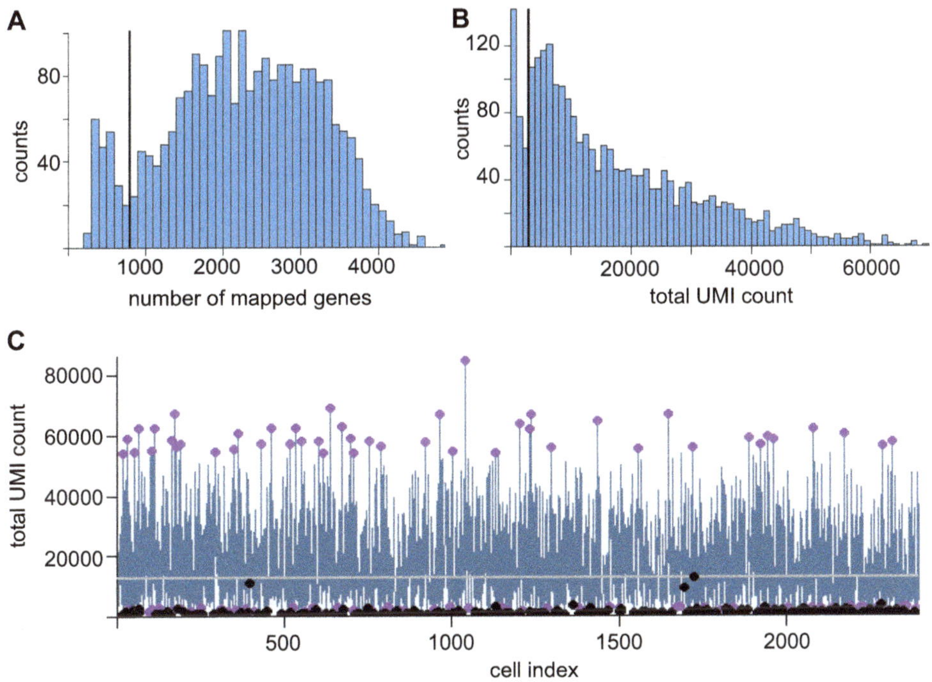

Fig. 1 Removing outlier cells. (**a**) Distribution of the number of detected genes per cell. The vertical line marks the chosen threshold. All samples with total number of genes lower than threshold are removed from further analysis. (**b**) Distribution of the total number of UMI counts per cell. The vertical line marks the chosen threshold. All samples with total UMI count lower than threshold are discarded from further analysis. (**c**) Total UMI count per cell. Purple dots mark cells excluded based on their total UMI count (too high or too low). The threshold for high outliers was Q3 + 1.5 · IQR. Black dots mark cells excluded based on the low total number of genes detected per cell. Horizontal gray line marks the overall sample median

the failure in RNA capture step. Again, the best way to decide on the threshold value is to visualize the data (Fig. 1b). A lot of these outliers will overlap with those detected in the previous step (Fig. 1a).

3. High total number of transcripts. Unexpectedly high numbers of transcripts could mean more than one cell was captured. The best way to detect outliers here is, again, to visualize the data. If there is no obvious threshold value, a Tukey Fences for outliers could be used, where values above the sum of the upper quartile (Q3) and 1.5 or 3 times the interquartile range (IQR) would be considered as outliers or far outliers, respectively.

An example of the outliers detected by the previously mentioned steps is shown in Fig. 1c. There are other ways of detecting suspicious samples (e.g., high percentage of mitochondrial transcripts, as used in [2]). These choices will depend on the biology of the sample, for example, if it is suspected that there may be rare populations of cells of interest, it would be a good idea to be less stringent in outlier removal.

3.3.3 Normalization

As the sequencing depth always varies between the samples, there is a need for appropriate normalization. Depending on the technology used, there might also be a need to normalize for gene length, as in the techniques not using unique barcoding (UMIs), there is a bias toward longer genes being assigned more counts. The initial standard methods used for normalization were RPM (reads per million mapped reads), where all the gene counts in the cell are divided by the same value (the so-called size factor) and RPKM (reads per kilobase per million mapped reads), where each gene in the cell is divided by a gene-specific value. Both these techniques are still used today. The drawback of these methods is that total read counts in cells are often dominated by a few highly expressed genes (e.g., actin), and this can distort the inferred relative level of expression of weakly expressed genes [8].

For deep sequenced samples acquired using the C1 approach [2, 3], we found that using the size factor from "DESeq" package in R [8] gives a good result. This method normalizes single cell read counts based on the geometrical mean over all transcripts (in the case of *Dictyostelium*, this means a geometrical mean of up to 14,000 different transcript counts). For the more shallow sequencing sample acquired using the 10× method used in [5], where over 4000 cells were sequenced, we used size factors calculated with "scran" package in R [9]. This method was developed alongside the advancement of single-cell RNA sequencing (scRNA-seq) and assumes a large and heterogenous population. It pre-clusters the cells based on their expression profile, calculating each cluster's size factor value, before deconvolving each individual cell's size factor.

For general reviews on normalization methods, we suggest [10, 11].

At the end of the analysis, one should always check whether the final observations are also valid in the unprocessed data and not an artefact introduced by preprocessing (*see* **Note 1**, Fig. 8).

3.4 Gene Expression Variability

A major advantage of the development of scRNA-seq is the ability to study the distribution of gene expression in a population of cells and how this leads to the emergence of different cell types. To explore the emergence of cellular heterogeneity during development, we analyzed individual transcriptomes of 433 cells from three different development timepoints [2]. Cells were collected at 0 h (undifferentiated cells), 3 h (starvation), and 6 h (aggregation), with a median of 3 million reads and 5600 mapped genes per cell. A typical way of showing gene expression variability in a cell population is plotting square coefficient of variation (CV^2) for each gene against its mean expression level (Fig. 2a). The variability in gene expression is negatively correlated with its mean expression level. We define a variable gene as a gene more variable in its expression than expected for a gene with that level of expression. The "expected" variance is described with a running median of CV^2 values and is depicted with a red curve in Fig. 2a. In this example, genes previously known to be variably expressed, contact site A (*csaA*) and discoidin I (*dscA*), and a gene known as homogeneously expressed, actin 5 (*act5*), are plotted on the same figure. We found that overall variability in the cell population increases over time (Fig. 2b) and that this increase affects genes at all levels of expression (Fig. 2c). This heterogeneity increased without clear separation of the population into different states [2].

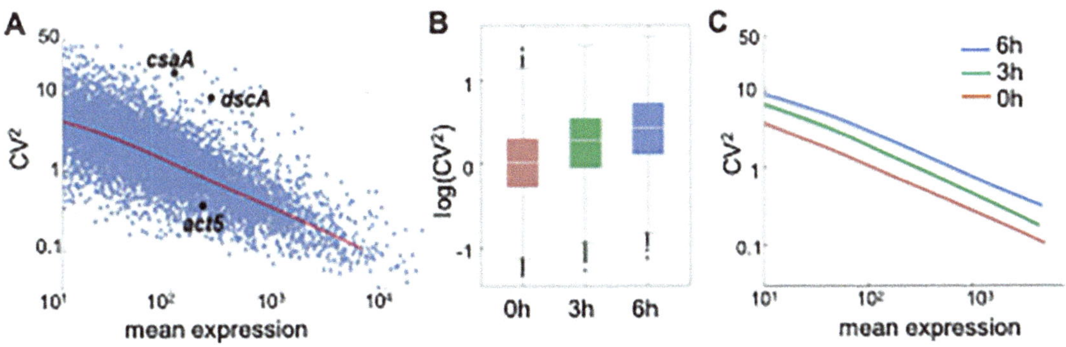

Fig. 2 Dynamics of gene expression heterogeneity during early development. (**a**) Transcript variability (expressed as CV^2) of each gene with mean expression >10 read counts plotted against the mean expression. The red line marks the running median. *csaA* and *dscA* are shown as examples of known variably expressed genes and *act5* as an example of a gene with more uniform expression in the population. (**b**) Overall transcript heterogeneity within a cell population increases during early development. Distribution of $\log_{10}(CV^2)$ values of genes is shown for populations at 0, 3, and 6 h of development. (**c**) The increase in variance occurs across all levels of expression. The running medians are plotted for all three timepoints. (Figure reproduced from [2])

Since many genes are downregulated during these initial stages of the development [2], we decided to examine how downregulation and upregulation of transcript levels affect transcript variability. This analysis used the distance from median (DM, [12]) as the measure of transcript variability. The first step for calculating DM is defining each transcript's variance as a distance between their CV^2 value and the value expected for the gene of such level of expression, as visualized by the running median in Fig. 2a. If read counts were extracted without UMIs, there is a need for another step, which also corrects for the effect of gene length [12]. As a result, we get a transcript variance measure that is independent of gene's length or its expression level (Fig. 3a). The DM method is now also a part of the "scran" package in R [9].

Comparing the variance between genes being up- and downregulated from undifferentiated cells (0 h) to the start of aggregation (6 h), we found that downregulated transcripts become more variable than upregulated transcripts. This effect is consistent across all levels of expression (Fig. 3a). In addition, the higher the fold-change expression threshold we used, the difference became even more apparent (*see* **Note 2**, Fig. 9). We tested whether the difference in transcript stability of up- and downregulated genes could be causing this effect on variability, by using the RNA turnover kinetic data from [13] and found no correlation between RNA turnover and gene expression variability (Fig. 3b). We then considered how regulation of transcription could influence the variability in transcript abundance between cells. We simulated transcription with a simple two-state model, where a gene is either in *off* or *on* state and switches between these states with the rates of k_{on} and k_{off}, respectively (Fig. 3c). This results in bursts of transcription, described by the burst size and the frequency of burst occurrence. When in the *on* state, the polymerase initiation event happens at the rate λ, and the transcripts have a degradation time τ. We ran the simulations using physiological parameter estimates [13–16]. This suggested that gene expression if regulated primarily by changing the burst frequency then transcription downregulation will result in a more variable distribution of the transcript abundance in the cell population (Fig. 3d), as observed in the data. This is best visualized in Fig. 3e: reduction of burst frequency results in a lower mean expression but a higher variance. In the opposite direction, the increase in frequency increases the expression but lowers the variance. Initially, we looked at the difference of variance at what we considered the end point—6 h of differentiation. But from the point of the model, there is no reason why one end of the process would be considered the beginning and the other the end. Following that logic, we would expect exactly the opposite characteristics of up- and downregulated genes at 0 h: genes to be

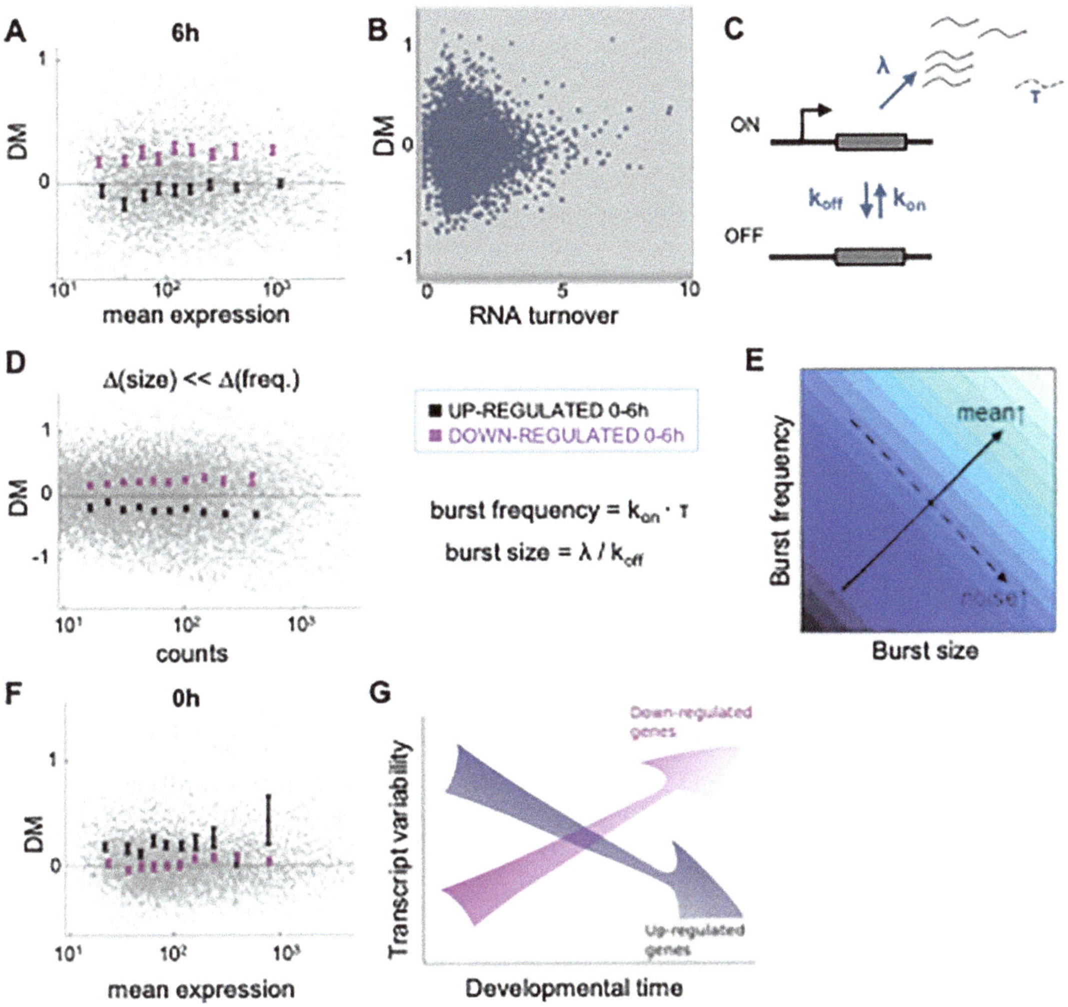

Fig. 3 Origins of gene expression heterogeneity. (**a**) Genes downregulated during development become more variable than upregulated genes. Transcript variability (expressed as DM) of each gene with mean expression >10 read counts in the population of aggregating cells (6 h) is plotted against the mean expression. Distribution of DM values for each bin is shown in purple and black, for down- and upregulated genes, respectively. Each bin contains 500 genes. The threshold for fold change (FC) is fivefold in both directions, that is, $|\log_2 FC| > 2.32$. (**b**) DM of each gene plotted against its RNA turnover. RNA turnover is measured as a ratio of the expression before and after 1 h actinomycin D treatment [13]. (**c**) Schematic of a two-state model of transcription. (**d**) Results of a transcription simulation presented in the same format as **a**. (**e**) Schematic of the influence of changing burst size and burst frequency on mean expression and variance. (**f**) Genes upregulated during development are more variable at the onset of starvation than downregulated genes. (**g**) Schematic of the findings. (Figure reproduced from [2])

developmentally upregulated would be more variable to start with, while the genes to be downregulated would be more homogeneous. This is exactly what we observe at the onset of aggregation (Fig. 3f). The summary of this findings is presented in Fig. 3g.

3.5 Linear Dimensionality Reduction

As each *Dictyostelium* cell is described with around 14,000 genes, it is impossible to visualize the data in its original form—each cell would be a point in the 14,000-dimensional space. In order to visualize, investigate and describe the data we need to use techniques of dimensionality reduction.

To be able to directly interpret the data in this reduced space, the first step is usually principal component analysis (PCA), a linear dimensionality reduction method. To remove experimental noise, we only used the genes with mean expression above the certain threshold. The choice of this threshold was based on the convention at the time of the study: we used a value where the coefficient of variation stops reaching a maximal value. For example, in Fig. 2a, the mean expression threshold value was ten read counts. Another standard approach (used, for example, in the Cell Atlas projects [17]) is to use only the most variably expressed genes, selecting all the genes more variable than average or only the 10% most variable genes (*see* **Note 3**, Fig. 10).

A useful example considers the mound phase of development [3]. For this phase, we captured 116 single cell transcriptomes with around 3 million reads and 5400 mapped genes per cell. PCA linearly transforms the data into new components, sorted in the way that the first component will capture most of the variance in the set, with each further component capturing less and less variance. The top-left panel in Fig. 4a shows the cells positioned based on their transcriptome profiles in principal component 1 (PC1) and PC2 space. To interpret this plot, we use loading analysis. As each component is a linear combination of original dimensions (i.e., genes), we can interpret each component based on the contributions of genes to its total variance. The higher the loading's absolute value an individual gene has in a principal component, the larger its contribution is.

We carried out the loading analysis in the following way. For each component, we sorted the genes based on their absolute loadings and took the top genes whose contribution summed to the 25% of the component's variance. We then used these genes for gene ontology enrichment analysis, using the PANTHER Classification System on the Gene Ontology website [18]. The resulting enrichments in biological processes are shown in the tables below the PCA panel for the PC1 and right of the panel for the PC2 in Fig. 4a. It is apparent from these results that PC1 and PC2 capture different aspects of development: PC1 captures translation being downregulated and development initiated, while PC2 captures cytokinesis, signalling, adhesion, and cAMP chemotaxis. If we overlay the known biological information on this plot, such as expression of developmentally upregulated and downregulated genes (Fig. 4b left and right panel, respectively, [19]), we can see the direction of development going from left to right (across PC1), with the tilt toward the positive PC2 value. Looking at the PC3,

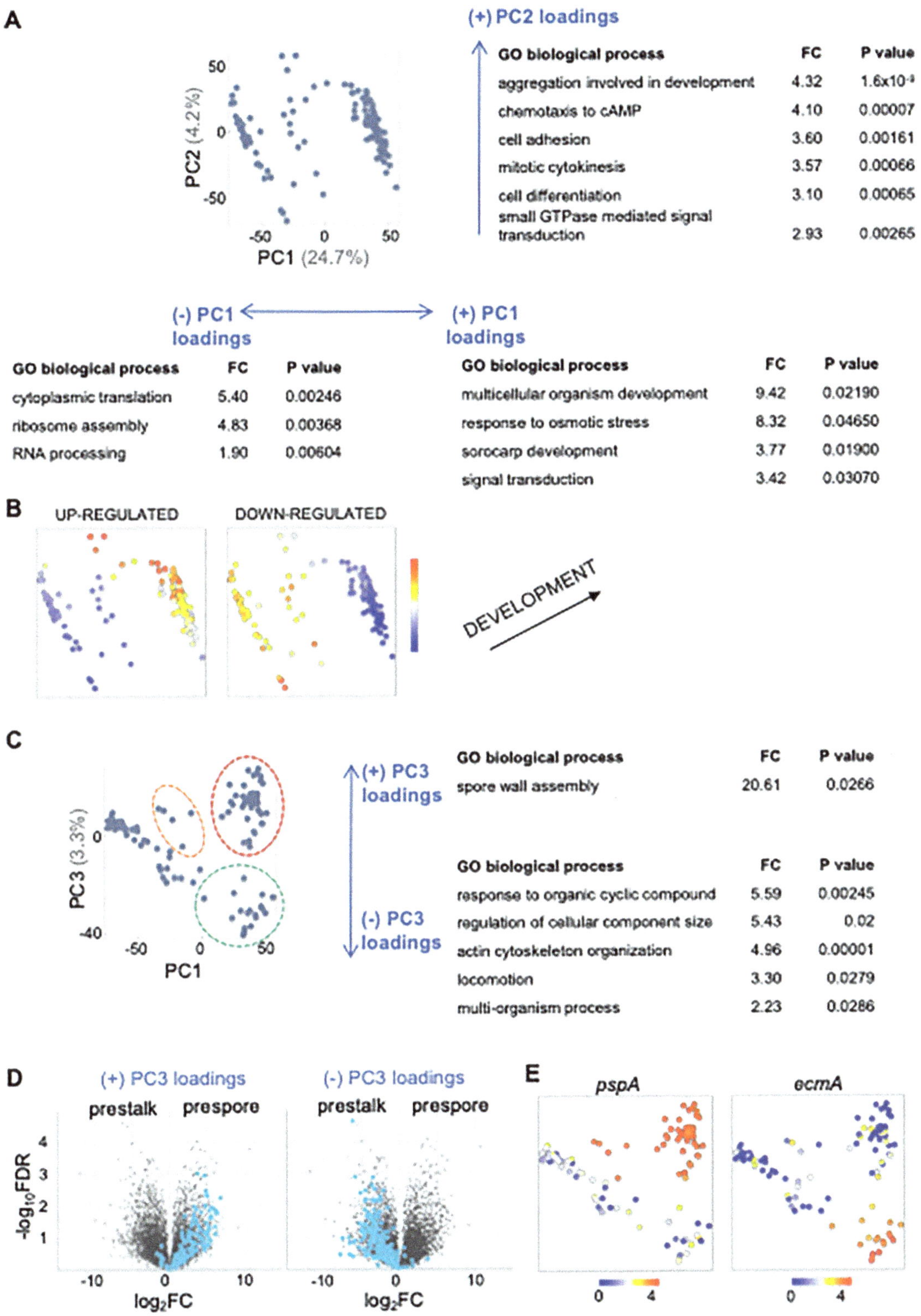

Fig. 4 Extracting biological information from PCA. (**a**) Cell positioned in PC1 and PC2 space based on their transcriptional profiles. Lower left panel: GO terms (biological processes) enriched in the negative loadings of PC1. Lower right panel: GO terms enriched in the positive loadings of PC1. Right panel: GO terms enriched in

while keeping PC1 as a proxy for development, we can observe the separation of cells in two separate clusters at the high values of PC1 (Fig. 4c, left panel). Loading analysis of PC3 gives easily interpretable information for the top cluster (spore wall assembly) but a more complex one for the bottom cluster (right panel in Fig. 4c). We then used the same set of genes to look at their enrichment in prestalk or prespore cells, using the information from [20] on the dictyExpress platform [21]. The result of this analysis is shown in Fig. 4d. PC3 captures the cell fate information, and we observe the obvious separation of prespore and prestalk cells across PC3. Overlaying the expression of known cell fate markers, *pspA* for spore and *ecmA* for stalk cells, illustrates this inference (Fig. 4e).

Following the same principle, we identified another cluster of cells along the PC4 axis [3]. This population appears as an intermediate between the early and differentiated cells, with a slight enrichment in prestalk markers. All identified clusters are shown in Fig. 5a.

3.6 Hierarchical Clustering

A simple and useful approach to explore a single cell transcriptomics dataset is two-way hierarchical clustering, where cells and genes are simultaneously ordered and clustered in an unbiased way based on their similarities (Fig. 5b). For this analysis, we selected genes based on the following conditions: their mean normalized read count was >10 and are correlated with at least ten other genes with Pearson's correlation >0.5. Over 5000 genes passed this criterion (almost half of the genome), which is a huge advantage of the C1 deep sequencing methods. With the similar criteria for data collected using 10× [5], we retrieved 957 genes.

The result of the hierarchical clustering is shown in Fig. 5b. Expression level of each gene is normalized, so each gene will have a mean of zero and a standard deviation of one. The colors in the heatmap represent the z-score for each gene across the cells, which reflect the up- or downregulation of the specific gene in the specific

Fig. 4 (continued) the positive loadings of PC2. (**b**) Same plot as in A but overlayed with the average expression of genes upregulated and downregulated during the mound stage. The arrow on the right shows inferred direction of developmental progression. (**c**) Cells positioned in PC1 and PC3 space based on their transcriptional profiles. PC1 here serves as a proxy for developmental time. The orange, red, and green ellipses mark early prespore, prespore, and prestalk populations, respectively, based on observations in **c**, **d**, and **e**. Upper right panel: GO terms enriched in the positive loadings of PC3. Lower right panel: GO terms enriched in the negative loadings of PC3. (**d**, **e**) PC3 describes the separation of cell population in the spore and stalk cells in the later mound stages. (**d**) Volcano plots of positive PC3 loadings (left panel, blue dots) and negative PC3 loadings (right panel, blue dots) show almost exclusive enrichment of spore and stalk markers, respectively. Log_2FC is a measure of fold change in prespore cells versus prestalk cells; $-log_{10}FDR$ is a measure of false discovery rate. Plots downloaded from dictyExpress platform based on data from [20]. (**e**) The same as plot in **c**, overlayed with the expression of *pspA* and *ecmA* as a spore and stalk marker, respectively. Figure partially reproduced from [3]

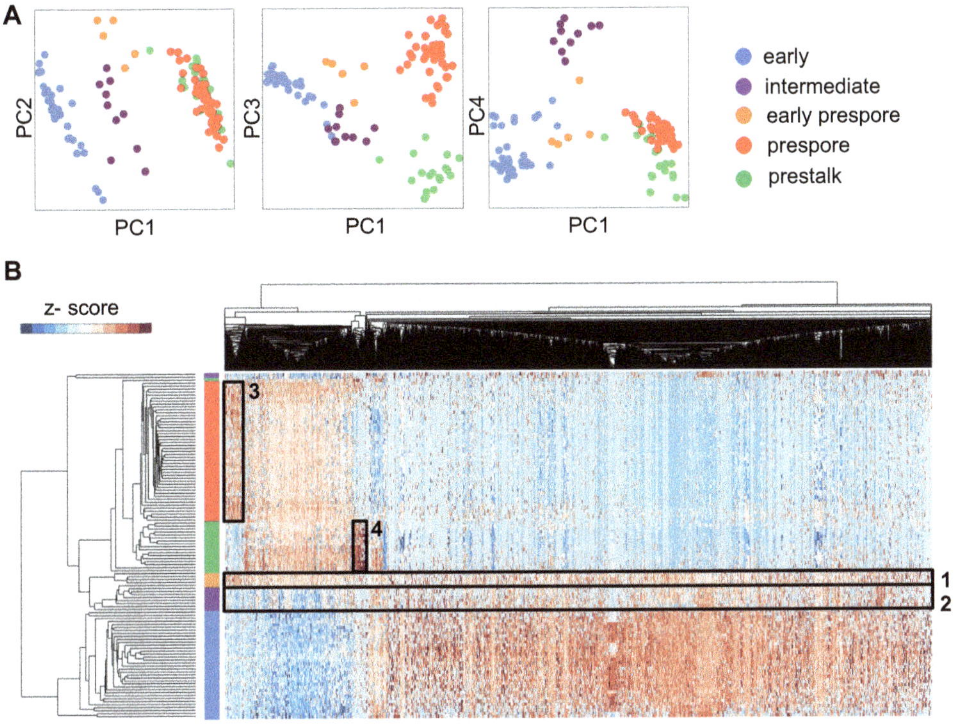

Fig. 5 Visualizing cell states. (**a**) Cells from the example in Fig. 4, positioned based on their transcriptome profiles in PC1 and PC2 (left), PC3 (middle), and PC4 (right panel). Color legend on the right describes all inferred cell states. (**b**) Two-way hierarchical clustering of cells and genes from the same experiment. Columns are genes and rows are cells. Colored boxes next to the cells' hierarchical tree represent same subpopulations marked in the PCA plots in **a**. Boxes marked 1 and 2 show early prespore and intermediate/early prestalk subpopulation, respectively. Boxes marked 3 and 4 show induced spore and stalk markers, respectively. (Figure adapted from [3])

cell. The colored boxes next to the cells' hierarchical tree mark cell clusters as in Fig. 5a. We observe a sudden shift in the transcriptome profiles as cells go through development within the initial multicellular structure (mound), with 80% of the genes being downregulated. Two intermediate populations are also clearly visible and seem to be captured at the moment of transition, with the "early" genes not yet being downregulated and "late" genes already showing activation (cell clusters marked with 1 and 2). Within the genes upregulated during this transition, there are clear clusters of cell fate-specific genes (marked with 3 and 4 for spore and stalk markers, respectively).

Overall, this analysis revealed key features of a developmental transition. Firstly, cells occupy discrete states during developmental progression. Secondly, these analyses confirm the existence of intermediates states that have been previously inferred from single gene analysis [22], showing that these states show mixed early and late gene expression, in addition to signatures of the ultimate fate. Thirdly, the major gene expression transition in the mound stage is dominated by repression, not activation, which contrasts the standard developmental narrative of cell differentiation being determined by what genes are "on" (rather than what genes are "off").

3.7 Nonlinear Dimensionality Reduction

More recently, we used the 10× method to extract individual transcriptomes from 4743 cells [5]. The sequencing depth here is much lower, with around 35,000 reads and 2300 mapped genes per cell. Although shallow sequencing is less useful for extracting detailed mechanistic information on gene expression, it enabled us to profile ten times more cells than in the previously mentioned experiments, which has provided very clear data on the transitions occurring from the undifferentiated state through to the mound [5].

Cells were captured while feeding on bacteria on agar plate, by taking a continuous streak of cells from the edge of the multicellular zone and into the layers of bacteria. This way we captured the cells in a mimic of their physiological environment, in all moments of development up to the tight mound stage, where differentiation of spore and stalk cells begins within the multicellular structure. This enabled us to capture a continuous time series of development in a very complex signaling environment, rather than the more standard timepoint sampled experiments where all cells are synchronously starved in homogeneous environment at uniform density. The early developmental process captured is shown in Fig. 6a. In Fig. 6b, 4743 cells described with 9698 genes (mean expression >0.01) are plotted in the space of PC1 and PC2. Overlaying the biological knowledge we have on the system (expression of genes active before the development begins, during the cell aggregation, and after the cell aggregate), we conclude that again, PC1 represents the axis of developmental time—the direction of development captured in this plot is schematically shown with the overlayed arrow in the last panel (Fig. 6b). Distribution of cell density across these two dimensions is shown in Fig. 6c. Using the higher-order principal components, a sudden change in transcriptome profiles becomes apparent which appears to correlate with the start of aggregation (Fig. 6d). PC5 captures the separation of the population toward two different cell fates at the end of development (Fig. 6e).

Although we are dealing with a relatively "simple" model organism, the information we are interested in spreads across at least five dimensions. To reduce dimensionality even further, we use nonlinear dimensionality reduction techniques. Many of these methods are often used in singe cell transcriptomes analysis (e.g., t-SNE, UMAP, diffusion maps, etc.). We decided on elastic

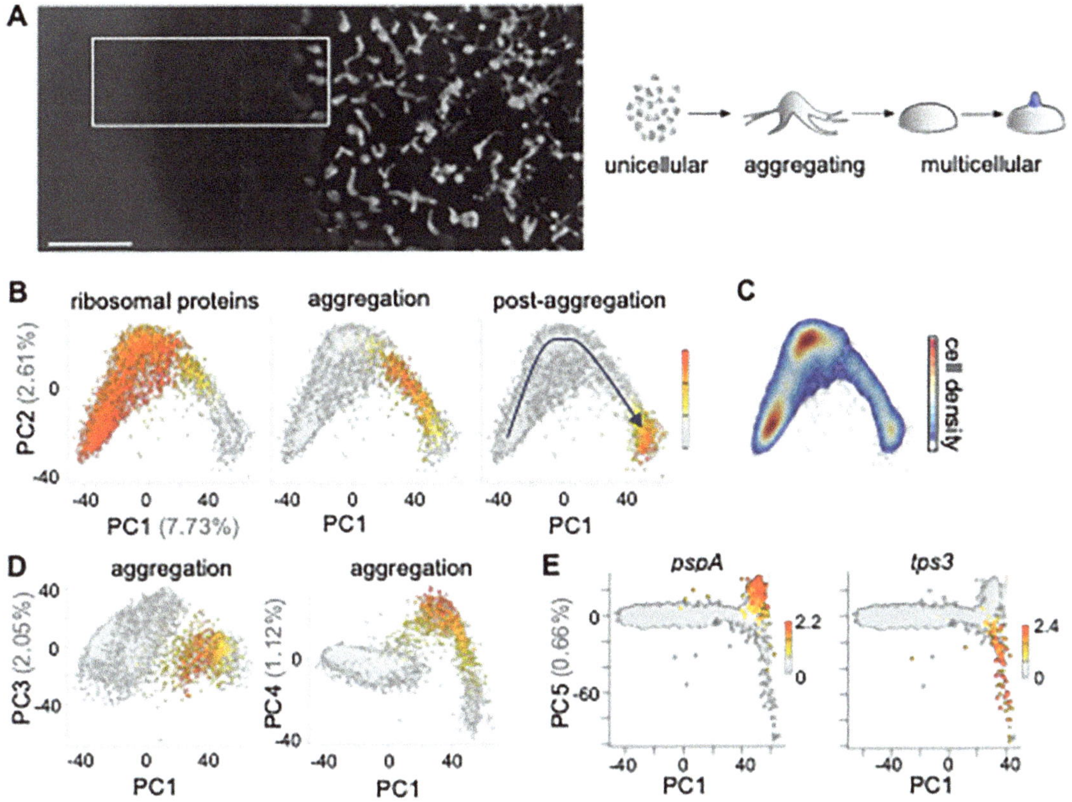

Fig. 6 Continuous sampling of developmental progression. (**a**) Image of cells plated on bacteria, with feeding front on the left and development progression to the right. The white box approximates the region sampled. Right panel: schematic of the developmental stages sampled. (**b**) Cell positioned in PC1 and PC2 space based on their transcriptional profiles. Overlayed is an expression of sets of genes expressed during early development, aggregation, and aggregate. The arrow in the right panel shows approximate direction of development. (**c**) Same as **b** but overlayed with relative cell density values. (**d**) Cells positioned in PC1 and PC3 space (left) and PC1 and PC4 space (right), overlayed with expression of aggregation genes. (**e**) Cells positioned in PC1 and PC5 space overlayed with expression of *pspA* and *tps3*, as spore and stalk markers, respectively. (Figure adapted from [5])

embedding (EE), which seems to preserve both local and global structures of the data [23]. For Fig. 7, we used EE on the first 11 PCs from the example in Fig. 6. Here, the visualization reveals more direct information: as cells transition from individual cells to streams (from northwest to southeast), there is a sudden substantial shift in the transcriptome profiles (the "jump"). Secondly, during tight mound formation, cell transcriptomes seem to converge to a similar expression profile (the "bottleneck"), before separating toward two different fates, stalks and spores (Fig. 7a, b). The developmental context of the jump is validated by overlaying expression of gene sets marking undifferentiated cells and streams from a separate study [24] (Fig. 7c), which also shows the sudden shift. The cell density plot in Fig. 7d shows the distribution of cells

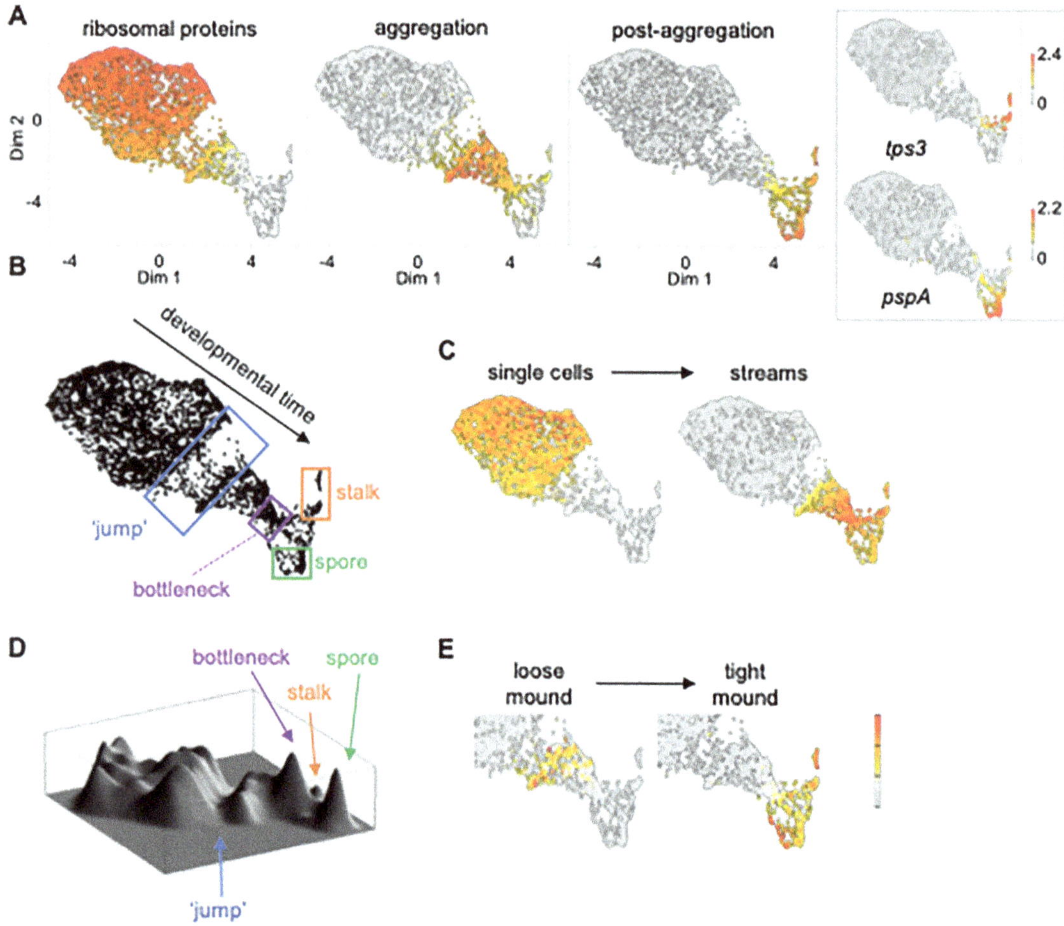

Fig. 7 Two-dimensional visualization of cell state transitions during development. (**a**) Cells positioned in two-dimensional space, after EE was performed on the first 11 PCs. From left to right: overlayed expression of genes expressed during early development, aggregation, and mound. Right panel: overlayed expression of *tps3* (stalk) and *pspA* (spore) markers. (**b**) Jump, bottleneck, stalk, and spore regions marked on the same plot as in **a**. The arrow shows the inferred direction of development. (**c**) The same as in **a**, overlayed with the average expression of genes down- and upregulated during the morphological transition from single cells to streams [24]. (**d**) Three-dimensional cell density plot, with the same regions marked as in **b**. (**e**) Zoom-in of the lower-right region of the plot in **a**, overlayed with the expression of genes down- and upregulated during the morphological transition from loose to tight mounds [24]. (Figure reproduced from [5])

across these two dimensions and shows a considerable heterogeneity just before the jump, with only a few cells caught in the moment of transition and an accumulation of cells at the bottleneck, before separating in two different cell types. Using again the data from [24], this bottleneck corresponds to the loose to tight mound transition (Fig. 7e).

3.8 Trajectory Inference

There are many trajectory inference methods developed over the last decade, aiming to determine the approximate order and direction of differentiation across the transcriptome space and to predict the trajectories by which individual cells might explore the possible cell states. We tried several of these but often found them inadequate for tracking *Dictyostelium* development for different reasons. For instance, using the scRNA-seq data from onset of development up to aggregation stage [2] and mound stage data [3], we found both Monocle [25] and Wishbone [26]-mapped 3 h starved cells branching into two clusters: aggregating and mound cells. When trying to infer the developmental progression trajectory on the continuously sampled development up to the mound stage [5] with DensityPath [23] and StemID [27], the heterogeneous region before the "jump" was interpreted as containing at least three clusters: two dead ends and only one continuing across the jump into the development that skips aggregation phase as another dead-end cluster. These observations seem unrealistic based on several decades of literature on *Dictyostelium* development.

For trajectory analysis, we also used RNA velocity [28], an algorithm which uses the information on spliced versus unspliced introns to predict the future state of each cell. We were concerned that *Dictyostelium* genes have too few introns for the algorithm to work, with other considerations that the introns are highly AT rich and therefore unlikely to be mapped effectively and also that RNA processing may be too efficient to leave substantial intron signatures in the data. Indeed, our analysis revealed very few introns in our reads, and these appeared limited to highly expressed genes with long introns, such as ribosome protein genes. However, we nevertheless attempted to apply RNA velocity to our data. Our initial analysis suggested cells in the early mound state "circulate" within this state, rather than progress forward. However, our velocity fields were weak and unconvincing. The initial data may warrant further study, perhaps using gene expression imaging [14] or more recent transcriptome inference methods [29]. We also suggest [30] for a detailed overview of different pseudotime methods. The possibilities for DNA barcode-based lineage tracing [31] may be limited for following cell histories during *Dictyostelium* development. Cells divide no more than once during development [32, 33], so the likelihood of sampling cells from the same division (to enable the construction of lineage trees) using the sequencing technologies we have used to date is expected to be small.

4 Notes

In this section, we briefly reflect on some key examples of how our exploratory data analysis works. As mentioned above, it is a good practice to always check whether the final observations are also valid

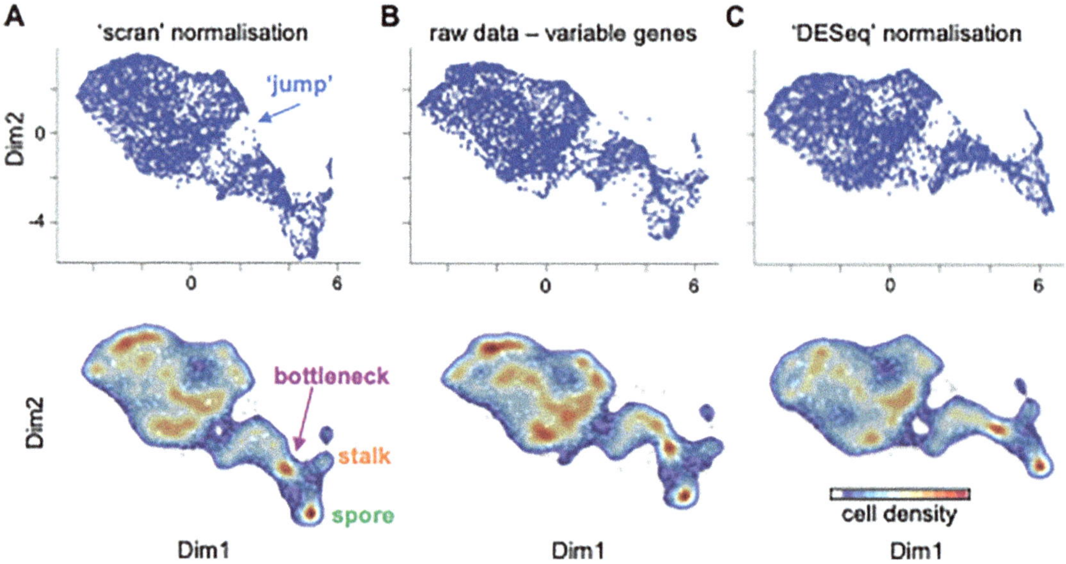

Fig. 8 Comparing different normalization approaches. (**a–c**) Cell positioned in two-dimensional space, based on their transcriptome profiles. (a) The same as in Fig. 7. (**b**) Same as in **a** but performed on the raw data (not normalized) and using only variable genes (4849, instead of 9698 genes). (**c**) Same as in **a** but normalized using the size factor from DESeq2 package in R. Lower panels show the same data as in upper panels but overlayed with relative cell density information. Jump, bottleneck, stalk, and spore regions are marked. (Data used are from [5])

in the unprocessed data or in the data processed in a slightly different manner. For instance, normalization is used to dampen the noise in a data. Different normalization methods could make a less clear signal but should not give a completely different conclusion.

1. Comparing Normalization Approaches.

   In Fig. 8, we show an example from [5] represented in Fig. 7. The top panels show EE plots of cells' transcriptome profiles, while the bottom plots show distribution of cell density. Figure 8a shows data normalized by pre-clustering and deconvolution method ("scran" package in R, [9]. In Fig. 8b, the data were not normalized, but only variable genes were included to reduce the noise. Figure 8c shows data normalized by the geometrical mean size factor from DESeq2 package in R [8]. The major observations are consistent in all three examples: high heterogeneity just before the jump, the jump, the decrease of heterogeneity in the bottleneck, and the separation of cell fates immediately after.
2. Binning, Threshold, and Variance.

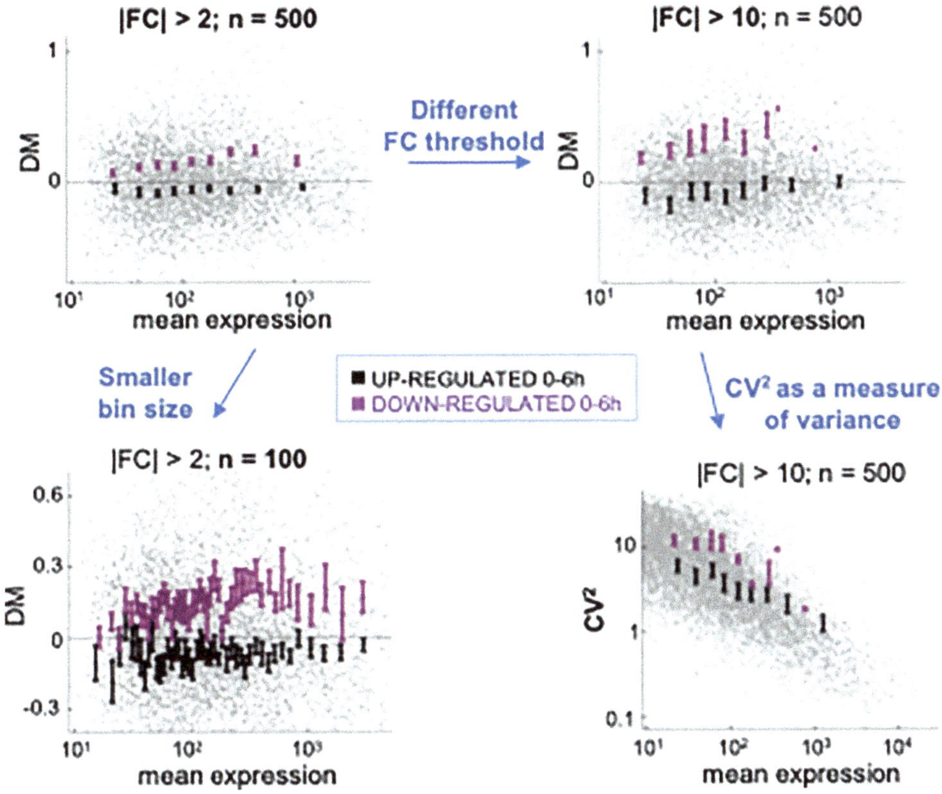

Fig. 9 Comparing different thresholds, bin sizes, and variance measures. Genes downregulated during development become more variable than upregulated genes. Transcript variability (expressed as DM) of each gene with mean expression >10 read counts in the population of aggregating cells (6 h) is plotted against the mean expression. Distribution of mean DM for each bin is shown in purple and black, for down- and upregulated genes, respectively. Each bin contains 500 genes. Same data as in Fig. 3, except for gene expression fold change threshold, is twofold in each direction ($|\log_2 FC| > 1$). Bottom panel: result of decreasing the bin size to 100 genes. Right panel: result of increasing the FC threshold to tenfold in each direction. Lower right: result of using CV^2 as a measure of variance (CV^2 does not correct for expression level or gene length). (Figure reproduced from [2])

In the next example (Fig. 9), we will refer to the example from gene expression variability study ([2] and Figs. 2 and 3 above). As in Fig. 3, the DM values for all genes are shown in gray. The transcript variabilities of genes up- and downregulated during development are marked in black and purple, respectively. The top-left panel shows the variance distribution for all genes with at least twofold change in expression, in any direction ($|\log_2 FC| > 1$). If we increase this threshold to at least tenfold (right panel), the difference in variance becomes even more apparent. In both these analyses, we compared the genes of the similar expression level by binning so that each bin contains 500 genes. Changing bin size (bottom-left panel), the overall difference between up- and downregulated genes is still clear. Finally, the bottom-right panel shows that overall conclusion is also retained if we use CV^2 values of genes rather than DM.

3. Deciding Which Genes to Include in the Analysis.

   This example is linked with example in Fig. 6, with PCA using 9698 genes (criteria being the mean expression >0.01; [5]). We can additionally simplify the data by selecting only variable genes, which is actually the most common trend in single cell transcriptomic analysis, notably in the studies focused on building cell atlases [17].

   In Fig. 10a, genes are plotted based on their variance (CV^2) and mean expression. The vertical line marks the gene expression threshold (>0.01). The genes marked yellow, black, and red are the top 50% variable genes. The top 25% most variable genes are marked black and red, while top 10% are marked only in red. Figures 10b–d show the result of PCA on the same sample as in Fig. 6 but using only 50%, 25%, or 10% most variable genes. We can see that the sudden transcriptome shift is already visible in PC1/PC2 space, when using only variable genes. In addition, fate specification already appears in PC3, rather than PC5 when using only top 10% variable genes. Finally, more variance is explained by single PCs: the variance captured by PC1 goes from 8% to 20%, when reducing the set of genes from 9698 genes with mean expression >0.01 to the 969 top 10% variable genes.

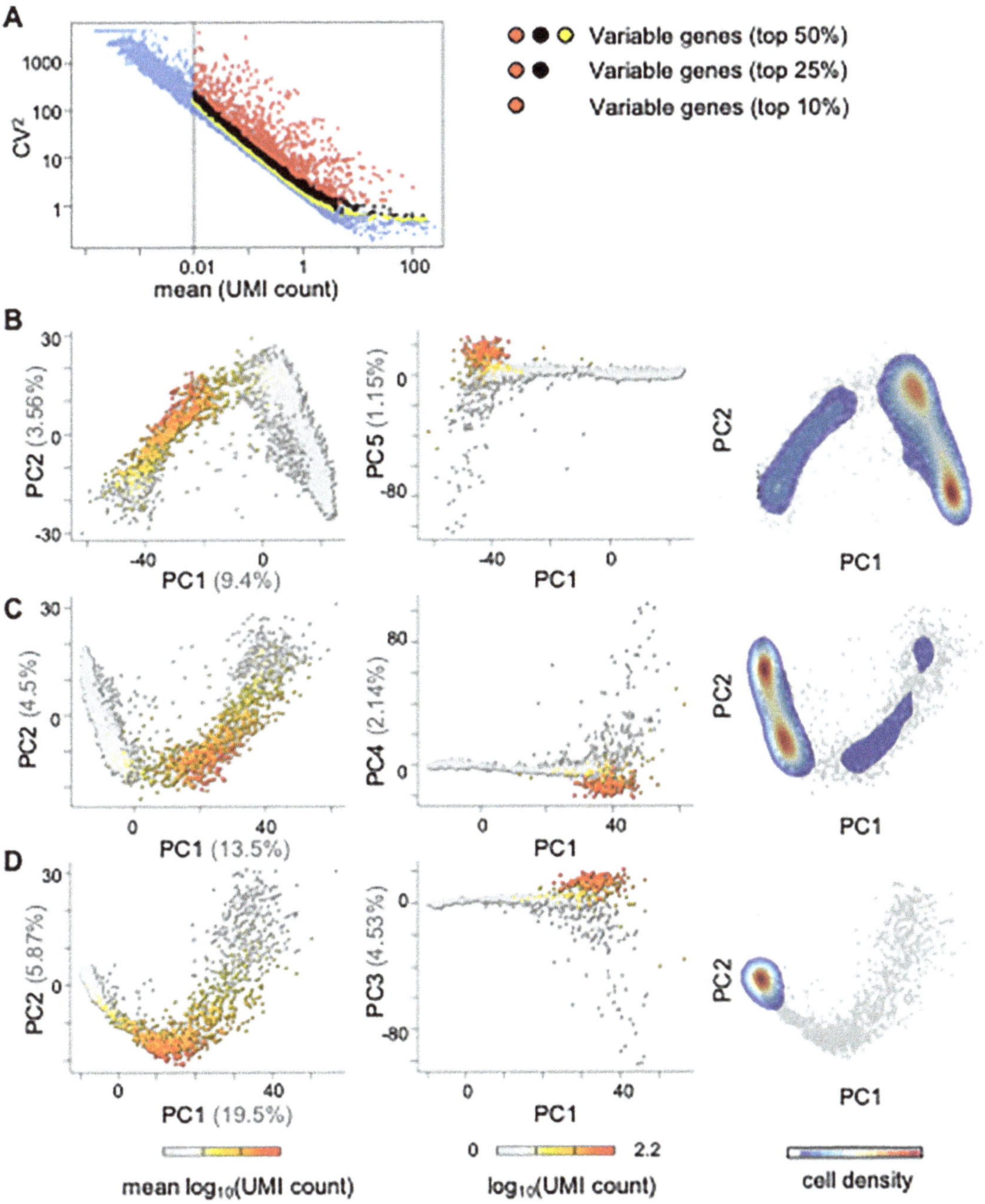

Fig. 10 Exploring effects of different numbers of genes on the analysis. (**a**) Transcript variability (CV^2) of each gene plotted against mean expression. The vertical line marks the gene expression threshold. Marked yellow, black, and red are genes more variable than average (50%, 4849 genes). Black and red are the top 25% variable genes (2424). Red only are the 10% top variable genes (969). (**b**) Same as in Fig. 6, but cells are described only with variable genes (top 50%). Left panel: cells in PC1 and PC2 space. Overlayed is average expression of aggregation genes. Middle: cells in PC1 and PC5 space. Overlayed is the expression of *pspA* (spore marker). Right: same as left panel but overlayed with relative cell density values. (**c**) The same as in **b**, but cells described only with top 25% variable genes. Middle panel is PC4 versus PC1. (**d**) The same as in **b**, but cells described only with top 10% variable genes. Middle panel is PC3 versus PC1. (Data from [5])

References

1. Eling N, Richard AC, Richardson S, Marioni JC, Vallejos CA (2018) Correcting the mean-variance dependency for differential variability testing using single-cell RNA sequencing data. Cell Syst 7:284–294 e12
2. Antolović V, Miermont A, Corrigan AM, Chubb JR (2017) Generation of single-cell transcript variability by repression. Curr Biol 27:1811–1817 e3
3. Antolović V, Lenn T, Miermont A, Chubb JR (2019) Transition state dynamics during a stochastic fate choice. Development 146: dev173740
4. Nichols JM, Antolović V, Reich JD, Brameyer S, Paschke P, Chubb JR (2020) Cell and molecular transitions during efficient dedifferentiation. elife 9:e55435
5. Westbrook ER, Lenn T, Chubb, JR, Antolović, V (2023) Collective signalling drives rapid jumping between cell states. bioRxiv 2023: 2023.05.03.539233
6. Miermont A, Antolović V, Lenn T, Nichols JME, Millward LJ, Chubb JR (2019) The fate of cells undergoing spontaneous DNA damage during development. Development 146: dev174268
7. Tunnacliffe E, Corrigan AM, Chubb JR (2018) Promoter-mediated diversification of transcriptional bursting dynamics following gene duplication. Proc Natl Acad Sci USA 15: 8364–8369
8. Anders S, Huber W (2010) Differential expression analysis for sequence count data. Genome Biol 11:R106
9. Lun AT, Bach K, Marioni JC (2016) Pooling across cells to normalize single-cell RNA sequencing data with many zero counts. Genome Biol 17:75
10. Vallejos CA, Risso D, Scialdone A, Dudoit S, Marioni JC (2017) Normalizing single-cell RNA sequencing data: challenges and opportunities. Nat Methods 14:565–571
11. Lytal N, Ran D, An L (2020) Normalization methods on single-cell RNA-seq data: an empirical survey. Front Genet 11:41
12. Kolodziejczyk AA, Kim JK, Svensson V, Marioni JC, Teichmann SA (2015) The technology and biology of single-cell RNA sequencing. Mol Cell 58:610–620
13. Muramoto T, Cannon D, Gierlinski M, Corrigan A, Barton GJ, Chubb JR (2012) Live imaging of nascent RNA dynamics reveals distinct types of transcriptional pulse regulation. Proc Natl Acad Sci USA 109:7350–7355
14. Chubb JR, Bloomfield G, Xu Q, Kaller M, Ivens A, Skelton J, Turner BM, Nellen W, Shaulsky G, Kay RR, Bickmore WA, Singer RH (2006) Developmental timing in Dictyostelium is regulated by the Set1 histone methyltransferase. Dev Biol 292:519–532
15. Stevense M, Muramoto T, Müller I, Chubb JR (2010) Digital nature of the immediate-early transcriptional response. Development 137: 579–584
16. Corrigan AM, Tunnacliffe E, Cannon D, Chubb JR (2016) A continuum model of transcriptional bursting. elife 5:e13051
17. HuBMAP Consortium (2019) The human body at cellular resolution: the NIH human biomolecular atlas program. Nature 574:187–192
18. Ashburner M, Ball CA, Blake JA, Botstein D, Butler H, Cherry JM, Davis AP, Dolinski K, Dwight SS, Eppig JT, Harris MA, Hill DP, Issel-Tarver L, Kasarskis A, Lewis S, Matese JC, Richardson JE, Ringwald M, Rubin GM, Sherlock G (2000) Gene ontology: tool for the unification of biology. The gene ontology consortium. Nat Genet 5:5–9
19. Rosengarten RD, Santhanam B, Fuller D, Katoh-Kurasawa M, Loomis WF, Zupan B, Shaulsky G (2015) Leaps and lulls in the developmental transcriptome of Dictyostelium discoideum. BMC Genomics 16:294
20. Parikh A, Miranda ER, Katoh-Kurasawa M, Fuller D, Rot G, Zagar L, Curk T, Sucgang R, Chen R, Zupan B, Loomis WF, Kuspa A, Shaulsky G (2010) Conserved developmental transcriptomes in evolutionarily divergent species. Genome Biol 11:R35
21. Stajdohar M, Rosengarten RD, Kokosar J, Jeran L, Blenkus D, Shaulsky G, Zupan B (2017) dictyExpress: a web-based platform for sequence data management and analytics in Dictyostelium and beyond. BMC Bioinformatics 18:291
22. Clay JL, Ammann RR, Gomer RH (1995) Initial cell-type choice in a simple eukaryote: cell-autonomous or morphogen-gradient dependent? Dev Biol 172:665–674
23. Chen Z, An S, Bai X, Gong F, Ma L, Wan L (2019) DensityPath: an algorithm to visualize and reconstruct cell state-transition path on density landscape for single-cell RNA sequencing data. Bioinformatics 35:2593–2601
24. Katoh-Kurasawa M, Hrovatin K, Hirose S, Webb A, Ho HI, Zupan B, Shaulsky G (2021) Transcriptional milestones in

Dictyostelium development. Genome Res 31: 1498–1511

25. Trapnell C, Cacchiarelli D, Grimsby J, Pokharel P, Li S, Morse M, Lennon NJ, Livak KJ, Mikkelsen TS, Rinn JL (2014) The dynamics and regulators of cell fate decisions are revealed by pseudotemporal ordering of single cells. Nat Biotechnol 32:381–386
26. Setty M, Tadmor MD, Reich-Zeliger S, Angel O, Salame TM, Kathail P, Choi K, Bendall S, Friedman N, Pe'er D (2016) Wishbone identifies bifurcating developmental trajectories from single-cell data. Nat Biotechnol 34:637–645
27. Grün D, Muraro MJ, Boisset JC, Wiebrands K, Lyubimova A, Dharmadhikari G, van den Born M, van Es J, Jansen E, Clevers H, de Koning EJP, van Oudenaarden A (2016) De Novo prediction of stem cell identity using single-cell transcriptome data. Cell Stem Cell 19:266–277
28. La Manno G, Soldatov R, Zeisel A, Braun E, Hochgerner H, Petukhov V, Lidschreiber K, Kastriti ME, Lönnerberg P, Furlan A, Fan J, Borm LE, Liu Z, van Bruggen D, Guo J, He X, Barker R, Sundström E, Castelo-Branco G, Cramer P, Adameyko I, Linnarsson S, Kharchenko PV (2018) RNA velocity of single cells. Nature 560:494–498
29. Maizels RJ, Snell DM, Briscoe J (2023) Deep dynamical modelling of developmental trajectories with temporal transcriptomics. bioRxiv. 2023.07.06.547989
30. Saelens W, Cannoodt R, Todorov H, Saeys Y (2019) A comparison of single-cell trajectory inference methods. Nat Biotechnol 37:547–554
31. Weinreb C, Rodriguez-Fraticelli A, Camargo FD, Klein AM (2020) Lineage tracing on transcriptional landscapes links state to fate during differentiation. Science 367:eaaw3381
32. Zimmerman W, Weijer CJ (1993) Analysis of cell cycle progression during the development of Dictyostelium and its relationship to differentiation. Dev Biol 160:178–185
33. Muramoto T, Chubb JR (2008) Live imaging of the Dictyostelium cell cycle reveals widespread S phase during development, a G2 bias in spore differentiation and a premitotic checkpoint. Development 135:1647–1657

Chapter 17

Proteomic Analysis of *Dictyostelium discoideum* by Mass Spectrometry

Ramesh Rijal and Richard H. Gomer

Abstract

The large-scale proteomic analysis of *Dictyostelium discoideum* has contributed to our understanding of intracellular as well as secreted proteins in this versatile model eukaryote. Mass spectrometry-based proteomic analysis is a robust, sensitive, and rapid analytical method for identification and characterization of proteins extracted from tissues, cells, cell fractions, or pull-down assays. The availability of core facilities which make proteomics inexpensive and easy to do has facilitated a wide range of research projects. In this chapter, we present a simple standard methodology to extract proteins and prepare samples from *D. discoideum* for mass spectrometry and methods to analyze the identified proteins.

Key words *Dictyostelium discoideum*, Proteomics, Mass spectrometry

1 Introduction

In a proteomics analysis, one can take cells or a cell fraction and identify thousands of specific proteins in the sample and determine the relative amounts of the different proteins [1, 2]. This can be used for basic discovery (what proteins are being secreted by this cell type? What proteins are present in this purified organelle?), comparative expression analysis of proteins from multiple protein samples (in this mutant cell type, or in cells treated with X, which proteins show a difference in abundance compared to wild type?), for protein-protein interactions studies, and for identification of posttranslational modification sites in proteins [3]. Mass spectrometry-based proteomics analysis involves extraction of protein from tissues, cells, cell fractions, or pull-down assays. The total protein in the sample is quantitated, and the proteins in the sample are then digested into small peptides by a protease such as trypsin that cuts at specific sites in proteins. The peptide masses, and peptide fragment masses, are then measured by mass spectrometry

Alan R. Kimmel (ed.), *Dictyostelium discoideum: Methods and Protocols*, Methods in Molecular Biology, vol. 2814, https://doi.org/10.1007/978-1-0716-3894-1_17,

[3]. For each detected peptide, the mass and fragment masses are compared to the predicted masses and fragment masses of all proteins predicted to be encoded in the genome of the corresponding cell, and a tally of these identified peptides, as well as a tally of peptides identified as coming from each protein, allows quantitation. The quantitation is accompanied by a measure of the confidence of the protein identification: Detecting eight peptides that come from protein Y strongly indicates that protein Y is present in the sample, while detecting only 1 peptide typically generates a lower confidence. Proteomics core facilities are found in many research institutions, and these cores typically perform digestion of proteins and the subsequent mass spectrometry. Using this method, one can quantify ~2000 to ~3000 proteins in a biological sample [4], with the core starting with a sample and returning a list of annotated proteins and the corresponding protein amounts. The analysis is relatively inexpensive: the University of Texas Southwestern Proteomics Core (https://proteomics.swmed.edu/wordpress/?page_id=44) charges $204, and the Chemistry MS facility at Texas A&M University (https://mass-spec.chem.tamu.edu/proteomics/proteomics-services.php) charges $85 per sample.

Mass spectrometry-based proteomics approach has been used in *D. discoideum* by multiple laboratories to study the transition of unicellular growth to multicellular development [4–11], proteins involved in host-pathogen interactions [12–26], secreted proteins [27–30], directed cell migration [31–35], and membrane trafficking [36].

As an example of what proteomics can do, we used proteomics to study the effect of exogenous polyphosphate on wild-type *D. discoideum* cells. Polyphosphate is a signal released by, and sensed by, growing cells [37]. The proteomics indicated that, unexpectedly, extracellular polyphosphate decreases levels of several proteasome proteins, and we then observed that polyphosphate inhibits proteasome activity in both *D. discoideum* and human leukemia cells [4]. The proteomics thus led to a new avenue of research.

2 Materials

Analytical grade reagents and ultrapure water with a resistivity of 18.2 MΩxcm at 25 °C should be used to prepare buffers and solutions. Buffers and solutions should be stored at room temperature and follow safety data sheet (or material safety data sheet) to safely use the reagents and dispose the waste materials generated. For ultrapure water, Synergy UV purification system (Millipore, Molsheim, France) may be used.

2.1 Sample Preparation

1. Ethanol and kimwipes for cleaning and latex gloves.
2. Hairnet or hood when working with small concentrations of sample.
3. Although not strictly necessary, a laminar flow hood will reduce sample contamination from dirt and skin cells in the air.
4. PBM [20 mM KH_2PO4, 0.01 mM $CaCl_2$, 1 mM $MgCl_2$ (pH 6.1) with KOH] [38].
5. 1.5 mL Eppendorf tubes.
6. Eppendorf tube centrifuge capable of running at both 500 × g and 18,000 × g.
7. Cell Lysis buffer: Mix 495 μL radioimmunoprecipitation assay (RIPA) lysis and extraction buffer (Ref 89,900, Thermo Scientific) with 5 μL of 100× protease and phosphatase inhibitors cocktail (Ref 1,861,281, Thermo Scientific).
8. Sonicator: Type 95563, Chicago Electric or the equivalent.
9. Determination of protein content: Total protein content is measured using Bradford reagent (B6916, Sigma) or (E530-1 L, Amresco). Bovine serum albumin (J65855, Alfa Aesar) is used as a standard. The staining of protein is measured with a Synergy Mx multimode microplate reader (BioTek) with a micro sample attachment for 5 μL samples.
10. 2× sample buffer: 100 mM Tris/HCl, pH 6.8, 20% glycerol, 4% SDS, and 0.04% Bromophenol blue.
11. 95 °C dry bath incubator (Fisher Scientific).

2.2 SDS-Page

1. Apparatus: Bio-Rad Mini-PROTEAN tetra cell single core (Cat# 1658005, Bio-Rad).
2. Gels: Mini-PROTEAN precast gels 8.6x6.7x0.1 cm (Cat#4561096, Bio-Rad).
3. SDS-PAGE running buffer: 1× buffer contains 25 mM Tris-HCl pH 8.3, 192 mM glycine, and 0.1% sodium lauryl sulfate (SDS), or use commercial 10x Tris/Glycine/SDS Buffer (Cat#161–0732, Bio-Rad).

2.3 Gel Staining

1. Staining and destaining container: Use a clear or translucent box that will hold the gel; a lid from a pipette tips box works well.
2. Plastic wrap or aluminum foil to cover the staining container.
3. Staining solution: 30% methanol, 10% acetic acid, and 0.05% Coomassie Brilliant Blue (CBB) G-250 (w/v) (Cat#161-0406, Bio-Rad). Alternatively, use ready-to-use Imperial Protein Stain (Prod# 24615, Thermo Scientific).
4. Destaining solution: 20% ethanol and 7% acetic acid in water.

2.4 Band Excision

1. Single edge industrial razor blade (surgical carbon steel) (Cat#55411-050, VWR) or the equivalent.
2. Light box (Model 2185, Mayline Co. Inc.) or the equivalent.

2.5 In-Gel Digestion and Mass Spectrometry

1. Performed at proteomics core facility such as at the University of Texas Southwestern (http://www.utsouthwestern.edu/research/corefacilities/proteomics-core.html).

2.6 Proteins' Annotation and Classification

1. Free online software: Gene Ontology database (http://geneontology.org/) and the PANTHER classification system. Alternatively, the biological DataBase network (bioDBnet) (https://biodbnet-abcc.ncifcrf.gov/db/db2db.php) can be used.

3 Methods

3.1 Sample Preparation

Prepare sample in a simple way to maximize reproducibility. Clean all surfaces with ethanol and kimwipes, and wear gloves (rinse powder off gloves after putting on gloves). For very small protein concentrations (such as a few nanograms of a purified protein excised from a gel), wear a lab coat and a hairnet or better yet a hood and a surgical mask to reduce the amounts of human keratin proteins in the sample from skin cells.

1. To extract at least 10 μg of proteins, collect ~10^6 *D. discoideum* cells in a 1.5 mL Eppendorf tube by centrifugation at 500 × g for 3 min (*see* **Note 1**).
2. Wash cells twice by resuspending in 500 μL of PBM and centrifugation.
3. Collect cell pellet, and add 40 μL RIPA buffer containing 1× protease and phosphatase inhibitors, and incubate on ice for 20 min (*see* **Note 2**).
4. Sonicate the cell lysate for 1 min (*see* **Note 3**), and centrifuge at 18,000 g at 4 °C for 10 min to pellet down the cell debris (*see* **Note 4**).
5. Transfer 30 μL of supernatant to a new 1.5 mL Eppendorf tube, and use 5 μL of this to measure total protein content using Bradford reagent and a microplate reader (*see* **Note 5**). Use known concentrations of bovine serum albumin as a standard protein to measure the concentration of the protein in the sample. Most proteomics core facilities require 10 μg of protein for analysis, which is equivalent to ~1.5×10^5 *D. discoideum* cells [39] (*see* **Note 6**).
6. Add 25 μL of 2× sample buffer to the 25 μL of the remaining cell lysate, and heat at 95 °C for 5 min in a dry bath incubator.

3.2 SDS-Page

1. Assemble electrophoresis gel in the electrophoresis apparatus and fill the tank with SDS-PAGE running buffer. Rinse the wells with the running buffer by gently pipetting, and load 10 μL of sample in each well. Leave an empty well as a gap between sample wells to prevent mixing of samples (*see* **Note 7**).
2. Run the gel at 80 V. When the Bromophenol blue dye front reaches 5 mm below the bottom of the wells (not to completion, this typically takes 20 minutes), turn off the power supply and disassemble the gel cassette on an ethanol-cleaned bench area, and carefully transfer the gel into a container precleaned with ethanol.

3.3 Gel Staining

1. Stain the gel with freshly made Coomassie Brilliant Blue staining solution or Imperial protein stain overnight with gentle shaking (40 rpm) (*see* **Note 8**).
2. After discarding the staining solution, add destaining solution and incubate for 30 minutes with gentle shaking. Replace destaining solution with water and shake overnight to get rid of the blue stain from the gel background.

3.4 Band Excision and Mass Spectrometric Analysis

1. Razor blades typically are very high carbon steel, which rusts easily. The blades are thus made with a thin film of oil to prevent rusting. Remove the oil by carefully wiping the blade with ethanol and kimwipes. Pour the water off the gel. On a light box in an ethanol-cleaned laminar flow hood (this reduces contamination with skin cells and other dirt) and of course wearing gloves, excise the entire protein band with a razor blade in the destaining container. Minimize the amount of unstained gel.
2. Dice the combined protein bands into approximately 2 mm × 2 mm pieces. Place the diced bands in an Eppendorf tube previously rinsed with ethanol and clean water, seal the tube with parafilm, and send to the proteomics core for trypsin digestion and mass spectrometric analysis (*see* **Note 9**).
3. Further instructions on sample preparation can be obtained from the core facility.

3.5 Characterization and Classification of Identified Proteins

1. The various proteomics cores send the list of identified proteins in an excel sheet, which includes protein IDs from UniProt, relative abundance, and confidence of the protein identification. A minimum of three biological replicates is needed for most work determining abundance between two or more conditions or genotypes.
2. Identified proteins can be classified using the Gene Ontology database (http://geneontology.org/) and the PANTHER

classification system into different categories based on biological processes, molecular functions, and cellular locations.

4 Notes

1. Protein samples can also be collected after immunoprecipitation and pull-down assays to identify interaction partners of a target protein.
2. Addition of protease and phosphatase inhibitors is critical to protein extraction to avoid loss of soluble protein and keep phosphorylated proteins phosphorylated.
3. Keep cells on ice during sonication to minimize heat generation and to prevent proteolysis of the protein sample. Although sonication is a harsher technique for *D. discoideum* cell disruption, other methods such as osmotic lysis, freeze thaw lysis, detergent lysis, and enzymatic lysis can also be used for cell lysis. Alternatively, to obtain cytoskeletal and other RIPA-insoluble proteins, cells can be directly mixed with SDS sample buffer and heated. When working with a new sample, one can try different methods to lyse the cells and measure the efficiency of lysis by quantifying total extracted proteins.
4. Cell debris in the lysate, which also includes sheared nucleic acids, should be removed by centrifugation. Purity of the sample is important for obtaining good results from mass spectrometry.
5. As an alternative to the Bradford assay, Lowry or bicinchoninic acid can be used assay to measure total protein concentration. Accurate protein quantification by Bradford detection is affected if the protein sample is contaminated with detergent. Since detergents are generally ionizable, contamination of protein samples with detergent can result in poor results in mass spectrometric data. Use clean gloves, and clean the workspace and gel apparatus with ethanol to minimize contamination of protein samples.
6. In-solution protein samples from immunoprecipitation or pull-down assays can be submitted directly to perform mass spectrometric analysis. If so, immediate use of the protein samples increases the accuracy in the results.
7. The use of precast PAGE gels prevents contamination from hair and skin that can occur during gel preparation.
8. A single protein band can be excised as previously described for tryptic digestion and MS analysis. A complex mixture of protein is resolved by SDS-PAGE and stained with Coomassie stain as previously described or with silver stain to detect

bands with low protein amounts. Coomassie detects as little as 10 ng of protein in a gel band, whereas silver stain detects less than 0.1 ng of protein in a gel band.

9. To study posttranslational modification of proteins such as methylation, acetylation, and phosphorylation, enrichment of peptides is done using commercially available kits. Most proteomics core facilities use affinity-based enrichment of modified peptides and have additional charges to do this enrichment before MS analysis.

Acknowledgments

We thank Sara Kirolos and Kristen Consalvo for helpful comments on the manuscript. This work was supported by NIGMS R35 GM139486.

References

1. Wilkins MR, Sanchez JC, Gooley AA, Appel RD, Humphery-Smith I, Hochstrasser DF, Williams KL (1996) Progress with proteome projects: why all proteins expressed by a genome should be identified and how to do it. Biotechnol Genet Eng Rev 13:19–50
2. Pandey A, Mann M (2000) Proteomics to study genes and genomes. Nature 405:837–846
3. Aebersold R, Mann M (2003) Mass spectrometry-based proteomics. Nature 422: 198–207
4. Suess PM, Watson J, Chen W, Gomer RH (2017) Extracellular polyphosphate signals through Ras and Akt to prime Dictyostelium discoideum cells for development. J Cell Sci 130:2394–2404
5. Kim WD, Yap SQ, Huber RJ (2021) A proteomics analysis of Calmodulin-binding proteins in Dictyostelium discoideum during the transition from unicellular growth to multicellular development. Int J Mol Sci 22
6. Srinivasan S, Alexander H, Alexander S (1999) The prespore vesicles of Dictyostelium discoideum. Purification, characterization, and developmental regulation. J Biol Chem 274: 35823–35831
7. Alexander S, Srinivasan S, Alexander H (2003) Proteomics opens doors to the mechanisms of developmentally regulated secretion. Mol Cell Proteomics 2:1156–1163
8. Roth U, Muller S, Hanisch FG (2006) Proteomic analysis of Dictyostelium discoideum. Methods Mol Biol 346:95–109
9. Czarna M, Mathy G, Mac'cord A, Dobson R, Jarmuszkiewicz W, Sluse-Goffart CM, Leprince P, De Pauw E, Sluse FE (2010) Dynamics of the Dictyostelium discoideum mitochondrial proteome during vegetative growth, starvation and early stages of development. Proteomics 10:6–22
10. Gonzalez-Velasco O, De Las Rivas J, Lacal J (2019) Proteomic and transcriptomic profiling identifies early developmentally regulated proteins in Dictyostelium Discoideum. Cells 8
11. Mazur M, Wojciechowska D, Sitkiewicz E, Malinowska A, Swiderska B, Kmita H, Wojtkowska M (2021) Mitochondrial processes during early development of Dictyostelium discoideum: from bioenergetic to proteomic studies. Genes (Basel) 12
12. Urwyler S, Brombacher E, Hilbi H (2009a) Endosomal and secretory markers of the legionella-containing vacuole. Commun Integr Biol 2:107–109
13. Urwyler S, Nyfeler Y, Ragaz C, Lee H, Mueller LN, Aebersold R, Hilbi H (2009b) Proteome analysis of legionella vacuoles purified by magnetic immunoseparation reveals secretory and endosomal GTPases. Traffic 10:76–87
14. Linares JF, Moreno R, Fajardo A, Martinez-Solano L, Escalante R, Rojo F, Martinez JL (2010) The global regulator Crc modulates metabolism, susceptibility to antibiotics and virulence in Pseudomonas aeruginosa. Environ Microbiol 12:3196–3212
15. Vial L, Groleau MC, Lamarche MG, Filion G, Castonguay-Vanier J, Dekimpe V, Daigle F,

Charette SJ, Deziel E (2010) Phase variation has a role in Burkholderia ambifaria niche adaptation. Isme J 4:49–60
16. Hoffmann C, Finsel I, Hilbi H (2012) Purification of pathogen vacuoles from legionella-infected phagocytes. J Vis Exp
17. Hoffmann C, Finsel I, Hilbi H (2013) Pathogen vacuole purification from legionella-infected amoeba and macrophages. Methods Mol Biol 954:309–321
18. Hoffmann C, Finsel I, Otto A, Pfaffinger G, Rothmeier E, Hecker M, Becher D, Hilbi H (2014) Functional analysis of novel Rab GTPases identified in the proteome of purified legionella-containing vacuoles from macrophages. Cell Microbiol 16:1034–1052
19. Hilbi H, Rothmeier E, Hoffmann C, Harrison CF (2014) Beyond Rab GTPases legionella activates the small GTPase ran to promote microtubule polymerization, pathogen vacuole motility, and infection. Small GTPases 5:1–6
20. Schmolders J, Manske C, Otto A, Hoffmann C, Steiner B, Welin A, Becher D, Hilbi H (2017) Comparative proteomics of purified pathogen vacuoles correlates intracellular replication of legionella pneumophila with the small GTPase Ras-related protein 1 (Rap1). Mol Cell Proteomics 16:622–641
21. Steiner B, Swart AL, Welin A, Weber S, Personnic N, Kaech A, Freyre C, Ziegler U, Klemm RW, Hilbi H (2017) Er remodeling by the large GTPase atlastin promotes vacuolar growth of legionella pneumophila. Embo Rep 18:1817–1836
22. Manske C, Finsel I, Hoffmann C, Hilbi H (2018) Analysis of legionella metabolism by pathogen vacuole proteomics. Methods Mol Biol 1841:59–76
23. Varas MA, Riquelme-Barrios S, Valenzuela C, Marcoleta AE, Berrios-Pasten C, Santiviago CA, Chavez FP (2018) Inorganic polyphosphate is essential for salmonella Typhimurium virulence and survival in Dictyostelium discoideum. Front Cell Infect Microbiol 8:8
24. Valenzuela C, Gil M, Urrutia IM, Sabag A, Enninga J, Santiviago CA (2021) SopB- and SifA-dependent shaping of the salmonella-containing vacuole proteome in the social amoeba Dictyostelium discoideum. Cell Microbiol 23:e13263
25. Shevchuk O, Batzilla C, Hagele S, Kusch H, Engelmann S, Hecker M, Haas A, Heuner K, Glockner G, Steinert M (2009) Proteomic analysis of legionella-containing phagosomes isolated from Dictyostelium. Int J Med Microbiol 299:489–508
26. Manske C, Hilbi H (2014) Metabolism of the vacuolar pathogen legionella and implications for virulence. Front Cell Infect Microbiol 4: 125
27. Srinivasan S, Traini M, Herbert B, Sexton D, Harry J, Alexander H, Williams KL, Alexander S (2001) Proteomic analysis of a developmentally regulated secretory vesicle. Proteomics 1: 1119–1127
28. Bakthavatsalam D, Gomer RH (2010) The secreted proteome profile of developing Dictyostelium discoideum cells. Proteomics 10: 2556–2559
29. Lavialle F, Deshayes S, Gonnet F, Larquet E, Kruglik SG, Boisset N, Daniel R, Alfsen A, Tatischeff I (2009) Nanovesicles released by Dictyostelium cells: a potential carrier for drug delivery. Int J Pharm 380:206–215
30. Huber RJ, O'day DH (2015) Proteomic profiling of the extracellular matrix (slime sheath) of Dictyostelium discoideum. Proteomics 15:3315–3319
31. Zhang P, Wang Y, Sesaki H, Iijima M (2010) Proteomic identification of phosphatidylinositol (3,4,5) triphosphate-binding proteins in Dictyostelium discoideum. Proc Natl Acad Sci Usa 107:11829–11834
32. Yan J, Mihaylov V, Xu X, Brzostowski JA, Li H, Liu L, Veenstra TD, Parent CA, Jin T (2012) A Gβγ effector, ElmoE, transduces Gpcr signaling to the actin network during chemotaxis. Dev Cell 22:92–103
33. Sobczyk GJ, Wang J, Weijer CJ (2014) Silac-based proteomic quantification of chemoattractant-induced cytoskeleton dynamics on a second to minute timescale. Nat Commun 5:3319
34. Lacal Romero J, Shen Z, Baumgardner K, Wei J, Briggs SP, Firtel RA (2018) The Dictyostelium GSK3 kinase GlkA coordinates signal relay and chemotaxis in response to growth conditions. Dev Biol 435:56–72
35. Kothari P, Srivastava V, Aggarwal V, Tchernyshyov I, Van Eyk JE, Ha T, Robinson DN (2019) Contractility kits promote assembly of the mechanoresponsive cytoskeletal network. J Cell Sci 132
36. Dieckmann R, Gopaldass N, Escalera C, Soldati T (2008) Monitoring time-dependent maturation changes in purified phagosomes from Dictyostelium discoideum. Methods Mol Biol 445:327–337
37. Suess PM, Gomer RH (2016) Extracellular polyphosphate inhibits proliferation in an autocrine negative feedback loop in Dictyostelium discoideum. J Biol Chem 291:20260–20269

38. Brock DA, Gomer RH (2005) A secreted factor represses cell proliferation in Dictyostelium. Development 132:4553–4562
39. Aubry L, Klein G (2006) Purification techniques of subcellular compartments for analytical and preparative purposes. In: Eichinger L, Rivero F (eds) Dictyostelium discoideum Protocols. Humana Press, Totowa

INDEX

A

Actin waves 21, 121, 136, 150, 158, 229
Activation membrane proteins 133–147
Activity assays 124, 164
APEX 120
Autophagic flux 102, 103
Autophagosomes 46, 97, 98
Autophagy 46, 56, 97, 98, 100–102

B

Bactericidal activity 92
Bioinformatics 224
Biotinylation 41, 120, 121, 127
BLAST 8, 215, 218, 219

C

Cathepsins 46, 56–58, 61, 65, 66, 70–72, 77
CDNA library complementation 2, 7
Cell migration 109, 195, 196, 248
Cellular signaling 128
Centrosomes 37, 38, 42
cGMP 177–193
CGMP-binding protein GbpC 177, 191–192
Chemically-induced dimerization 107, 108
Chemotaxis 163, 164, 167, 170, 177, 195, 196, 205, 232
Confocal fluorescence microscopy 195–205
Contact guidance 149, 150
CRISPR 4, 11, 13, 129, 219
Cytokinesis 1, 2, 5–8, 10–17, 21, 22, 210, 232
Cytoskeletal network 17
Cytosolic-RFP 135, 136, 143

D

Developments 4, 8, 56, 58, 76, 77, 82, 85, 92, 97, 110, 119, 120, 128, 146, 149, 153, 167, 172, 195, 209, 210, 218, 223–242, 248
Dextrans 46, 48, 50–52, 82, 85, 87
Dictyostelium discoideum 2–5, 7–13, 15, 17, 19–21, 25, 46, 48–52, 56, 58, 60, 76, 77, 82, 83, 85, 90, 91, 98–100, 102, 109, 115, 116, 129, 155, 195–197, 247–253
Digitonin 46–50, 121, 122, 124

E

Endolysosomes 46, 48–50
Engineered ascorbic peroxidase 120
Enzymes 5, 45, 55, 89, 119, 177, 210
Expansion microscopy (ExM) 29, 32, 37

F

FKBPs 107, 108, 110, 115
Förster resonance energy transfer (FRET) 134, 196, 197, 199–205
FRB 107, 108, 110

G

GFP-sensor 134, 135, 144–146
Glycyl-L-phenylalanine 2-naphthylamide (GPN) 46–49, 52
G protein-coupled receptors (GPCRs) 115, 121, 195
GTPases 109, 163
Guanylyl cyclases 177, 180, 182, 183, 188–190

H

Heat-killed yeast 83–85
Hydrolases 56, 60, 64, 66–70, 73
Hyperosmotic shock 50

I

Imaging 2, 5–6, 9, 12–19, 21, 22, 24, 29, 34, 48–51, 85, 111, 115, 116, 124, 127, 139, 151–154, 157, 158, 163–167, 170–172, 198–200, 202, 203, 205, 239
1,4,5-Inositol trisphosphate (IP3) 196

Alan R. Kimmel (ed.), *Dictyostelium discoideum: Methods and Protocols*, Methods in Molecular Biology, vol. 2814, https://doi.org/10.1007/978-1-0716-3894-1,

K

Klebsiella aerogenes 5, 14, 74, 91, 92, 94, 109, 116, 136, 182

L

Lamin 30
Latex-bead 81–83, 86
L-leucyl-L-leucine methyl ester (LLOMe) 46–52
Lov domain 109
Lysosensor 46, 48, 50, 51
Lysosomes 55, 56, 89, 91, 97, 130, 182, 192

M

Macropinocytosis 81, 82, 85, 86
Mannosidases 89
Mass spectrometry (MS) 2, 119, 121, 123, 127, 129, 247–253
Mechanoresponsive 22
Membranes 45–50, 56, 58, 81, 90, 91, 99, 101, 104, 108, 112, 113, 115, 116, 126, 127, 133–147, 165, 169, 170, 175, 178, 188, 195, 196, 248
Microfabrication 150, 152
Model system 56, 109, 209, 211

N

Neutrophils 195, 196

O

Optogenetics 108, 109, 115

P

Phagocytosis 81, 83–86, 91, 93, 94, 163
Phagosome maturation 81, 89, 94
Phosphodiesterases 177, 178, 182, 190, 191
Phospholipase C (PLC) 195, 196
PIP_3 157
Propidium iodide (PI) 47, 48, 50, 108, 112, 113, 134–136, 178
Proteases 31, 33, 46, 56, 61, 70–73, 75, 99, 101, 104, 130, 166, 167, 178, 180–182, 247, 249, 250, 252
Protein dynamics 2, 20
Protein protein interaction 2, 8, 17, 109, 178, 200, 247
Proteomics 120, 121, 247, 248, 250, 251, 253
Proximity labeling 119–121
Pull-down 130, 163–170, 172, 247, 252

R

Rap 163–175
Rap1 .... 136, 163–166, 172, 174
Rapamycin 107, 108, 112, 113, 116
Ras 134–136, 141, 143–145, 163–175, 178, 191, 193, 196
Ras binding domain (RBD) 163–166, 172, 173
Restriction enzyme-mediated integration (REMI) 209–219
RNA sequencing 228

S

Single cell transcriptomics 223, 234, 242
Superresolution microscopy 31

T

Thioesterase 56, 58, 64, 73
Transcription bursting 230

GPSR Compliance
The European Union's (EU) General Product Safety Regulation (GPSR) is a set of rules that requires consumer products to be safe and our obligations to ensure this.

If you have any concerns about our products, you can contact us on

ProductSafety@springernature.com

In case Publisher is established outside the EU, the EU authorized representative is:

Springer Nature Customer Service Center GmbH
Europaplatz 3
69115 Heidelberg, Germany

www.ingramcontent.com/pod-product-compliance
Ingram Content Group UK Ltd.
Pitfield, Milton Keynes, MK11 3LW, UK
UKHW052004051026
15624UKWH00003B/562
* 9 7 8 1 0 7 1 6 3 8 9 3 4 *